ELEMENTS
OF
STOCHASTIC PROCESS
SIMULATION

ELEMENTS
OF
STOCHASTIC PROCESS
SIMULATION

Byron S. Gottfried, Ph.D.

University of Pittsburgh

PRENTICE-HALL, INC., Englewood Cliffs, New Jersey 07632

Library of Congress Cataloging in Publication Data

Gottfried, Byron S. (date)
   Elements of stochastic process simulation.

   Includes index.
   1. Digital computer simulation.   2. Stochastic
processes.   I. Title.
QA76.9.C65G67 1984   0.01.4'34   83-3251
ISBN   0-13-272500-2

Editorial/production supervision and
   interior design: Bette Kurtz, Elizabeth Athorn, Natalie Krivanek
Cover design: Edsal Enterprizes
Manufacturing buyer: Anthony Caruso

Printed in the United States of America

10 9 8 7 6 5 4 3 2 1

ISBN 0-13-272500-2

Prentice-Hall, International, Inc., *London*
Prentice-Hall of Australia Pty. Limited, *Sydney*
Editora Prentice-Hall do Brasil, Ltda., *Rio de Janeiro*
Prentice-Hall Canada Inc., *Toronto*
Prentice-Hall of India Private Limited, *New Delhi*
Prentice-Hall of Japan, Inc., *Tokyo*
Prentice-Hall of Southeast Asia Pte. Ltd., *Singapore*
Whitehall Books Limited, *Wellington, New Zealand*

To my parents,
Faye and Sidney Gottfried,
who contributed to this book in many different ways.

CONTENTS

3 SOME ELEMENTARY SIMULATION PROBLEMS 48

4 NONUNIFORM RANDOM VARIATES 76

5 INDUSTRIAL AND BUSINESS APPLICATIONS 112

6 CONDUCTING A COMPLETE SIMULATION STUDY 154

7 QUEUING APPLICATIONS 184

8 SPECIAL-PURPOSE SIMULATION LANGUAGES 229

APPENDIX A. Selected Fortran Programs and Subprograms   259

APPENDIX B. Answers to Selected Problems   287

Index   297

PREFACE

Of the many analytical techniques that comprise the so-called management sciences, few are used more frequently than stochastic process simulation. This is due to the applicability of this technique to a wide variety of problems. Yet the content of this important field cannot be clearly established, since it is a potpourri of individual topics drawn from the fields of business, computer science, statistics, engineering, and mathematics. Thus the subject can be approached from many different points of view, both in the classroom and in professional practice.

The lack of uniformity associated with this subject is clearly illustrated by the large variation in currently available textbooks. Some are very qualitative in nature and are therefore unable to provide the necessary skills for building and solving actual simulation models. On the other hand, there are some highly mathematical texts exclusively devoted to certain specialized topics in simulation. Such books are inappropriate for beginning students because they do not impart an overall understanding of the subject.

This book is intended as an introductory-level text for a one-semester course in stochastic process simulation. I have therefore attempted to maintain some reasonable balance between model building, computer implementation, and analysis and interpretation of results. The prospective audience is reasonably diverse, as it includes undergraduate students enrolled in industrial engineering, computer science, and information science programs. In addition, the material may be appropriate for a simulation course in an MBA program or a quantitatively oriented undergraduate business curriculum.

The reader is expected to have some background in computer programming and in probability and statistics. A high level of proficiency is not required in either area, although the reader should be able to translate detailed flowcharts into working computer programs using a general-purpose language such as Fortran, Pascal, or BASIC.

The entire text can easily be covered within a fifteen-week semester. Normally the material can best be covered by proceeding through the book sequentially, though the choice of topics can be varied to meet individual needs. In particular, some of the material in Chapter 2 dealing with random-number generation and statistical tests for randomness can be skipped. Also, the elementary simulation problems presented in Chapter 3 may not be of interest to more advanced students, though I have found that most of my undergraduate engineering students become very "turned on" by the simulation of gambling games.

The book includes a large number of numerical examples and computational flowcharts. These items facilitate the implementation of the material described in the text. The flowcharts are developed for a general-purpose programming language, such as Fortran. I have chosen to emphasize general-purpose languages for model development because they provide deep insights into the actual logical intricacies of the models. Many of the models can, however, be described more easily via special-purpose simulation languages, such as GPSS, SLAM, and SIMULA. Several of these languages are briefly described in Chapter 8. The use of one of these languages as a supplement to a general-purpose language is highly recommended.

Each chapter is followed by an extensive list of problems. Students should be encouraged to solve as many of these problems as time will allow. This is especially true of the more comprehensive simulation problems at the end of Chapters 3, 5, 7, and 8. In order to assist in the solution of these problems, I have included a number of Fortran programs in Appendix A. Most of these are utility-type subprograms that generate random variates, group data, and calculate statistical parameters. However, I have also included a few complete programs that solve some of the more complicated models.

I cannot overemphasize the importance of having students develop and solve their own simulation models on a computer, making proper use of the methodology presented in the text. The solutions should then be examined critically and refined whenever necessary. Ultimately, this is the only way that simulation can really be learned.

Finally, I wish to express my gratitude to Maria-Victoria Tobon, who assisted me in writing this book by solving many of the problems, writing some of the Fortran subprograms, and tabulating the results of countless simulation programs over a period of several years. I also wish to thank José Sepulveda for writing the GASP program presented in Chapter 8, John Wizzard for assisting me with some of the problems in Chapter 7 and writing the SIMULA program in Chapter 8, and Penny Brown for typing most of the manuscript.

Byron S. Gottfried

ELEMENTS
OF
STOCHASTIC PROCESS
SIMULATION

1

INTRODUCTION

One of the most important tasks in running a business organization is the establishment of efficient operating policies. For example, a bank manager must determine how many teller windows to operate during a typical busy day. Similarly, the manager of a manufacturing facility would like to know how best to schedule a variety of different types of orders. A hospital administrator must decide upon appropriate inventory levels of expensive, perishable substances such as blood, drugs, and food, and must schedule the use of the hospital's facilities as efficiently as possible. In each of these examples, the choice of a particular operating policy can determine the difference between a profitable versus unprofitable, or, an efficient versus inefficient operation.

A complicating factor in the choice of a suitable operating policy is the uncertainty that is inherent in most business operations, since a manager rarely knows in advance exactly what will occur on any given day. Therefore, for a particular operating policy to be truly effective, it must be able to accommodate the unexpected as well as the commonplace without undue disruption or unreasonable expense.

One way to establish a suitable operating policy is to experiment with the actual system. Thus an effective mode of operation can usually be found by trial and error, given a reasonable amount of patience and common sense. The trouble with this method is that it may be impractical, and perhaps impossible, to tamper with the actual system. The hospital administrator, for example, cannot tolerate shortages or unreasonable delays when dealing with critically ill patients; the bank manager will probably lose many customers if they are forced to wait in long, slowly moving lines, and so on. A different method must therefore be found for determining effective operating policies.

Computer simulation offers a practical solution to this problem. By experimenting with a computer simulation model, the analyst can try a number of different operating policies without disturbing the actual system. These policies can then be compared and the best one selected.

This book is concerned with the development of computer simulation models for a variety of problems that arise in business and industry. Since all of these problems involve some significant degree of uncertainty, we will be considering techniques for simulating the effects of randomness on system behavior. In particular, we will see how random events can be generated by a computer in the same manner that they occur in an actual problem situation. These computer-generated events can then be analyzed to draw conclusions about the behavior of the real system.

1.1 BASIC TERMINOLOGY

There are certain basic terms, such as system, state, model, and operating policy, that are used repeatedly in any discussion of stochastic-process simulation. We must, therefore, consider the meaning of these terms before we can begin studying actual simulation techniques.

System

There are many different ways to define a *system*, depending upon the context within which the term is used. For our purposes, it is most appropriate to think of a system as *an isolated collection of interacting components*. The nature of the components and the degree of interaction between them may vary substantially from one system to another. It is this interaction, however, that prevents the individual system components from being analyzed as separate, independent entities.

A jet aircraft is an excellent example of a complex system consisting of numerous mechanical, electronic, chemical, and human components. A major corporation, together with its customers and its suppliers, represents another example of a system containing complex, interacting components. Finally, a national economy can be represented and studied as a system, since it is comprised of many interacting elements.

From a systems analysis standpoint there are two general types of systems—*deterministic* and *stochastic* (or *probabilistic*). In a deterministic system, the individual system components always behave in a well-defined, predictable manner. Stochastic systems, on the other hand, involve the occurrence of random events. Such systems are encountered when analyzing many realistic problems, such as games of chance, sales forecasts, financial acquisitions, equipment maintenance, inventory control, networks, and situations involving queues (waiting lines). This book is concerned exclusively with this latter type of system.

Stochastic systems can be further categorized as being either *static* or *dynamic*. In a static system the occurrence of random events is independent of the passage of time. Such systems are relatively easy to simulate. On the other hand, the random events in a dynamic system must occur sequentially with respect to time (for example, a customer cannot enter a service area until the previous customer has departed). Thus, dynamic systems are relatively complicated and, therefore, more difficult to simulate. As a rule, gambling games, financial problems, and network problems (for example, CPM and PERT) are static in nature, whereas equipment maintenance problems, inventory control problems, and queuing problems are dynamic.

State

The *state* of a system can be thought of as the totality of all relevant system characteristics. Usually, a system can be characterized by a specific set of attributes. The state of the system will then be determined by assigning a particular value to each of these attributes.

In the case of a jet aircraft, for example, the state of the system would be determined by such factors as the aircraft's speed, altitude, direction of travel, weather conditions, number of passengers, amount of fuel remaining, and operating status. Notice that some of these factors will remain constant, whereas others will vary with time. Thus we see that the state of a system can (and often does) change with time. Moreover, some of these factors are, for all practical purposes, deterministic, whereas others, such as weather conditions and operating status, are stochastic.

State Variables

If each of the attributes that characterize the state of a system can be quantified, then a unique variable can be used to represent each attribute. These variables are known as *state variables*. If a system has m state variables, we can express them mathematically as $(s_1, s_2, \ldots, s_m)$ or, in vectorial form, simply as **S**.

When considering a system that involves discrete, random events, each state variable may be a vector (that is, a list) whose components represent discrete events of a particular type. Each individual event will be represented by a *random variable*. For example, if we were analyzing a supermarket checkout counter, one of the state variables would contain the arrival times of the various customers; another would contain the corresponding departure times. In such cases we will express the (vectorial) state variables as $(\mathbf{S}_1, \mathbf{S}_2, \ldots, \mathbf{S}_m)$. Also, the entire set of state variables will be expressed collectively as $\mathbf{S}$.

In order to obtain values for the state variables, it will be necessary to carry out a number of calculations (that is, evaluate a *mathematical model*), which may be rather complicated for certain types of problems. Specific values will have to be assigned to the random variables. Additional information (that is, values of the *decision variables* and the *system parameters*) will also be required in order to carry out these calculations.

Decision Variables

In most situations there are certain variables whose values can be specified by the analyst (or the "decision maker") at the beginning of a problem, independent of any other considerations. These variables are known as *decision variables*. Mathematically, we can represent the decision variables as $(x_1, x_2, \ldots, x_n)$, or, using vector notation, as $\mathbf{X}$.

Normally the values chosen for the decision variables will affect the state of the system. Thus the state variables will be dependent upon the decision variables. We can therefore think of the state variables as *dependent variables*, and the decision variables as *independent variables*.

System Parameters

System parameters are similar to decision variables in the sense that their values can be specified *a priori*. These quantities usually represent physical constants, design parameters, constants of proportionality, etc., over which the decision maker has little or no control. Therefore the values of the system parameters may not change from one problem situation to another, whereas the decision variables will take on different values. We will refer to the system parameters as $(c_1, c_2, \ldots, c_k)$ or, in vector notation, simply as $\mathbf{C}$.

The state variables will be dependent upon the choice of system parameters as well as the choice of decision variables. We can express this dependence symbolically as

$$\mathbf{S} = f(\mathbf{C}, \mathbf{X}) \tag{1.1}$$

where f may represent a specific single equation or a detailed set of equations that describe the actual system behavior.

Cause-and-Effect Relationships

All systems are governed by certain relationships that describe the interaction between state variables, decision variables, and system parameters. These relationships may represent physical laws, economic principles, statistical correlations, etc. We will refer to them in general terms as *cause-and-effect relationships*.

For a given system, each cause-and-effect relationship can be expressed symbolically as

$$\varphi\,(\mathbf{S},\,\mathbf{X},\,\mathbf{C}) = 0 \tag{1.2}$$

Several such relationships will be required in order to obtain an accurate representation of a typical, realistic system.

Model

A *model* is used to provide some type of description of an actual system. Some models are used to provide a *physical* description of a real system. Model airplanes, architectural models, etc., fall into this category. There is also another, more abstract kind of model, called a *mathematical model*, that is used to describe the *behavior* of an actual system. Such models are comprised of a set of equations that represent the underlying cause-and-effect relationships within the system. We will be concerned with this latter type of model in this book.

Suppose that l cause-and-effect relationships are required to obtain an accurate description of a given system. These relationships can be expressed in symbolic form as

$$\left.\begin{aligned}
\varphi_1\,(\mathbf{S},\,\mathbf{X},\,\mathbf{C}) &= 0 \\
\varphi_2\,(\mathbf{S},\,\mathbf{X},\,\mathbf{C}) &= 0 \\
&\;\;\vdots \\
\varphi_l\,(\mathbf{S},\,\mathbf{X},\,\mathbf{C}) &= 0
\end{aligned}\right\} \tag{1.3}$$

These equations constitute a mathematical model for the system. The model is used to evaluate both the state of the system and some particular quantity that is representative of system performance, as discussed later. Thus, for a given set of conditions (that is, a given set of values for $\mathbf{C}$ and $\mathbf{X}$), the model will provide us with a quantitative measure of overall system behavior as well as a set of values for the state variables. Moreover, by specifying different sets of conditions and evaluating the model repeatedly for each case, we can see how the system behaves in

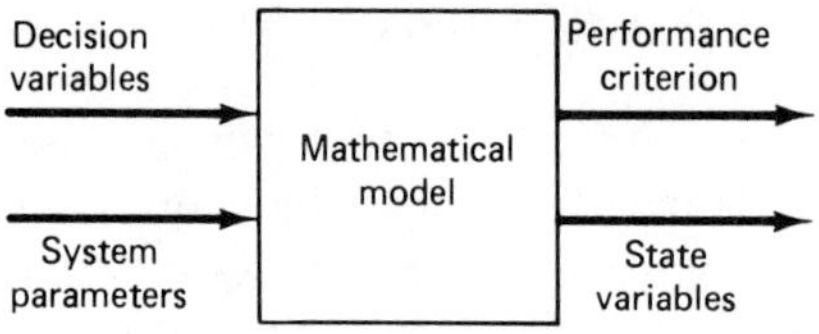

Decision variables → | Mathematical model | → Performance criterion

System parameters → | Mathematical model | → State variables

Figure 1.1

response to changes in various system parameters or decision variables. Figure 1.1 illustrates the role of the model in providing a description of system behavior.

Since a computer is normally used to solve the mathematical model, the terms *computer model* and *simulation model* are frequently used. It should be understood, however, that the actual description of the system is provided by the cause-and-effect relationships that make up the mathematical model. The computer is merely a device that is used to evaluate the model numerically.

System Performance Criterion

We have already mentioned the fact that some specific criterion is required as a measure of system performance. This is called the *system performance criterion*. Mathematically, this function is dependent upon the state variables as well as the decision variables and the system parameters. The performance criterion can therefore be expressed symbolically as

$$Y = f(\mathbf{S}, \mathbf{X}, \mathbf{C}) \tag{1.4}$$

Physically (or economically), the performance criterion may represent profit, cost, production level, quality, waiting time, queue length, etc.

Operating Policy

Recall that the choice of the decision variables affects the state of the system, and hence the manner in which the system behaves. We therefore refer to a given set of values for the decision variables as an *operating policy*. Whenever a different value is assigned to one or more of the decision variables, a new operating policy is obtained. A change in the model (for example, an alteration of one or more cause-and-effect relationships) also results in a different operating policy.

Our overall objective in carrying out a simulation study is to determine the best possible operating policy, relative to some particular measure of system performance. In practice, however, the amount of time and effort required to find the "best possible" policy may be excessive. Therefore, in reality, we often settle for an operating policy that is reasonably satisfactory, even though it may not be the very best that is attainable.

Example 1.1

An enterprising group of students has opened up a hot dog stand adjacent to their college campus. They plan to charge $0.85 for each hot dog, which will cost them $0.50. In addition to the cost of the hot dogs, the students must pay a fixed fee of $300 per month for rent, electricity, taxes, etc.

The expected number of hot dogs sold each day will depend upon the selling price. At $0.85 per hot dog, the students expect to sell about 200 hot dogs each day. Figure 1.2 shows the expected number of hot dogs sold per day as a function of selling price. It should be understood, however, that the actual number of hot dogs sold each day will fluctuate randomly about the expected value.

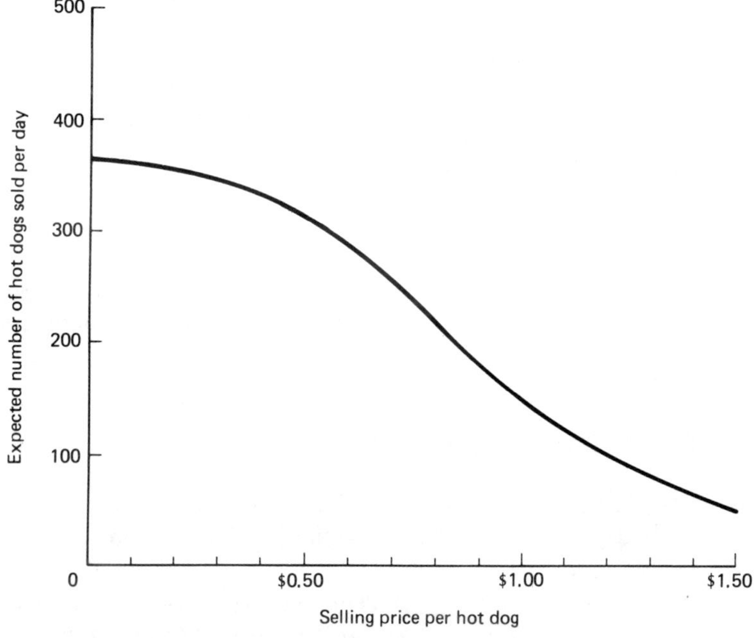

Figure 1.2

Suppose that we wish to determine the actual profit resulting from 1 year of operation. We can then compare this figure with the expected profit, and see how much variation there is between the actual and the expected values. A large variation would suggest that there is a high degree of risk associated with the proposed venture.

Let us assume that the hot dog stand will be open every day of the year. The yearly profit can then be expressed as

$$YP = (\$0.85 - \$0.50) \times (n_1 + n_2 + \ldots + n_{365}) - (\$300 \times 12)$$

where

$$n_1 \;=\; \text{number of hot dogs sold during the first day}$$
$$n_2 \;=\; \text{number of hot dogs sold during the second day}$$

etc.

In this example the system is the financial operation of the hot dog stand, including the financial transactions associated with the customers and the suppliers. The state of the system is determined by the daily sales. Thus, the system can be characterized by one state variable, **n**, whose components $(n_1, n_2, \ldots, n_{365})$ are the daily sales volumes. Each component will be represented by a random variable. (We will discuss methods for generating random variables elsewhere in this book.) The selling price of each hot dot ($0.85) is a decision variable, whereas the cost per hot dog ($0.50) and the monthly overhead ($300) are system parameters. (Note that the students have some control over the selling price of the hot dogs, but the cost of the hot dogs and the monthly overhead are fixed.)

Figure 1.2 represents a cause-and-effect relationship. A complete mathematical model will consist of this relationship, plus additional information which describes the random variation of the daily sales volumes. The yearly profit will provide a measure of system performance, and the decision to sell the hot dogs for $0.85 each represents an operating policy.

If we were actually going to simulate this system, we would probably carry out several different simulations, each with a different selling price, in order to determine the effect of selling price on yearly profit. (Figure 1.2 would be used for this purpose.) By carrying out such a study, the most profitable selling price could be established. Thus, the stated price of $0.85 per hot dog might only be a trial value for one part of a more comprehensive study.

1.2 WHAT IS SIMULATION?

Although we have already made casual reference to the term "simulation" in this book, we have not defined exactly what is meant by this term. We will do so now.

Simulation is an activity whereby one can draw conclusions about the behavior of a given system by studying the behavior of a corresponding model whose cause-and-effect relationships are the same as (or similar to) those of the original system. Hence, simulation is concerned with the development and use of models that realistically describe system performance. We have already seen that there are different kinds of models that can be used to represent a system, depending on the type of description that is desired. In this book we will be concerned only with the use of mathematical models to represent the behavior of actual systems.

Stochastic-process simulation (also called *discrete-event simulation* or *Monte-Carlo simulation*) refers to the use of mathematical models to study systems that are characterized by the occurrence of discrete, random events. These individual events are represented by random variables whose values are generated by a computer. The randomness that is encountered in a real system can therefore be synthesized, allowing the behavior of the original system to be reproduced artificially. Such studies allow us to assess both the expected behavior of the system and the amount of random variation that will be encountered. The latter information is indicative of the degree of risk that is associated with the given system.

In most problems of practical interest, a complete simulation study will require that the system be simulated repeatedly under varying conditions (that is, using a different set of values for the decision variables and system parameters during each simulation). The impact of various operating policies and different imposed conditions can be assessed in this manner. Since each simulation requires a separate evaluation of a mathematical model, a comprehensive simulation study is actually a form of numerical experimentation in which the mathematical model takes the place of an experimental facility.

This book presents a methodology for simulating various stochastic processes that arise in business and industry. Two main themes predominate. One is the formulation of appropriate mathematical models; the other is the solution of these models, using a computer to generate the required random events. Attention is also given to the analysis and interpretation of the simulated results. Readers who have mastered this material should be capable of carrying out their own simulation studies for a variety of realistic problem situations.

1.3 A SYSTEMS APPROACH

It is sometimes helpful to visualize a simulation program from a *systems viewpoint*, in which the mathematical model resides within a "black box" whose contents cannot be seen from the outside. Known information (that is, *input data*) is supplied to the black box, and the corresponding desired information (*output data*) emerges from the black box. Thus the black box causes the input data to be *transformed* into the output data. The procedure is illustrated schematically in Fig. 1.3.

We have already seen that the input data will consist of a set of values for the decision variables and the system parameters, and that the output data will contain the values of the state variables and the system performance criterion. We have not said anything, however, about the *form* of the performance criterion. In many problems the performance criterion is expressed as a statistical distribution function rather than as a single number. Under such circumstances the output data should include *at least* the *expected value* (that is, the *mean*) and the *standard*

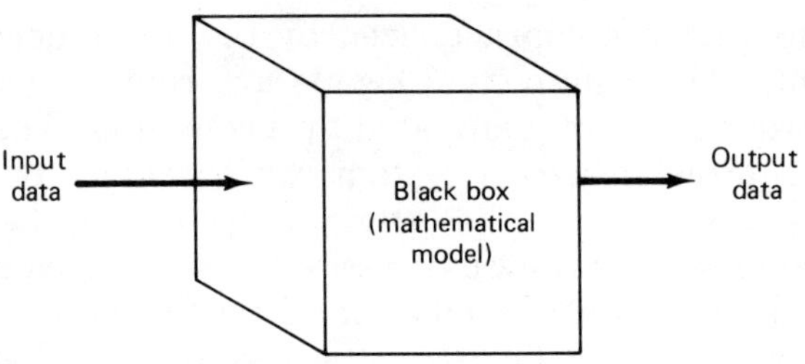

Figure 1.3

deviation of the performance criterion. It is usually desirable to include additional information also, such as the actual distribution data.

For most problems, the mean value of the system performance criterion, $\overline{Y}$, is determined as

$$\overline{Y} = \frac{1}{n} \sum_{i=1}^{n} Y_i \tag{1.5}$$

where the Y_i's represent individual values of the performance criterion, each corresponding to a particular set of values of **S**, **X**, and **C**. (Remember that the numerical values for the state variables will be dependent upon the values obtained for certain randomly generated events.)

The corresponding standard deviation, s, is determined as

$$s = \frac{1}{n} \sum_{i=1}^{n} \left[Y_i - \overline{Y} \right]^2 \tag{1.6}$$

or equivalently, as

$$s = \left[\frac{1}{n} \sum_{i=1}^{n} Y_i^2 \right] - (\overline{Y})^2 \tag{1.7}$$

(Note that Eq. (1.6) is often written with $(n-1)$ rather than n in the denominator. The numerical difference between these two expressions for $\overline{Y}$ becomes negligible, however, when n is large, as is typically the case when carrying out a computer-based simulation study.)

Sometimes the computed values of Y_i are *grouped* into successive intervals. A set of *relative frequencies* can be obtained from these grouped data as follows. Let n_j be the number of values of Y_i falling into the jth interval (where j is an interval index which ranges from 1 to m). Then the relative frequency for the jth interval, f_j, is simply

$$f_j = n_j/n \tag{1.8}$$

where n is the total number of Y_i's generated, that is,

$$n = \sum_{j=1}^{m} n_j \tag{1.9}$$

Note that the relative frequencies are required, by definition, to sum to unity. Thus,

$$\sum_{j=1}^{m} f_j = \frac{1}{n} \sum_{j=1}^{m} n_j = 1 \tag{1.10}$$

as expected.

Finally, a cumulative distribution can be obtained from the relative frequencies as

$$Y_1 = f_1$$
$$Y_2 = f_1 + f_2$$
$$\cdot$$
$$\cdot$$
$$\cdot$$
$$Y_j = f_1 + f_2 + \ldots + f_j \tag{1.11}$$
$$\cdot$$
$$\cdot$$
$$\cdot$$
$$Y_m = \sum_{j=1}^{m} f_j = 1$$

The relative frequencies and the cumulative distribution can, of course, be expressed either as fractions, as in the above expressions, or as percentages. In the latter case, the fractional values are simply multiplied by 100. Either form is acceptable, though there may be some bias toward the use of percentages within a business environment.

We will make repeated use of these simple relationships in the succeeding chapters of this book.

Example 1.2

In Ex. 1.1 we determined the profit resulting from the operation of a hot dog stand for 1 year (that is, for 365 consecutive days). This yearly profit (YP) can be thought of as the sum of 365 different values, where each value represents the daily profit. Thus,

$$YP = P_1 + P_2 + \ldots + P_{365}$$

where P_i, the profit for the ith day, can be calculated as

$$P_i = (\$0.85 - \$0.50)\, n_i - \$300\,(12/365)$$

Since n_i (the number of hot dogs sold during the ith day) will vary randomly from 1 day to the next, the corresponding values for P_i will also exhibit random day-to-day variation. These quantities will be represented by random variables, whose values will be assigned by the computer during the process of solving the mathematical model.

We can categorize the flow of information for this problem:

Input data:

1. Selling price of each hot dog ($0.85)
2. Cost of each hot dog ($0.50)
3. Monthly overhead ($300)

Output data:

1. Yearly profit (YP)
2. Expected daily profit

$$\bar{P} = [P_1 + P_2 + \ldots + P_{365}]/365$$

3. Standard deviation of the daily profit

$$s = \sqrt{\frac{(P_1 - \bar{P})^2 + (P_2 - \bar{P})^2 + \ldots + (P_{365} - \bar{P})^2}{365}}$$

4. Grouped data for the P_i's (that is, the percentage of P_i's that fall within each of several adjacent intervals)
5. It may also be desirable to present the mean value, the standard deviation, and the grouped data for the n_i's, though this information will be less essential than the profit data.

For example, suppose that the following data were obtained for 1 simulated year of operation:

Day	Daily sales	Daily profit
1	183	$54.19
2	212	64.34
3	195	58.39
.	.	.
.	.	.
.	.	.
364	208	62.94
365	223	68.19

These values result in a total yearly profit of $22,706.65, which corresponds to an expected daily profit of $62.21 and a standard deviation of $6.00. In addition, the daily profit values can be grouped:

Interval ($)	Percentage of values falling within interval	Percentage of cumulative distribution
45–50	2.2	2.2
50–55	10.6	12.8
55–60	17.4	30.2
60–65	41.5	71.7
65–70	19.0	90.7
70–75	7.6	98.3
75–80	1.7	100.0
	100.0	

The grouped profit data (particularly the cumulative distribution) can be used to determine the likelihood of various levels of success or failure. This information indicates the degree of risk that is associated with the proposed venture. We will see how to interpret the grouped data in order to obtain this kind of information in later chapters of this book.

1.4 THE ROLE OF SIMULATION

Before we become involved with the details of building and solving simulation models, let us pause briefly and examine the pros and cons associated with the use of discrete-event simulation.

The principal advantage in using simulation is its flexibility. Practically any problem involving risk can be represented with a reasonable degree of accuracy by means of a simulation model. There are, of course, other ways to approach problems which involve risk. In particular, models based upon the use of linear programming or queuing theory are very well known, although these models are applicable only to certain narrow classes of problems whose characteristics are relatively uncommon in actual practice. Simulation, on the other hand, has no such restrictions. Thus, simulation can be used to analyze many problem situations that cannot be studied otherwise.

We have already discussed another advantage of the use of simulation, namely, the fact that a simulation study will provide full statistical information about the system performance criterion. This allows us to assess not only the expected system behavior but also the probability that the system behavior may be significantly different. In other words, simulation provides an indication of the risk associated with a particular operating policy as well as a measure of expected system performance.

Simulation also has its drawbacks. The most serious is the fact that large blocks of computer time may be required to carry out a meaningful simulation study. Since computer time can be expensive, particularly when working with a large scientific computer, the simulation of a realistic system can become a very costly undertaking. This is especially true of comprehensive simulation studies that involve a comparison of several different operating policies.

Another difficulty that arises with simulation studies is the fact that erroneous conclusions can be drawn about the behavior of a system, even though the model is accurate and is solved correctly. This is usually caused by a sample size that is too small (that is, too few simulated events) to provide accurate statistical information about the system. This problem can be overcome, however, using techniques that will be presented later in this book.

On balance, the advantages associated with the use of simulation outweigh the disadvantages. This is why simulation, together with linear programming and certain statistical techniques, forms the cornerstone of current management science practice. In the succeeding chapters of this book we will learn to use this flexible, effective tool for analyzing a variety of problems that typically arise in business and industrial situations.

PROBLEMS

For each of the problem situations described in Probs. 1.1 through 1.6, identify
(a) the system parameters
(b) the decision variables
(c) the state variables
(d) the cause-and-effect relationships (in general, qualitative terms)
(e) the system performance criterion

1.1. A company that has just developed a new product wishes to assess its profitability. The company estimates that there is a 35 percent chance of selling between 40,000 and 60,000 units a year, a 40 percent chance of selling between 60,000 and 80,000 units a year, and a 25 percent chance of selling between 80,000 and 100,000 units a year. Based upon current market conditions, it appears unlikely that the company will sell fewer than 40,000 or more than 100,000 units a year.

The cost of manufacturing and distributing this product also appears somewhat uncertain. There is a 20 percent chance that the cost will be between \$60 and \$70 per unit, a 35 percent chance that it will be between \$70 and \$80 per unit, a 30 percent chance that it will be between \$80 and \$90 per unit, and a 15 percent chance that it will be between \$90 and \$100 per unit. It is unlikely that the cost will be less than \$60 or greater than \$100 per unit.

The company wishes to determine the expected yearly profit corresponding to some specified selling price. (Most likely, the profitability associated with several different selling prices will eventually be investigated.) In addition, the company would like to estimate the chances that the yearly profit will be much higher or much lower than the expected value. This latter information will provide some indication of the degree of risk associated with the proposed venture.

1.2. Roulette is played with a wheel containing 38 different squares along its circumference. Two of these squares, numbered 0 and 00, are green; 18 squares are red, and 18 are black. The red and black squares alternate in color, and are randomly numbered 1 through 36. A small marble is spun within the wheel, which eventually comes to rest within one of the squares. The game is played by betting on the outcome of each spin, in any one of the following ways:

(a) By selecting a single red or black square, at 35-to-1 odds. Thus a player betting $1.00 and winning, would receive $36.00 (the original $1.00 plus an additional $35.00).

(b) By selecting a color (either red or black) at 1-to-1 odds. Thus a player choosing red on a $1.00 bet would receive $2.00 if the marble came to rest in any red square.

(c) By selecting either the odd or the even numbers (excluding 0 and 00) at 1-to-1 odds.

(d) By selecting either the low 18 or the high 18 numbers at 1-to-1 odds. The player will automatically lose if the marble comes to rest in one of the green squares (0 or 00).

Computer simulation will be used to determine some typical player histories. These results will be used to estimate the likelihood of winning at roulette.

1.3. A printing company currently owns three presses, which are operated on a full-time basis. Overflow work is subcontracted to another printer, on a break-even basis, in order to maintain customer goodwill. The company is now considering the purchase of a fourth press, at a cost of $150,000. Management would like some estimate of the time required to recover this initial investment (neglecting the time-value of money). A simulation model has been proposed for this purpose.

The demand for the fourth press (that is, the time between arrivals of successive overflow orders) is assumed to be exponentially distributed, with a mean of 1.5 days. The service times (that is, the times required to process the orders) are believed to be normally distributed, with a mean of 1 day and a standard deviation of 0.4 day. In addition, there is a constant setup time of 0.1 day required for each order. A net profit of $50 per hour ($400 per day) can be realized when the press is in operation.

For simplicity, assume that any new order that comes in while *all* presses are in operation (including the new press) is still subcontracted to another printer. Thus there will never be a backlog of orders waiting to be processed.

1.4. A company is considering the desirability of the following proposed venture:

Initial investment: $15,000,000
Sales price: distributed as

Price per unit ($)	Relative frequency
20–25	0.08
25–30	0.17
30–35	0.22
35–40	0.27
40–45	0.13
45–50	0.09
50–55	0.04

Sales volume: normally distributed, with a mean of 200,000 units per year and a standard deviation of 50,000 units per year.

Annual costs: fixed cost of $100,000 per year, plus a unit cost which is distributed:

Cost per unit ($)	Relative frequency
10–12	0.12
12–14	0.22
14–16	0.36
16–18	0.20
18–20	0.10

Tax rate: 48%
Depreciation: straight line, over a 10-year period.
Interest rate: 12% per year, compounded annually.

Computer simulation will be used to determine the present worth of the proposed venture.

1.5. The manager of a supermarket wishes to study the operation of a checkout counter. The arrival times of customers to the checkout counter are known to be exponentially distributed, with a mean interarrival time of 2.3 minutes. Only one customer can go through the checkout counter at any given time. Thus, if the checkout counter is busy when a new customer arrives, that customer must wait in line until the checkout counter becomes free. If several customers are waiting, each customer will enter the checkout counter on a first-come, first-served basis. The checkout times are known to be normally distributed, with a mean of 1.8 minutes and a standard deviation of 0.5 minutes.

The checkout operation is to be simulated in order to determine some representative customer waiting times and representative waiting-line lengths. Also of interest is the fraction of time that the checkout counter is not in use.

1.6. An airline check-in counter in a major airport is open from 6:00 A.M. to 11:00 P.M. every weekday. At least one check-in window will always be

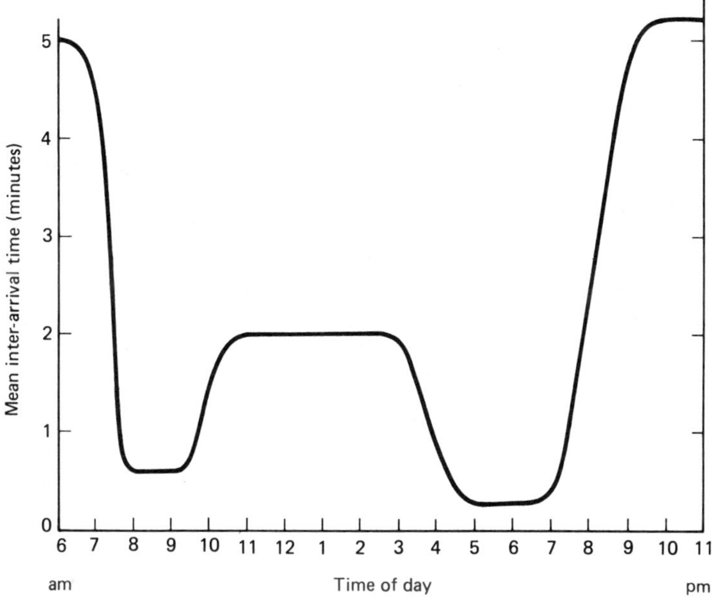

Figure 1.4

open during these hours. Additional check-in windows will be opened as the need arises, during busy periods of the day. As many as six windows can be open at any time. Each window will have its own waiting line.

Suppose that the passenger interarrival times are exponentially distributed, with mean values that fluctuate during the day as indicated in Fig. 1.4. Also, suppose the check-in times (service times) are distributed in accordance with the following empirical distribution.

Time interval (min.)	Relative frequency
0– 0.5	0.159
0.5– 1.0	0.238
1.0– 1.5	0.212
1.5– 2.0	0.132
2.0– 2.5	0.095
2.5– 3.0	0.063
3.0– 4.0	0.042
4.0– 5.0	0.032
5.0– 7.0	0.016
7.0–10.0	0.011
	1.000

We wish to simulate the behavior of the system for various operating policies and recommend a satisfactory operating policy based upon the simulated results. (An *operating policy* in this situation will be a schedule indicating how many windows will be open at different times during the day).

1.7. Refer to several other simulation textbooks and obtain a definition of simulation from each textbook. Compare these definitions with each other and with the definition given in this text. Are these definitions compatible with one another? Discuss thoroughly.

1.8. Consider the problem situation described in Prob. 1.1. Describe the manner in which the yearly profit will be presented, and explain the reasons for expressing the profit in this manner. Why is it inadequate to simply present a single value (that is, the expected value) for the yearly profit?

2

UNIFORMLY DISTRIBUTED RANDOM VARIATES

The key to simulating discrete, random events is the ability to generate random numbers on a computer. A great many random numbers will be required for a typical simulation study. It is therefore essential that they be generated as quickly and efficiently as possible.

This chapter presents methods for generating such random numbers. We will see that these numbers are not really random, since they are generated in reproducible sequences by deterministic techniques. The numbers *appear* random, however, and are able to satisfy a variety of statistical tests for randomness. For all practical purposes, then, we can assume that these numbers are truly random. Such numbers are referred to as *pseudorandom numbers*.

In this chapter we will consider only pseudorandom numbers that are *uniformly distributed within the unit interval (0, 1)*, that is, numbers that fall within the interval (0, 1) with equal likelihood. Such numbers can be used directly to simulate random events, as shown in Chap. 3. (When used to represent random events on a one-to-one basis, the random numbers are known as *random variates*.) Moreover, in Chap. 4 we will see that these uniformly distributed random numbers are used to generate other random numbers which are governed by different (nonuniform) types of distributions. Thus, uniformly distributed random numbers provide a basis for generating the random variates required in a wide variety of realistic simulation problems.

2.1 DESIRABLE PROPERTIES OF RANDOM NUMBER GENERATORS

Ideally, a pseudorandom number generator should possess all of the following desirable characteristics.

1. *Randomness.* First and foremost, it is essential that the generated sequence of pseudorandom numbers exhibits the same properties as truly random numbers. We have already mentioned the fact that such random behavior is determined by a number of different statistical tests. Some of these tests will be described elsewhere in this chapter (see Sec. 2.7).

2. *Large period.* Since all pseudorandom number generators are based upon the use of precise, deterministic formulas, every pseudorandom number sequence will eventually begin to repeat itself. The size of the nonrepeating sequence is called the *period*. We would like the period to be as large as possible. From a practical viewpoint, the period should at least be sufficiently large so that the random numbers do not repeat themselves during any single simulation.

3. *Reproducibility.* When debugging a simulation program or carrying out a parametric study (that is, varying the input data), it may be desirable to generate the exact same sequence of random numbers during each simulation. There are other situations, however, in which different sequences of random numbers are required in a given simulation study. Therefore, the random number generator should be capable of providing both repeated and distinct random number sequences, in accordance with the wishes of the analyst.

4. *Computational Efficiency.* Since a typical simulation study will require that a great many random numbers be generated, the random number generator should provide these numbers using as little computer time as possible. Moreover, the random number generator should not require extensive computer memory.

In actual practice, the realization of all four of these properties is quite difficult to achieve.

2.2 EARLY RANDOM NUMBER GENERATORS

The earliest attempts to generate random numbers on a computer were based upon the use of random number tables. Typically, a table of random numbers would be stored on a memory device, such as a magnetic tape. The memory device would then be read, in a random fashion, whenever a random number was required.

Unfortunately, such table look-up methods were found to be unsatisfactory, for several reasons. To begin with, it is much more difficult than one might imagine to obtain a table of numbers exhibiting truly random behavior, although a few such tables do exist. (Perhaps the best known is a book published by the RAND Corporation in 1955, entitled "One Million Random Digits.") Second, these tables are finite and comparatively limited in size. Table look-up methods are also very slow and hence computationally inefficient. Finally, most of the methods used to sample such tables are themselves more or less random, resulting in random number sequences that are not reproducible. Because of these inadequacies table look-up methods were eventually abandoned, and attention was given to the development of techniques for calculating random numbers directly as they are needed.

The first method for actually *generating* pseudorandom numbers is the *center-square* method, which was proposed by John Von Neumann in 1946. In this method a new random number is obtained by squaring the old random number and retaining the middle digits. Thus,

$$n_{i+1} = \left\{ \text{middle digits of } n_i^2 \right\} \tag{2.1}$$

where n_i and n_{i+1} represent the old and the new random numbers, respectively. If n_i consists of d digits, then n_i^2 will contain either $2d$ or $(2d\text{-}1)$ digits. The centermost d digits are then selected for n_{i+1}.

Example 2.1

Use the center-square method to calculate several consecutive random quantities, beginning with $n_1 = 25073$.

The calculations are summarized below:

$$n_1 = 25073$$
$$n_1^2 = 628\underbrace{65532}9$$
$$n_2$$

$$n_2 = 86553$$
$$n_2^2 = 74\underbrace{91421}809$$
$$n_3$$

$$n_3 = 14218$$
$$n_3^2 = 20\underbrace{21515}24$$
$$n_4$$

$$n_4 = 21515$$

$$n_4^2 = 462\underbrace{8952}_{n_5}25$$

$$n_5 = 28952$$

and so on.

The center-square method is simple but ineffective, since the generated sequence of numbers may not have the desired random behavior. In fact, the sequence will frequently degenerate into a distinct, repeated pattern, or a sequence of zeros. Moreover, even if such degeneration does not occur, the sequence will eventually begin to repeat itself. The period is unpredictable, as it is highly dependent upon the initial value (that is, the value chosen for n_1). Many initial values result in periods that are unacceptably small.

A variation of the center-square method which is somewhat more satisfactory is based upon forming the product of the *previous two* random numbers. The next number is then obtained by extracting the middle digits of this product. Thus,

$$n_{i+1} = \left\{ \text{middle digits of } n_i \times n_{i-1} \right\} \tag{2.2}$$

Example 2.2

Use the above variation of the center-square method to calculate several consecutive random quantities, beginning with $n_1 = 60455$ and $n_2 = 71287$.

The calculations are summarized below:

$$n_1 = 60455$$

$$n_2 = 71287$$

$$n_2 \times n_1 = 43\underbrace{09655}_{n_3}585$$

$$n_3 = 96555$$

$$n_3 \times n_2 = 68\underbrace{83116}_{n_4}285$$

$$n_4 = 31162$$

$$n_4 \times n_3 = 300\underbrace{88469}_{n_5}10$$

$$n_5 = 88469$$

and so on.

This method is susceptible to the same difficulties experienced with the original center-square method, even though the number sequences tend to be more random and have larger periods. In practice neither method is used, since better methods are now available.

2.3 METHODS BASED UPON CONGRUENT NUMBERS

Most modern random number generators are based upon the use of *congruent numbers*. Suppose we are given some positive integer, m, which is called a *modulus*. We say that two numbers, a and b, are *congruent modulo m* if $(a - b)$ is some integral multiple of m, that is, if

$$(a - b) = km \tag{2.3}$$

where k is an integer. This relationship is usually written as

$$a \equiv b \,(\text{mod } m) \tag{2.4}$$

where the symbol "$\equiv$" denotes congruence.

Example 2.3

Suppose that we are given a value of $m = 8$. The following pairs of numbers are congruent modulo 8:

$$
\begin{array}{ll}
(8, 0) & [\text{because } (8 - 0) \quad\; = 1 \times 8] \\
(11, 3) & [\text{because } (11 - 3) \quad = 1 \times 8] \\
(6, -2) & [\text{because } [6 - (-2)] \;\; = 1 \times 8] \\
(35, 3) & [\text{because } (35 - 3) \quad = 4 \times 8] \\
(4, -12) & [\text{because } [4 - (-12)] = 2 \times 8] \\
(100, 140) & [\text{because } (100 - 140) = -5 \times 8]
\end{array}
$$

These congruence relationships can be expressed as

$$
\begin{aligned}
8 &\equiv 0 \,(\text{mod } 8) \\
11 &\equiv 3 \,(\text{mod } 8) \\
6 &\equiv -2 \,(\text{mod } 8) \\
35 &\equiv 3 \,(\text{mod } 8) \\
4 &\equiv -12 \,(\text{mod } 8) \\
100 &\equiv 140 \,(\text{mod } 8)
\end{aligned}
$$

Suppose that

$$a \equiv b \,(\text{mod } m) \tag{2.5}$$

and

$$b \equiv c \,(\text{mod } m) \tag{2.6}$$

Then it is easy to establish that

$$a \equiv c \pmod{m} \tag{2.7}$$

Residues

For a given integer a, the smallest nonnegative integer, r, that is congruent to a (mod m) is called a *residue* modulo m. Thus,

$$r \equiv a \pmod{m}, \qquad 0 \le r < m \tag{2.8}$$

Each value of a will have its own corresponding residue. There are, however, only m distinct residues, whose values range from 0 to $(m-1)$.

Example 2.4

Suppose that $m = 8$. Determine the residue corresponding to a value of $a = 13$.

The desired residue is $r = 5$, since $5 < 8$ and $5 \equiv 13 \pmod 8$.

Since $m = 8$, there are a total of 8 distinct residues. They are $r = 0, 1, 2, 3, 4, 5, 6,$ and 7.

Power Residues

A *power residue* is the smallest nonnegative integer, p_i, that is congruent to a^i (mod m), where a is a given integer and i is a positive integer exponent. Thus,

$$p_i = a^i \pmod{m} \tag{2.9}$$

There will be many different power residues corresponding to a given set of values for a and m, since different exponents (that is, different values for i) will result in different power residues. Eventually, however, the sequence of power residues will begin to repeat itself.

Example 2.5

Determine the power residues corresponding to $m = 32$ and $a = 13$.

$p_1 = 13$, since $13 < 32$ and $13 \equiv 13 \pmod{32}$

$p_2 = 9$, since $9 < 32$ and $9 \equiv 13^2 \pmod{32}$
     [Note that $(13^2 - 9) = 160 = 5 \times 32$]

$p_3 = 21$, since $21 < 32$ and $21 \equiv 13^3 \pmod{32}$
     [Note that $(13^3 - 21) = 2176 = 68 \times 32$]

$p_4 = 17$, since $17 < 32$ and $17 \equiv 13^4 \pmod{32}$
     [Note that $(13^4 - 17) = 28{,}544 = 892 \times 32$]

$p_5 = 29$, since $29 < 32$ and $29 \equiv 13^5 \pmod{32}$
     [Note that $(13^5 - 29) = 371{,}264 = 11{,}602 \times 32$]

$p_6 = 25$, since $25 < 32$ and $25 \equiv 13^6 \pmod{32}$
      [Note that $(13^6 - 25) = 4{,}826{,}784 = 150{,}837 \times 32$]

$p_7 = 5$, since $5 < 32$ and $5 \equiv 13^7 \pmod{32}$
      [Note that $(13^7 - 5) = 62{,}748{,}512 = 1{,}960{,}891 \times 32$]

$p_8 = 1$, since $1 < 32$ and $1 \equiv 13^8 \pmod{32}$
      [Note that $(13^8 - 1) = 815{,}730{,}720 = 25{,}491{,}585 \times 32$]

Beyond this point the power residues will begin to cycle, that is,

$$
\begin{array}{ll}
p_9 = 13 \qquad & p_{17} = 13 \\
p_{10} = 9 \qquad & p_{18} = 9 \\
p_{11} = 21 \qquad & p_{19} = 21 \\
p_{12} = 17 \qquad & p_{20} = 17 \\
p_{13} = 29 \qquad & p_{21} = 29 \\
p_{14} = 25 \qquad & p_{22} = 25 \\
p_{15} = 5 \qquad & p_{23} = 5 \\
p_{16} = 1 \qquad & p_{24} = 1 \\
\end{array}
$$

and so on.

The Power Residue Method

We now consider a simple, popular random number generator that is based upon the use of power residues. Hence, the method is known as the *power residue method* (it is also called the *multiplicative congruential method*). The method makes use of the following recursive congruential relationship:

$$n_i \equiv a n_{i-1} \pmod{m} \tag{2.10}$$

where n_i and n_{i-1} are successive random integers, and a (the multiplier) and m (the modulus) are specified.

If we begin with some known integer constant, n_0, then the repeated use of Eq. (2.10) results in

$$
\begin{aligned}
n_1 &\equiv a n_0 \pmod{m} \\
n_2 &\equiv a^2 n_0 \pmod{m} \\
&\cdot \\
&\cdot \\
&\cdot \\
n_i &\equiv a^i n_0 \pmod{m}
\end{aligned}
\tag{2.11}
$$

Thus, the successive random integers are related to the power residues of a.

In order that the calculated n_i's exhibit acceptable random behavior, it is essential that the values for m, n_0, and a be chosen in accordance with a carefully developed set of rules. One such set of rules, which is quite commonly used, is:

1. The modulus, m, should be chosen as large as possible in order to maximize the period of the random number sequence. When the method is implemented on a computer having w bits per word, the modulus is selected as

$$m = 2^{w-1} \tag{2.12}$$

2. The multiplier, a, must be chosen in such a manner that the correlation between successive n_i's is minimized, while at the same time obtaining the largest possible period. This can be accomplished provided the multiplier satisfies the following two conditions:

$$a \cong 2^{w/2} \tag{2.13}$$

$$a \equiv \pm\, 3 \;(\text{mod } 8) \tag{2.14}$$

3. The initial value, n_0 (which is often referred to as the *seed*), can be any positive, odd integer whose value is less than m. Note that different seeds can be used to generate different sequences of random numbers, even though the modulus and the multiplier remain the same.

When the values for m, a, and n_0 are chosen in this manner, each resulting sequence of random numbers will have a period equal to $m/4$.

The values of the n_i's obtained in this manner will range from 1 to $(m - 1)$. In order to obtain the desired uniformly distributed random variates, $u_1, u_2, u_3, \ldots$, where $0 < u_i < 1$, we simply divide each of the n_i's by m. Thus,

$$u_i = n_i/m \tag{2.15}$$

Example 2.6

The power residue method is to be implemented on a computer having a 12-bit word. Determine an appropriate set of values for m, a, and n_0, and evaluate the first few u_i's.

For a 12-bit computer, $w = 12$. Hence from Eq. (2.12), $m = 2^{11} = 2048$. Moreover Eqs. (2.13) and (2.14) are both satisfied by a value of $a = 67$, since 67 is approximately equal to 2^6, and $67 \equiv 3 \;(\text{mod } 8)$. Finally, we arbitrarily select the value $n_0 = 129$, since n_0 can be assigned any positive, odd integer value. To summarize, then,

$$m \; = 2048$$
$$a \; = \quad 67$$
$$n_0 \; = \quad 129$$

If we apply Eqs. (2.10) and (2.15) recursively, we obtain

$$n_1 \equiv 67 \times 129 \;(\text{mod } 2048) = 451$$
$$u_1 = 451/2048 = 0.220215$$
$$n_2 \equiv 67 \times 451 \;(\text{mod } 2048) = 1545$$
$$u_2 = 1545/2048 = 0.754395$$
$$n_3 \equiv 67 \times 1545 \;(\text{mod } 2048) = 1115$$
$$u_3 = 1115/2048 = 0.544434$$
$$n_4 \equiv 67 \times 1115 \;(\text{mod } 2048) = 977$$
$$u_4 = 977/2048 = 0.477051$$

.

.

.

etc.

There will be a total of $2048/4 = 512$ distinct n_i's. The entire sequence will begin to repeat itself once $i = 513$ (thus $n_{513} = n_1 = 451$, $n_{514} = n_2 = 1545$, $n_{515} = n_3 = 1115$, and so on).

The power residue method, with the parameters chosen in the manner indicated above, is very widely used, both for instructional purposes and for solving actual simulation problems in business and industry. There are two reasons for the method's popularity. First, the method is able to satisfy most statistical tests for randomness (see Sec. 2.6); and second, it can very easily be implemented in a high-level programming language such as FORTRAN (see Sec. 2.4).

Close scrutiny reveals, however, that the power residue method produces triples of numbers (that is, groups of three numbers) that are in some sense correlated (that is, related to one another) when the parameters are chosen in accordance with the above rules. Fishman (1973) discusses this problem in considerable detail. In a practical sense, it is difficult to assess the importance of this shortcoming.

The above problem can be avoided if some other set of rules is used to evaluate m, a, and n_0, though the power residue method is more difficult to implement under such conditions. Some alternative sets of rules are presented by Fishman (1973), Knuth (1981), and by Naylor, Balintfy, Burdick, and Chu (1966). These rules are utilized in certain sophisticated applications, particularly the special-purpose simulation languages discussed in Chap. 8. Nevertheless the present version of the power residue method appears to be adequate for most classroom situations.

The Mixed Congruential Method

The mixed congruential method is based upon the use of the following congruential relationship:

$$n_i \equiv a\, n_{i-1} + b \;\;(\mathrm{mod}\; m) \tag{2.16}$$

where n_i and n_{i-1} are successive random integers, and the parameters a, b, and m are specified integer constants.

Suppose we start with some known initial constant, n_0. The repeated use of Eq. (2.16) then results in

$$
\begin{aligned}
n_1 &\equiv a n_0 + b & (\mathrm{mod}\; m) \\[2mm]
n_2 &\equiv a^2 n_0 + \frac{b\,(a^2 - 1)}{(a - 1)} & (\mathrm{mod}\; m) \\[2mm]
&\;\;\cdot \\
&\;\;\cdot \\
&\;\;\cdot \\[2mm]
n_i &\equiv a^i n_0 + \frac{b\,(a^i - 1)}{(a - 1)} & (\mathrm{mod}\; m)
\end{aligned}
\tag{2.17}
$$

If we select values for the n_i's that are residues of the right-hand side, then each of the n_i's will be less than m, that is, $0 \leq n_i < m$. We can then obtain the desired uniformly distributed random variates, $0 \leq u_i < 1$, simply as

$$u_i = n_i/m \tag{2.18}$$

as before.

There is still the matter of selecting appropriate values for m, a, b, and n_0. The following set of rules is frequently used when the method is implemented on a computer:

1. Choose the modulus, m, as

$$m = 2^{w-1} \tag{2.19}$$

    where w represents the number of bits per word.

2. The multiplier, a, should satisfy the conditions

$$a \cong 2^{w/2} \tag{2.20}$$

$$a \equiv 1 \;(\mathrm{mod}\; 4) \tag{2.21}$$

3. The additive constant, b, and the initial value, n_0, can be any positive, odd integers whose values are less than m.

The reasons for these particular choices are similar to those that apply to the power residue method. However, the mixed congruential method will result in sequences of random numbers whose period is equal to m (in contrast to the power residue method, whose period equals $m/4$ when the constants are selected as described previously.) The mixed congruential method is therefore known as a *full-period* method.

Example 2.7

The mixed congruential method is to be implemented on a computer having a 12-bit word. Determine an appropriate set of values for m, a, b, and n_0, and evaluate the first few u_i's.

Let us arbitrarily select $b = 1$ and $n_0 = 129$ (note that both are positive, odd integers). The modulus will be selected in accordance with Eq. (2.19) as $m = 2^{11} = 2048$, since $w = 12$. Finally, Eqs. (2.20) and (2.21) are both satisfied by a value of $a = 65$, since 65 is approximately equal to 2^6, and $65 \equiv 1 \pmod 4$. Summarizing we obtain

$$m = 2048$$
$$a = 65$$
$$b = 1$$
$$n_0 = 129$$

If we now apply Eqs. (2.17) and (2.18) recursively, we obtain

$$n_1 \equiv 65 \times 129 + 1 \pmod{2048} \qquad = 194$$
$$u_1 = 194/2048 = 0.094727$$
$$n_2 \equiv 65 \times 194 + 1 \pmod{2048} \qquad = 323$$
$$u_2 = 323/2048 = 0.157715$$
$$n_3 \equiv 65 \times 323 + 1 \pmod{2048} \qquad = 516$$
$$u_3 = 516/2048 = 0.251953$$
$$n_4 \equiv 65 \times 516 + 1 \pmod{2048} \qquad = 773$$
$$u_4 = 773/2048 = 0.377441$$

.

.

.

.

etc.

The first 2048 n_i's will be distinct, since this method has a period equal to m. Once $i = 2049$, however, the sequence will begin to repeat itself, that is, $n_{2049} = n_1 = 194$, $n_{2050} = n_2 = 323$, etc.

The mixed congruential method has been found to generate number sequences that are statistically less random than those produced by the power residue method. Moreover, the method is slower than the power residue method. For these reasons the mixed congruential method is used less often than the power residue method, despite the fact that its period may be larger.

2.4 A FORTRAN-BASED RANDOM NUMBER GENERATOR

We have already learned that the power residue (multiplicative congruential) method can easily be implemented in a high-level programming language, such as FORTRAN. This is one of the factors that contributes to the method's popularity.

Presented below is a commonly used FORTRAN subprogram for generating uniformly distributed random variates, $0 < u_i < 1$, which is based upon the use of the power residue method.

```
      FUNCTION RAND(KX)
C*********************************************************************************
C                                                                               *
C THIS FUNCTION GENERATES A UNIFORMLY DISTRIBUTED RANDOM NUMBER                  *
C BETWEEN ZERO AND ONE, USING THE POWER RESIDUE METHOD.                         *
C                                                                               *
C KX SHOULD BE ASSIGNED A POSITIVE, ODD, INTEGER VALUE, NOT EXCEEDING           *
C 2147483647, THE FIRST TIME THE FUNCTION IS CALLED.                            *
C THEREAFTER, KX SHOULD BE ASSIGNED A VALUE OF ZERO.                            *
C                                                                               *
C NOTE: THIS FUNCTION IS VALID ONLY FOR A COMPUTER HAVING A 32-BIT WORD.        *
C                                                                               *
C*********************************************************************************
      IF (KX.GT.0) IX=KX
      IY=65539*IX
      IF (IY.LT.0) IY=IY+2147483647+1
      RAND=FLOAT(IY)/2147483647
      IX=IY
      RETURN
      END
```

Figure 2.1

The constant 65539 represents the multiplier (that is, $a = 65539$), and the quantity 2147483647 is one less than the modulus (that is, $m - 1 = 2147483647$).

The crux of the routine is the statement

$$IY = 65539*IX$$

This statement generates a deliberate overflow and then retains the low-order bits, thus causing the value of IY to be congruent to IX (modulo 2147483648). The next statement assures that IY have a positive value, since a negative value could have been obtained as a result of the overflow. Finally, the statement

RAND = FLOAT(IY)/2147483647

generates the desired random number — a decimal quantity that is uniformly distributed within the interval (0, 1). A different random number will be generated each time the function is accessed.

In order to see this routine, a positive, odd integer (that is, a seed) must first be provided for KX. This initializes the random number generator. (Typically, this value for KX will be an input quantity.) Whenever the routine is accessed thereafter, a value of zero should be provided for KX. The routine will then return a uniformly distributed random variate. The procedure is illustrated in the example below.

Example 2.8
Shown below are portions of a FORTRAN program that utilizes the above random number generator. The program first initializes the random number generator. It then generates and prints out 100 uniformly distributed (0, 1) random variates.

```
100   FORMAT (I9)
200   FORMAT (F8.6)
         .
         .
         .
      READ (5,100) KX
         .
         .
         .
      DUMMY = RAND (KX)
         .
         .
         .
      DO 50 I = 1,100
      X = RAND (0)
50    WRITE (6,200) X
```

Note that the variable DUMMY is not used elsewhere in the program. Its only purpose is to initialize the random number generator. The actual random numbers are generated within the DO loop, by accessing the function RAND.

A positive, odd integer (a seed) must be read into the computer when the program is executed.

The reader should understand that *the version of RAND which is presented above is valid only for a computer having a 32-bit word.* In order to use this routine with a computer having a different word size, the constants 65539 and 2147483647 must be replaced by $2^{w/2} + 3$ and $2^{w-1} - 1$, respectively, where w is the number of bits per word. Table 2.1 provides the proper values for several common word sizes.

TABLE 2.1 Values for Constants in Random Number Generator

Word size (w)	Multiplier (a)	Modulus, less one $(m - 1)$
8 bits	19	127
12	67	2047
16	259	32767
32	65539	2147483647
36	262147	34359738367
64	4294967299	9223372036854775807
w	$(2^{w/2} + 3)$	$(2^{w-1} - 1)$

There are some versions of FORTRAN that include a random number generator as a standard library function. In such cases an external routine, such as RAND, need not be supplied. Specific information on the availability and use of various library functions is usually available at each particular computer installation.

2.5 GENERATING RANDOM NUMBERS IN BASIC

BASIC is another general-purpose programming language that is commonly available and frequently used for many business and technical applications. It is similar to FORTRAN, though easier to use. A random number generator is included in the language as a standard library function. Use of the random number generator is illustrated in the example below.

Example 2.9

Shown below are portions of a BASIC program that first initializes the random number generator, and then generates and prints out 100 uniformly distributed (0, 1) random variates.

```
10  RANDOMIZE
    .
    .
    .
50  FOR I = 1 TO 100
60  LET X = RND
70  PRINT X
80  NEXT I
```

The RANDOMIZE statement initializes the random number generator. (Note that the user need not provide a specific value for the seed.) The actual random numbers are generated within the FØR - TØ loop (statements 50 through 80) by accessing the library function RND. Notice that an argument is not required when accessing this function.

2.6 OTHER UNIFORMLY DISTRIBUTED RANDOM VARIATES

Once a routine is available for generating uniformly distributed random numbers within the interval (0, 1), it is very easy to generate other types of uniformly distributed random variates.

Continuous Random Variates

Suppose that X is a continuous random variate, uniformly distributed within the interval (a,b), where $a < b$. Let U represent a continuous random variate that is uniformly distributed over the interval (0, 1). From simple proportionality, we can write

$$\left(\frac{X - a}{b - a}\right) = \left(\frac{U - 0}{1 - 0}\right)$$

or

$$X = a + (b - a)\, U \tag{2.22}$$

Thus it is very simple to generate X from a given U provided a and b are known.

Example 2.10

A FORTRAN statement that will generate a continuous, uniformly distributed random variate within the interval (1, 4) is presented below.

    X = 1. + 3. * RAND (0)

This statement makes use of the FORTRAN random number generator RAND, presented in Sec. 2.4

Discrete Random Variates

Now suppose that a and b are integer quantities, $a < b$, and X is a discrete, integer-valued random variate that is uniformly distributed within the interval (a, b). Thus X can take on the values a, $a + 1$, $a + 2, \ldots, b - 1, b$. If U is continuous and uniformly distributed within the interval (0, 1), as before, then

$$X = a + \text{INT}\left\{(b - a + 1)\, U\right\} \tag{2.23}$$

where INT denotes truncation (that is, dropping the decimals and thus retaining only the integer portion of the given quantity).

In order to understand the basis for Eq. (2.23), note that, if $0 \leq U < 1$, then $0 \leq (b - a + 1)\, U < (b - a + 1)$. Therefore the quantity

$$\text{INT}\left\{(b - a + 1)\, U\right\}$$

will take on the integer values $0, 1, 2, \ldots, (b - a)$, and hence X will assume the values $a, a + 1, a + 2, \ldots, b$, with equal likelihood.

Example 2.11

Presented below are three FORTRAN statements that simulate one throw of a pair of dice.

```
INTEGER X, X1, X2
    .
    .
    .
    X1 = 1 + INT (6.*RAND (0))
    X2 = 1 + INT (6.*RAND (0))
    X  = X1 + X2
```

Notice that we have utilized the random number generator RAND, presented in Sec. 2.4, in order to generate two continuous, uniformly distributed random numbers within the interval $(0, 1)$. Also, note that INT is a standard FORTRAN library function for carrying out truncation.

It is interesting to note that the last three statements can easily be combined. Thus,

```
INTEGER X
    .
    .
    .
    X = 2 + INT (6.*RAND (0)) + INT (6.*RAND (0))
```

Several applications of uniformly distributed random variates will be presented in the next chapter.

2.7 STATISTICAL TESTS FOR RANDOMNESS

Numerous statistical tests have been devised in order to assess the randomness of pseudorandom number sequences. A few of the more commonly used tests are presented in this section.

The Chi-Square Statistic

The chi-square (χ^2) statistic is used to determine how well a set of observations can be represented by a given distribution, provided each observation falls into one of k different categories. If the number of observed events (O_i) and the expected number of events (E_i) are known for each category, then the χ^2 statistic can be determined as

$$\chi^2 = \frac{(O_1 - E_1)^2}{E_1} + \frac{(O_2 - E_2)^2}{E_2} + \ldots + \frac{(O_k - E_k)^2}{E_k}$$

$$= \sum_{i=1}^{k} \frac{(O_i - E_i)^2}{E_i}$$

(2.24)

This statistic is used in conjunction with a χ^2 table, as shown in Table 2.2. The top row of the table indicates the *rejection probability* (p), which is the probability, in percent, of incorrectly rejecting the assumed distribution. (Some χ^2 tables give the *level of significance*, 100-p, rather than the rejection probability.) The left-hand column represents the number of degrees of freedom (v), which is one less than the number of categories (that is, $v = k - 1$).

When carrying out a statistical test, we are essentially testing the hypothesis that the observed results can be represented by the given distribution. The chance that this hypothesis is incorrect (that is, that the distribution is inappropriate) increases as the calculated value for χ^2 increases. Hence, the likelihood of *incorrectly rejecting* the assumed distribution *decreases*.

In practice, the hypothesis is rejected if the calculated χ^2 value exceeds the tabulated value for some reasonably small rejection probability (say $p = 5\%$ or $p = 1\%$), since it would be highly unlikely that the observed results would differ so greatly from the expected results if the hypothesis were valid. Furthermore, the hypothesis is usually rejected if the calculated χ^2 value is smaller than the tabulated value for some fairly large rejection probability (for example, $p = 95\%$ or $p = 99\%$); in this case, it would be highly unlikely that the observed results would fit the given distribution so perfectly. Hence, the hypothesis is accepted if the calculated χ^2 value falls within the interval that is formed by the tabulated values corresponding to the two extreme rejection probabilities.

Example 2.12

A set of observations has been obtained for a statistical experiment involving six categories. The number of observations falling into each category are summarized below, along with the expected number of observations for each category (based upon an assumed distribution).

Category	Number of observations	Expected number of observations
1	3	6
2	7	8
3	10	10
4	13	10
5	10	8
6	5	6

TABLE 2.2 Selected Values of the Chi-Square Distribution

Degrees of freedom	$p = 99\%$	$p = 95\%$	$p = 75\%$	$p = 50\%$	$p = 25\%$	$p = 5\%$	$p = 1\%$
$\nu = 1$	0.00016	0.00393	0.1015	0.4549	1.323	3.841	6.635
$\nu = 2$	0.00201	0.1026	0.5753	1.386	2.773	5.991	9.210
$\nu = 3$	0.1148	0.3518	1.213	2.366	4.108	7.815	11.34
$\nu = 4$	0.2971	0.7107	1.923	3.357	5.385	9.488	13.28
$\nu = 5$	0.5543	1.1455	2.675	4.351	6.626	11.07	15.09
$\nu = 6$	0.8720	1.635	3.455	5.348	7.841	12.59	16.81
$\nu = 7$	1.239	2.167	4.255	6.346	9.037	14.07	18.48
$\nu = 8$	1.646	2.733	5.071	7.344	10.22	15.51	20.09
$\nu = 9$	2.088	3.325	5.899	8.343	11.39	16.92	21.67
$\nu = 10$	2.558	3.940	6.737	9.342	12.55	18.31	23.21
$\nu = 11$	3.053	4.575	7.584	10.34	13.70	19.68	24.73
$\nu = 12$	3.571	5.226	8.438	11.34	14.84	21.03	26.22
$\nu = 15$	5.229	7.261	11.04	14.34	18.25	25.00	30.58
$\nu = 20$	8.260	10.85	15.45	19.34	23.83	31.41	37.57
$\nu = 30$	14.95	18.49	24.48	29.34	34.80	43.77	50.89
$\nu = 50$	29.71	34.76	42.94	49.33	56.33	67.50	76.15

Calculate a χ^2 statistic for this experiment and determine whether to accept or reject the hypothesis that the assumed distribution can be used to represent the data, based upon a 5% rejection probability.

The χ^2 statistic is easily determined as

$$\chi^2 = \frac{(3-6)^2}{6} + \frac{(7-8)^2}{8} + \frac{(10-10)^2}{10} + \frac{(13-10)^2}{10} + \frac{(10-8)^2}{8} + \frac{(5-6)^2}{6} = 3.19$$

Since there are six categories, $k = 6$ and hence $\nu = 5$. The tabulated χ^2 value corresponding to $\nu = 5$ and $p = 5\%$ is 11.07. Since the calculated value does not exceed this quantity, we accept the hypothesis. Moreover, the tabulated χ^2 value corresponding to $\nu = 5$ and $p = 95\%$ is 1.1455. Since the calculated value is greater than this quantity, we have an additional justification for accepting the hypothesis.

It should be noted that the χ^2 distribution of the statistic given by Eq. (2.24) is only approximate. The accuracy of the approximation increases as E_i, the expected number of events per category, increases. Normally, it is recommended that E_i exceed 5 when using Eq. (2.24).

The Frequency Test

The frequency test is perhaps the simplest and most common of all statistical tests used with random numbers, though the test actually measures uniformity rather than randomness. In this test the interval (0, 1) is subdivided into k subintervals of equal width. The number of random variates falling into each subinterval is then determined and a χ^2 statistic is calculated, with $\nu = k - 1$. The success or failure of the test is determined by the value obtained for the χ^2 statistic.

Example 2.13

Consider the following 50 uniformly distributed pseudorandom numbers.

0.923	0.060	0.497	0.710	0.807
0.406	0.852	0.632	0.953	0.903
0.562	0.526	0.358	0.756	0.195
0.478	0.262	0.318	0.720	0.837
0.017	0.513	0.199	0.083	0.335
0.233	0.035	0.116	0.848	0.432
0.133	0.451	0.081	0.467	0.249
0.299	0.575	0.600	0.695	0.380
0.979	0.849	0.999	0.892	0.580
0.544	0.378	0.872	0.427	0.548

We now determine how many of these quantities fall into each of the subintervals 0–0.099, 0.1–0.199, . . . , 0.9–0.999. The results are summarized below.

$$
\begin{array}{ll}
0\text{-}0.099: & 5 \\
0.1\text{-}0.199: & 4 \\
0.2\text{-}0.299: & 4 \\
0.3\text{-}0.399: & 5 \\
0.4\text{-}0.499: & 7 \\
0.5\text{-}0.599: & 7 \\
0.6\text{-}0.699: & 3 \\
0.7\text{-}0.799: & 3 \\
0.8\text{-}0.899: & 7 \\
0.9\text{-}0.999: & \underline{5} \\
& 50
\end{array}
$$

Theoretically we would expect to find five quantities in each subinterval. Therefore, we can calculate a χ^2 statistic as

$$
\chi^2 = \frac{(5-5)^2}{5} + \frac{(4-5)^2}{5} + \frac{(4-5)^2}{5} + \frac{(5-5)^2}{5} + \frac{(7-5)^2}{5} + \frac{(7-5)^2}{5} + \frac{(3-5)^2}{5}
$$

$$
+ \frac{(3-5)^2}{5} + \frac{(7-5)^2}{5} + \frac{(5-5)^2}{5} = 4.4
$$

From Table 2.2, we obtain a tabulated χ^2 value of 16.92 for $\nu = 9$ and a 5% rejection probability. Since $4.4 < 16.92$, the test is successful (that is, we accept the hypothesis that the numbers are uniformly distributed). Moreover we find a tabulated χ^2 value of 3.325 for $\nu = 9$ and a 95% rejection probability. Since $4.4 > 3.325$, we again conclude that the test is successful.

Fishman (1973) and Naylor, Balintfy, Burdick, and Chu (1966) explain how the frequency test can be applied to multiple sequences of pseudorandom numbers. Basically this is accomplished by determining an overall χ^2 statistic from the individual χ^2 statistics.

Finally, the reader should recognize the distinction between randomness and uniformity. Consider, for example, the number sequence 0.05, 0.10, 0.15, 0.20, . . ., 0.95, 1.00. This sequence is obviously not random, though it is perfectly uniform. The frequency test does examine randomness, in a sense, by rejecting a number sequence that is too uniform (that is, a number sequence whose calculated χ^2 value is less than the tabulated value for a high rejection probability).

The Gap Test

The gap test is used to test the ordering of a sequence of pseudorandom digits. In this test we count the number of intervening digits between the ith and the $(i + 1)$st occurrence of some particular digit, d. If n intervening digits are present, then we have a *gap* of length n. All of the gaps are determined for each d, where d is allowed to take on all 10 values (that is, $d = 0, 1, 2, \ldots, 9$). We then tabulate the total number of occurrences of each gap length.

The expected number of gaps of length n is given by

$$E_n = (0.9)^n \, (0.1) \, N \qquad\qquad (2.25)$$

where N is the total number of gaps. This equation can be used, in conjunction with Eq. (2.24), to compute a χ^2 statistic for the gap test.

Example 2.14

Consider the following sequence of 48 consecutive random digits:
923060497710807406852632953903562526358756195478.
We first determine the gaps that are associated with each digit. The results are summarized below.

d	Gap length (n)
0	1, 5, 1, 2, 11
1	31
2	18, 2, 8, 1
3	19, 3, 2, 6
4	8, 29
5	5, 4, 2, 3, 2, 3
6	12, 3, 9, 3, 5
7	0, 4, 24, 6
8	5, 19, 8
9	6, 16, 2, 15

A total of 38 gaps have been found, ranging in length from $n = 0$ to $n = 31$ (note that the largest possible gap is $n = 46$).

The observed and expected number of gaps of each length are tabulated below (O_n and E_n represent the observed and expected number of gaps of length n, respectively).

n	O_n	E_n	n	O_n	E_n
0	1	3.80	24	1	0.30
1	3	3.42	25	0	0.27
2	6	3.08	26	0	0.25
3	5	2.77	27	0	0.22
4	2	2.49	28	0	0.20
5	4	2.24	29	1	0.18
6	3	2.02	30	0	0.16
7	0	1.82	31	1	0.14
8	3	1.64	32	0	0.13
9	1	1.47	33	0	0.12
10	0	1.32	34	0	0.11
11	1	1.19	35	0	0.10
12	1	1.07	36	0	0.09
13	0	0.97	37	0	0.08
14	0	0.87	38	0	0.07
15	1	0.78	39	0	0.06
16	1	0.70	40	0	0.06
17	0	0.63	41	0	0.05
18	1	0.57	42	0	0.05
19	2	0.51	43	0	0.04
20	0	0.46	44	0	0.04
21	0	0.42	45	0	0.03
22	0	0.37	46	0	0.03
23	0	0.34			

In order to apply the χ^2 test we must combine some of the categories so that $E_i > 5$ for each new category. Thus we obtain

i	n	O_i	E_i
1	0–1	4	7.22
2	2–3	11	5.85
3	4–6	9	6.75
4	7–10	4	6.25
5	11–16	4	5.58
6	17–46	6	6.08

We can now calculate a value for the χ^2 statistic as

$$\chi^2 = \frac{(4 - 7.22)^2}{7.22} + \frac{(11 - 5.85)^2}{5.85} + \frac{(9 - 6.75)^2}{6.75} + \frac{(4 - 6.25)^2}{6.25} + \frac{(4 - 5.58)^2}{5.58}$$

$$+ \frac{(6 - 6.08)^2}{6.08} = 7.98$$

From Tab. 2.2, we obtain a value of 11.07 for $v = 5$ and $p = 5\%$. Since this value exceeds the calculated value, we accept the hypothesis that the given sequence of digits is randomly ordered. A similar conclusion is obtained from the tabulated data at the other end of the distribution (for example, at $p = 95$ percent).

The gap test can also be applied to multiple sequences of pseudorandom digits, using essentially the same procedure as that used for the frequency test.

Increasing and Decreasing Runs

A *run* is a succession of similar events, preceded and followed by different events. In this particular test a succession of continually increasing or continually decreasing pseudorandom numbers will constitute a run.

The procedure is to count the total number of increasing and decreasing runs, and also the number of runs of length n, where $n = 1, 2, 3, \ldots$, etc. The observed number of runs can then be compared with the expected number of runs, where the latter values are obtained from the following expressions.

1. Total number of runs:

$$E_{\text{TOT}} = (2N - 1)/3 \tag{2.26}$$

where N is the total number of pseudorandom variates.

2. Runs of length n:

$$E_n = 2\,[(n^2 + 3n + 1)\,N - (n^3 + 3n^2 - n - 4)]/(n + 3)! \tag{2.27}$$

for

$$n = 1, 2, 3, \ldots, N - 2$$

and

$$E_{N-1} = 2/N! \tag{2.28}$$

The success or failure of the test can be determined by calculating a value for the χ^2 statistic based upon the runs of length n.

Example 2.15

Let us apply the increasing and decreasing runs test to the 50 pseudorandom numbers presented in Ex. 2.13. These numbers are repeated below. Reading from left to right, we place a "+" beneath each number that is greater than its predecessor, and a "−" beneath each number that is less.

0.923	0.060	0.497	0.710	0.807
	−	+	+	+
0.406	0.852	0.632	0.953	0.903
−	+	−	+	−
0.562	0.526	0.358	0.756	0.195
−	−	−	+	−
0.478	0.262	0.318	0.720	0.837
+	−	+	+	+
0.017	0.513	0.199	0.083	0.335
−	+	−	−	+
0.233	0.035	0.116	0.848	0.432
−	−	+	+	−
0.133	0.451	0.081	0.467	0.249
−	+	−	+	−
0.299	0.575	0.600	0.695	0.380
+	+	+	+	−
0.979	0.849	0.999	0.892	0.580
+	−	+	−	−
0.544	0.378	0.872	0.427	0.548
−	−	+	−	+

An increasing or decreasing run can now be identified as a sequence of like signs. We find a total of 32 runs (16 positive and 16 negative). The expected number of runs, based upon $N = 50$, is obtained from Eq. (2.26) as $E_{TOT} = 33$. The results obtained for runs of length n are summarized below.

n	O_n	E_n
1	23	20.92
2	4	8.93
3	2	2.51
4	3	0.53
5-49	0	0.11

Regrouping the data so that $E_i > 5$ for each new category, we obtain

i	n	O_i	E_i
1	1	23	20.92
2	2-49	9	12.08

An χ^2 statistic can now be calculated as

$$\chi^2 = \frac{(23 - 20.92)^2}{20.92} + \frac{(9 - 12.08)^2}{12.08} = 0.99$$

Tab. 2.2 indicates a value of 3.841 for $\nu = 1$ and $p = 5\%$. Since this value exceeds the calculated value, we conclude that the given pseudorandom numbers are sequenced randomly. This conclusion is further supported by the tabulated χ^2 value of 0.00393, corresponding to $\nu = 1$ and $p = 95\%$.

Another commonly used runs test examines runs above and below the mean. The procedure is similar to the one outlined above.

Other Tests

The purpose of this section has been simply to familiarize the reader with the concepts of uniformity and randomness, in the statistical sense, by describing a few of the more commonly used statistical tests. Numerous other statistical tests for randomness, uniformity, and independence have been devised. The reader should recognize the existence of these tests, and should appreciate the effort that may be involved in establishing the validity of a given random number generator.

2.8 SOME CLOSING REMARKS

In most practical situations a person carrying out a simulation study will make use of whatever random number generator may happen to be readily available. The analyst should recognize, however, that the generation of pseudorandom numbers is a rather complex procedure, with many opportunities for error. In particular, a given routine must produce results that are statistically valid, and it must employ numerical values that are appropriate for the word size of the particular computer being used. The analyst should seek assurances that these matters have been adequately considered before using an available random number generator for a simulation study.

PROBLEMS

2.1. Use the center-square method and an electronic calculator to generate 12 consecutive, 4-digit pseudorandom numbers, beginning with $n_1 = 7308$. (If n_i^2 contains 8 digits, drop the 2 right digits and the 2 left digits to obtain n_{i+1}; if n_i^2 contains only 7 digits, drop the 2 right digits and the 1 left digit.)

2.2. Write a computer program that will generate k 4-digit, pseudorandom numbers using the center-square method (note that k will be an input parameter). Convert each 4-digit value to a uniformly distributed (0, 1) random variate by dividing each value of n_i by 9999. Extract the center digits from n_{i+1}^2 by shifting the decimal point and truncating.

2.3. Use the program written in Prob. 2.2 to generate 200 pseudorandom numbers. Print out both the 4-digit numbers and the corresponding (0, 1) values, in 2 separate columns. Calculate the mean and the standard deviation of the $(0, 1)$ values and compare with the theoretically expected values ($\mu = 1/2$ and $\sigma = \sqrt{1/12}$).

2.4. Use the modification of the center-square method given at the end of Sec. 2.2 to generate 12 consecutive, 4-digit pseudorandom numbers. Carry out the calculations with an electronic calculator, beginning with $n_1 = 2520$ and $n_2 = 5473$. Extract each value of n_{i+1} from the product $n_i \times n_{i-1}$ using the method described in Prob. 2.1.

2.5. Repeat Prob. 2.2 using the modified center-square method described at the end of Sec. 2.2.

2.6. Use the program written in Prob. 2.5 to generate 200 pseudorandom numbers. Print out both the 4-digit numbers and the corresponding $(0, 1)$ values, in 2 separate columns. Calculate the mean and the standard deviation of the $(0, 1)$ values and compare with the theoretically expected values ($\mu = 1/2$ and $\sigma = \sqrt{1/12}$).

2.7. How many residues correspond to a value of $m = 256$ (where m represents the modulus)? What are their values?

2.8. Determine the power residues corresponding to $m = 32$ and $a = 8$. How many unique power residues are there? (This problem should be solved by hand, using an electronic calculator.)

2.9. Repeat Prob. 2.8 for $m = 32$ and $a = 9$. Compare with the results obtained in Prob. 2.8.

2.10. Repeat Prob. 2.8 for $m = 32$ and $a = 19$. Compare with the results obtained in Probs. 2.8 and 2.9.

2.11. Write a FORTRAN program that will generate a sequence of uniformly distributed pseudorandom integers using the power residue method. Use the following FORTRAN statement to generate the required congruent numbers.

```
N (I) = M * (FLOAT (A * N (I − 1))/M − A * N (I − 1)/M)
```

where all variables are assumed to represent integers. (Notice that this is a *different* method from that used in subprogram RAND, which is presented in Sec. 2.4.)

Use the program to generate 600 consecutive values for n_i, based upon $m = 2^9 = 512$, $a = 35$, and $n_o = 373$. Print out all 600 pseudorandom numbers and then answer the following questions.

(a) How large is the period (that is, how many numbers are generated before the sequence begins to repeat)?

(b) What is the last nonrepeated number in the sequence?

(c) Can you identify an obvious subset of numbers that is excluded from the sequence?

2.12. Write a FORTRAN program that will generate a sequence of uniformly distributed pseudorandom integers using the mixed congruential method. Use the following FORTRAN statements to generate the required congruent numbers.

```
K = A * N (I − 1) + B
N (I) = M * (FLOAT(K)/M − K/M)
```

where all variables are assumed to represent integers. (Notice that this is a *different* method from that used in subprogram RAND, which is presented in Sec. 2.4).

Use the program to generate 600 consecutive values for n_i, based upon $m = 2^9 = 512, a = 33, b = 1$, and $n_0 = 373$. Print out all 600 pseudorandom numbers and then answer the following questions.

(a) How large is the period (that is, how many numbers are generated before the sequence begins to repeat)?

(b) What is the last nonrepeated number in the sequence?

(c) Can you identify an obvious subset of numbers that is excluded from the sequence?

Compare the results obtained in this problem with those obtained in Prob. 2.11.

2.13. Use the FORTRAN subprogram RAND, presented in Sec. 2.4, to generate 200 uniformly distributed (0, 1) pseudorandom numbers. Calculate the mean and the standard deviation, and compare with the theoretically expected values ($\mu = 1/2$ and $\sigma = \sqrt{1/12}$). Print out all 200 numbers and examine carefully.

2.14. Repeat Prob. 2.13 using an *even* value for the multiplier, obtained by subtracting 1 from the proper value shown in Table 2.1. Compare with the results obtained in Prob. 2.13.

2.15. Repeat Prob. 2.13 using an *even*-valued seed. Compare with the results obtained in Prob. 2.13.

2.16. Write a complete BASIC program that will generate a sequence of N uniformly distributed (0, 1) pseudorandom numbers, making use of the library function RND. Use the program to generate and print out 200 random values. Calculate the mean and the standard deviation, and compare with the theoretically expected values ($\mu = 1/2$ and $\sigma = \sqrt{1/12}$).

2.17. Use the FORTRAN subprogram RAND, presented in Sec. 2.4, to generate 200 uniformly distributed (0, 1) pseudorandom numbers. Arrange the numbers into 10 groups of 20 numbers each. Calculate and print a mean value for each group. Can you draw any conclusions about the manner in which the mean values are distributed? (*Hint:* consider the central-limit theorem.) Does this suggest a method for generating random variates that are governed by some distribution other than the uniform distribution?

2.18. Write a computer program that will generate a sequence of N uniformly distributed pseudorandom numbers within the interval (a, b). Use the program to generate and print out 100 random values for each of the following cases.

(a) $a = 2, b = 7$

(b) $a = -5, b = 3$

Calculate the mean and the standard deviation for each case and compare with the theoretically expected values.

2.19. Write a single FORTRAN statement that will generate the integers 1, 3, 5, 7, and 9 randomly and uniformly. Include a reference to the FORTRAN random number generator RAND, presented in Sec. 2.4.

2.20. A biased coin has a 60 percent probability of landing with a "head" when it is tossed. Write a complete computer program that will simulate a sequence of N consecutive tosses and print the outcome of each toss. Use the program to simulate 300 consecutive tosses. Then answer the following questions, based upon the computed results.

(a) What is the likelihood of obtaining three consecutive heads?
(b) What is the likelihood of obtaining four consecutive heads?
(c) What is the likelihood of obtaining three consecutive tails?
(d) What is the likelihood of obtaining three consecutive heads followed by two consecutive tails?

Compare the results obtained from the simulation with the answers obtained using elementary probability theory.

2.21. Write a complete computer program that will simulate a sequence of N consecutive throws of three dice and print the outcome of each throw. Use the program to simulate 300 consecutive throws. Then answer the following questions, based upon the computed results.

(a) What is the likelihood that a single throw will result in a score of 10?
(b) What is the likelihood of obtaining three successive 10's?
(c) What is the likelihood that a single throw will result in a score of 7?

Compare the results obtained from the simulation with the answers obtained using elementary probability theory.

2.22. A set of observations has been obtained for a statistical experiment involving eight categories. The number of observations falling into each category are summarized below, along with the expected number of observations for each category (based upon an assumed distribution).

Category	Number of observations	Expected number of observations
1	23	20
2	28	30
3	55	45
4	73	65
5	50	60
6	32	40
7	30	25
8	9	15
	300	300

Calculate a χ^2 statistic for this experiment and determine whether to accept or reject the hypothesis that the assumed distribution can be used to represent the data. Base the conclusions upon

(a) a 25% rejection probability
(b) a 5% rejection probability
(c) a 1% rejection probability

2.23. Consider the following set of 50 random numbers.

0.945	0.489	0.971	0.598
0.402	0.183	0.552	0.728
0.129	0.423	0.846	0.896
0.508	0.552	0.992	0.109
0.525	0.067	0.009	0.203
0.045	0.133	0.333	0.603
0.834	0.974	0.488	0.875
0.633	0.652	0.436	0.277
0.110	0.613	0.287	0.357
0.698	0.574	0.339	0.529
0.797	0.793	0.940	0.008
0.708	0.519	0.262	
0.468	0.180	0.351	

Carry out a frequency test on these numbers, using 10 equal subintervals (0–0.099, 0.1–0.199, . . ., 0.9–0.999) and a 5 percent rejection probability.

2.24. Form a string of 48 consecutive random digits from the first 16 random numbers given in Prob. 2.23 (that is, 945402129 . . . 183423). Carry out a gap test on this sequence, using a 5 percent rejection probability.

2.25. Apply the increasing and decreasing runs test to the 50 random numbers given in Prob. 2.23. Use a 5 percent rejection probability with the χ^2 statistic.

2.26. Examine the random numbers given in Prob. 2.23 for runs above and below the mean (where $\mu = 0.50$). Tabulate the total number of runs and the number of runs of each length. Carry out the usual χ^2 test, using the following expressions for the expected number of runs.

$$E_{\text{TOT}} = (N + 1)/2$$

and

$$E_n = (N - n + 3)/2^{n+1}$$

where N is the total number of random values and n is the run length.

3

SOME ELEMENTARY SIMULATION PROBLEMS

Most realistic simulation problems that arise in business and industry require the use of random variates that are nonuniformly distributed. We will see how such nonuniform random variates can be generated in the next chapter. For now, however, let us consider some well-known, elementary simulation problems that require only uniformly distributed random variates for their solution. These problems provide useful insights into the procedures that are used to build simulation models, run them on a computer, and extract useful information from the results.

The essential features of each problem will be presented through the use of flowcharts. Thus, our presentation will not favor any particular programming language. The level of detail will correspond, however, to the commonly used, general-purpose languages such as FORTRAN, BASIC, or PL/1. Therefore, it should be quite easy for the reader to construct a working computer program from these flowcharts, using the language of the reader's choice. We will retain this method of presentation throughout the remainder of this book.

3.1 GAMES OF CHANCE

Games of chance (that is, gambling games) typically refer to card games, dice games, etc. Many of these games are based upon the outcome of uniformly distributed random events. In other cases the events that occur are not uniformly distributed, but their outcome can readily be synthesized using uniformly distributed random variates.

The latter situation is illustrated by the throw of a pair of dice, where the outcome for each die is uniformly distributed (see Ex. 2.11, page 34). The outcome for the *pair* of dice is not uniform, however, because of the different ways that the individual dies can combine to form pairs (for example, there is only one way to form a 2, namely, $1 + 1$; in contrast, there are six different ways to form a 7: $1 + 6$, $2 + 5$, $3 + 4$, $4 + 3$, $5 + 2$, and $6 + 1$).

If the individual events in a gambling game can be simulated, then it is easy to build a simulation model of the entire game. Such a model can be used to determine the results that might be experienced if the game were played repeatedly, over a long period of time. The example below illustrates an application of this type.

Example 3.1—Shooting Craps

Let us now consider the game of "craps," in which a player throws a pair of dice one or more times until either a win or a loss is experienced. There are two ways to win in craps. The player can either throw the dice once and obtain a score of either 7 or 11; or obtain a 4, 5, 6, 8, 9, or 10 on the first throw and then repeat this same score on a subsequent throw before obtaining a 7. Conversely there are two ways to lose. The player can either throw the dice once and obtain a score of 2, 3, or 12; or obtain a 4, 5, 6, 8, 9, or 10 on the first throw and then obtain a 7 on a subsequent throw before repeating the original score.

In order to simulate this game, we will make use of a uniformly distributed random number generator and two subprograms. The first subprogram will simulate one throw of a pair of dice, using the method shown in Ex. 2.11. The second subprogram will simulate one complete play, that is, the dice will be thrown as many times as required in order to establish either a win or a loss. In addition, a main program will be required. This program will read input data, simulate multiple plays (by repeatedly accessing the subprograms), keep track of the number of wins and losses, and then write the necessary output data after the plays have been completed.

Figure 3.1 illustrates the first subprogram (SUB1), which simulates one throw of the dice. The routine is based upon the method presented in Ex. 2.11. Thus, $X1$ and $X2$ represent the values of the first and second dies, respectively, and X represents the value obtained for the pair of dice. Notice that the subprogram requires a uniformly distributed, (0, 1) random number generator (RAND), such as that presented in Section 2.4.

Figure 3.2 illustrates the second subprogram (SUB2), which is used to simulate one complete play. The outcome of the play is indicated by SCORE, which is set equal to 1 in the event of a win and 0 for a loss. Note that the value of X, which represents the outcome of one throw of the dice, is provided by SUB1.

The main program is outlined in Fig. 3.3. The program begins by reading in, and then writing out, the following input parameters:

N   (the total number of plays to be simulated)

KX   (a seed for the random number generator)

IOUT (an output indicator, which causes the outcome of each individual play to be printed if IOUT = 1)

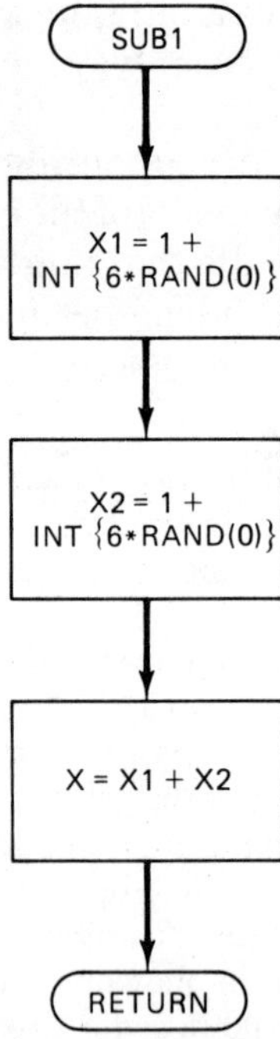

Figure 3.1

The program then initializes the random number generator (by accessing SUB1), and assigns an initial value of zero to SUM.

The individual plays are simulated within a loop, which is indicated by the dashed lines in Fig. 3.3. Each pass through the loop causes SUB2 to be accessed, thus simulating one play. The resulting value for SCORE is then added to SUM. (Remember that SCORE = 1 whenever there is a win, and SCORE = 0 for a loss.) Thus, after N passes through the loop, SUM will represent the total number of wins. This value is printed after the loop has been completed.

The program also includes an optional feature that causes the result of each play to be printed. This feature is activated only if IOUT has been assigned a value of 1. By printing the outcome of each play in this manner, the analyst can observe the progress of the simulated game, just as though it were actually taking place.

The complete program can be used in at least two different ways. First, it can be used to simulate a large numbers of plays—say 1000 or more—to determine the *expected* outcome of the game, that is, the expected fraction of plays that are wins. (This value can be established as 0.493, using the theory of probability.)

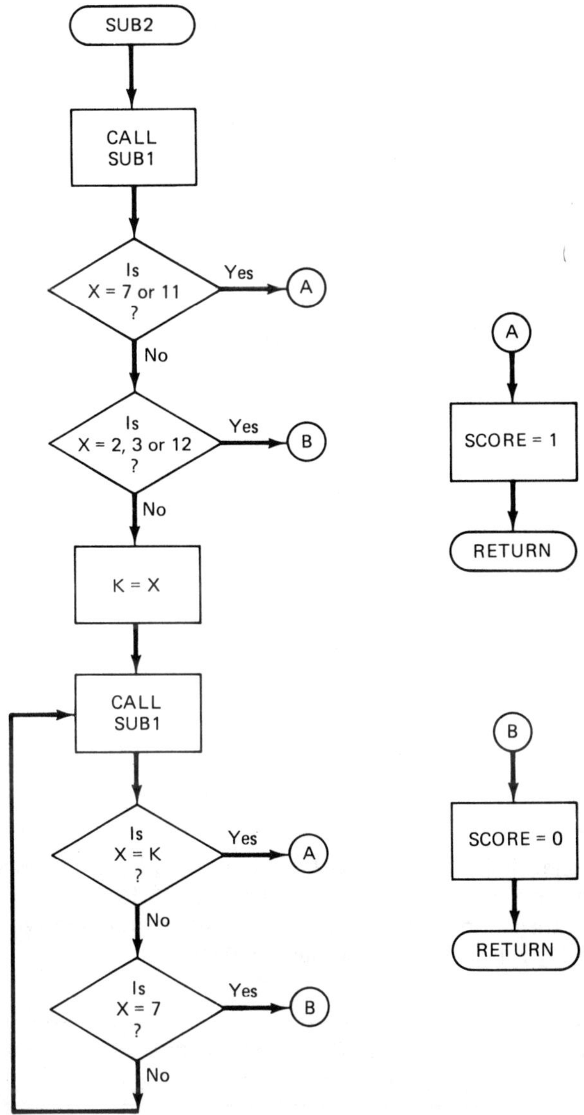

Figure 3.2

Another use of the program is to simulate several typical gambling sessions, each consisting of perhaps 25 to 50 plays. The outcome of each session can be recorded, resulting in a distribution of wins and losses for the multiple sessions. From this distribution we can obtain not only the expected outcome of the game but also the odds (that is, the probability) of "winning big" or of a disastrous loss.

Suppose, for example, that 1000 plays had been simulated and the results had been tabulated in blocks of 50, as shown below.

Block number	Runs	Number of wins
1	1- 50	22
2	51-100	20
3	101-150	28
4	151-200	24
5	201-250	19
6	251-300	26
7	301-350	23
8	351-400	30
9	401-450	22
10	451-500	25
11	501-550	18
12	551-600	26
13	601-650	26
14	651-700	29
15	701-750	21
16	751-800	24
17	801-850	20
18	851-900	24
19	901-950	29
20	951-1000	25
		481

We see that there have been 481 wins in 1000 plays, from which we might conclude that the expected outcome of the game is 0.481.

We can also group the blocked data to illustrate the relative frequencies of obtaining closely related scores, as shown below.

Wins per block	Number of blocks	Relative frequency	Cumulative frequency
18-19	2	0.10	0.10
20-21	3	0.15	0.25
22-23	3	0.15	0.40
24-25	5	0.25	0.65
26-27	3	0.15	0.80
28-29	3	0.15	0.95
30-31	1	0.05	1.00
	20	1.00	

(Note that the relative frequencies were obtained by dividing each of the frequencies by 20, that is, the total number of blocks.)

Once the relative frequencies have been determined, it is easy to obtain the *cumulative* frequencies, as tabulated in the last column above. Each value represents a partial sum of the current and previous relative frequencies. Thus the second value, 0.25, is obtained as 0.10 + 0.15; the third value, 0.40, is obtained as 0.10 + 0.15 + 0.15, and so on. We normally refer to this type of data as a *cumulative distribution.*

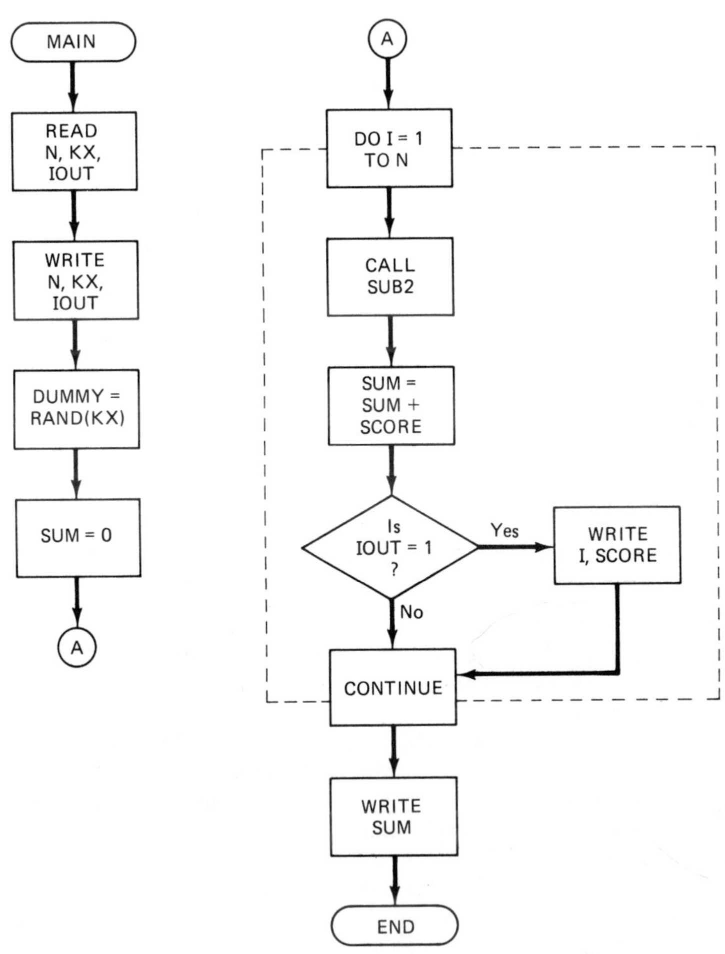

Figure 3.3

Figure 3.4 shows the cumulative distribution for the above data, with a broken curve drawn through the upper right corner of each rectangle. The odds of various win/loss situations can be estimated using this curve. Thus, there is

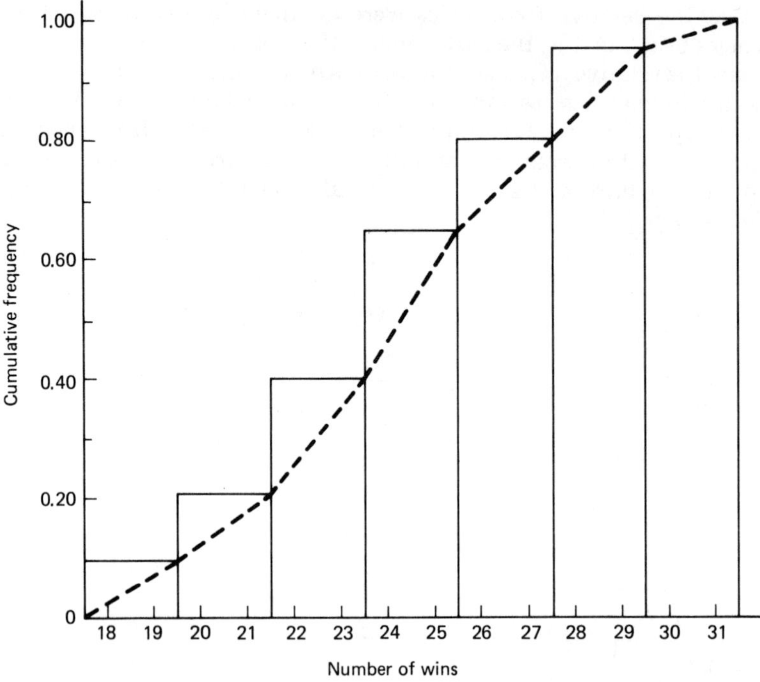

Figure 3.4

roughly a 20 percent chance of experiencing fewer than 22 wins in 50 plays. Similarly, there is a 20 percent chance of winning at least 28 out of 50 plays.

Example 3.2—Blackjack

Many games of chance involve playing cards. These games differ from dice games in the sense that the chances of various events occurring change as cards are withdrawn from the deck.

In order to see how a shrinking deck can be handled, let us consider the popular card game "blackjack." The object of blackjack is to obtain the highest score possible without exceeding 21. Picture cards (that is, Jack, Queen, King) count as 10 points, an Ace can be either a 1 or an 11, and all other cards are equal to their face value.

Blackjack is always played against the dealer. Initially each player is dealt two cards. Any player who receives an Ace and a picture card or an Ace and a 10 has blackjack (21 points with two cards) and therefore wins. A player who is originally dealt a low score may request one or more additional cards from the dealer. The player will win with a final score that is higher than the dealer's, but that does not exceed 21.

In this example we will simulate only the initial portion of the game, namely, dealing the first two cards to each player. We will do so by first dealing one card to each of the players, and then dealing a second card to each player. We will, of course, have to make certain that the same card is not dealt more than once.

Let us represent the deck of cards as a one-dimensional array called C. Initially the array will contain 52 elements [that is, $C(I)$, $I = 1, 2, \ldots 52$], arranged as follows:

I	$C(I)$
1	1.1 (Ace of clubs)
2	2.1 (2 of clubs)
.	.
.	.
.	.
10	10.1 (10 of clubs)
11	11.1 (Jack of clubs)
12	12.1 (Queen of clubs)
13	13.1 (King of clubs)
14	1.2 (Ace of diamonds)
.	.
.	.
.	.
26	13.2 (King of diamonds)
27	1.3 (Ace of hearts)
.	.
.	.
.	.
39	13.3 (King of hearts)
40	1.4 (Ace of spades)
.	.
.	.
.	.
52	13.4 (King of spades)

The numerical value of each card can be obtained from $C(I)$ by truncation, that is,

$$V = \text{INT}\left\{C(I)\right\}$$

with the additional restrictions that

$$V = 10 \quad \text{if INT}\left\{C(I)\right\} > 10$$

and

$$V = 11 \quad \text{if INT}\left\{C(I)\right\} = 1$$

(Note that an Ace will always be assigned a value of 11 during this initial portion of the game.) Also each card's suit can be determined as

$$S = 10 \left\{ C(I) - \text{INT} \left[C(I) \right] \right\}$$

Thus, $S = 1$ will correspond to a club, $S = 2$ to a diamond, etc.

The cards will be selected by generating a random value for I, that is,

$$I = 1 + \text{INT} \left\{ NI * \text{RAND}(0) \right\}$$

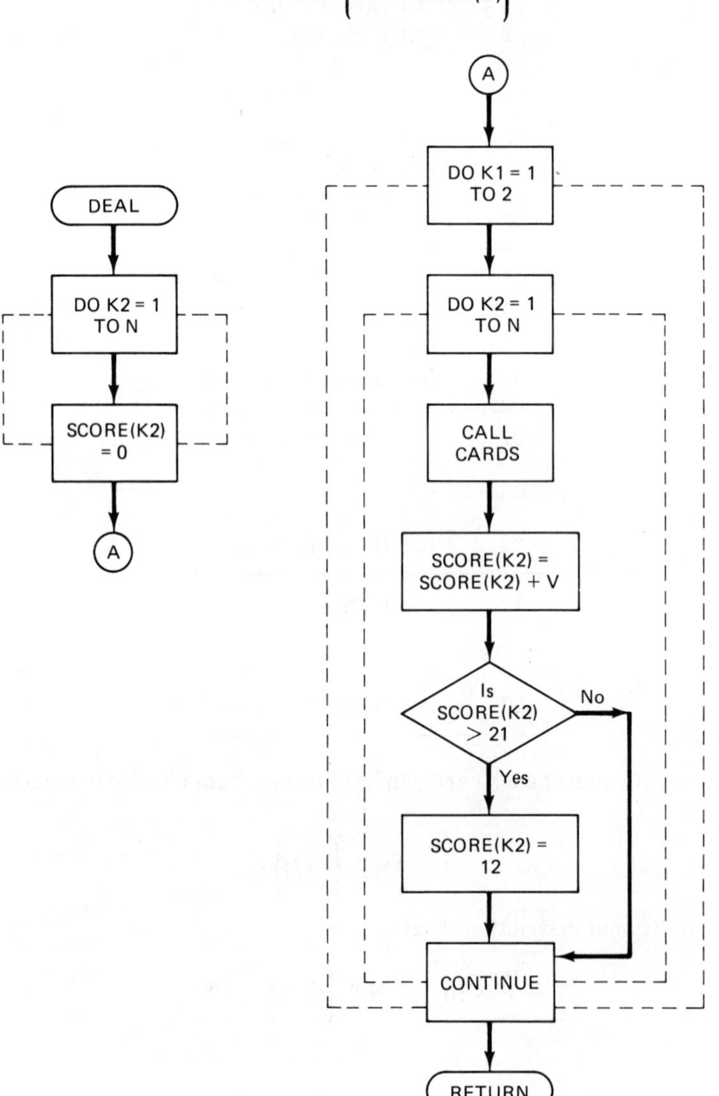

Figure 3.5

where NI is the number of cards remaining in the deck (initially, NI = 52). Once a card has been chosen and "dealt" it will be deleted from the array, thus preventing the same card from being dealt more than once. Every such deletion will cause the array to become smaller by one element.

In this example we will develop two subprograms. The first (DEAL) will deal two cards to each player; the second (CARDS) will select a card at random, determine the value of that card, and then eliminate that card from the array. Thus, the first subprogram will access the second each time a different card is dealt.

Figure 3.5 presents a flowchart for subprogram DEAL. Notice that two loops are included. The inner loop deals one card to each of N players, and the outer loop repeats the process a second time, thus dealing the second card.

Figure 3.6 presents a flowchart of subprogram CARDS. The first part of the flowchart (to entry point A) causes a card to be selected at random, and a corresponding numerical value to be calculated. The second part of the flowchart (after entry point A) causes a compression in the array, thus eliminating the card that has just been selected.

In order to complete the simulator a main program will be required which will read input data, access subprogram DEAL, calculate and record the total score for each player, and print the required output data. The detailed structure of this program is left to the reader.

Once the entire program has been written it can be used to generate the distribution of scores that are dealt to the various players. Suppose, for example, that 25 games have been simulated, with four players per game, resulting in the following initial scores (the individual cards are shown in parentheses after each complete score).

Game	Player 1	Player 2	Player 3	Player 4
1	12 (5.3 + 7.4)	7 (3.1 + 4.2)	11 (7.2 + 4.1)	21 (12.3 + 1.4)
2	21 (1.3 + 10.4)	14 (4.1 + 11.2)	20 (11.4 + 12.3)	20 (13.1 + 10.1)
3	10 (3.3 + 7.3)	15 (7.1 + 8.2)	11 (6.2 + 5.1)	4 (2.4 + 2.2)
4	12 (10.2 + 2.3)	8 (2.4 + 6.1)	7 (3.4 + 4.3)	18 (8.1 + 13.4)
5	11 (5.4 + 6.2)	16 (9.2 + 7.4)	13 (3.4 + 10.2)	16 (6.1 + 11.2)
6	18 (8.4 + 11.1)	13 (3.4 + 11.4)	15 (13.3 + 5.1)	12 (2.3 + 13.4)
7	12 (2.4 + 11.2)	8 (2.3 + 6.3)	17 (1.2 + 6.2)	21 (1.3 + 11.4)
8	13 (2.2 + 1.4)	6 (4.1 + 2.1)	20 (12.3 + 11.3)	17 (7.4 + 13.3)
9	15 (13.3 + 5.2)	18 (8.4 + 11.2)	18 (7.3 + 1.1)	9 (4.3 + 5.3)
10	11 (5.2 + 6.4)	10 (6.2 + 4.1)	13 (13.2 + 3.4)	21 (11.4 + 1.4)
11	12 (5.2 + 7.4)	19 (13.1 + 9.1)	14 (4.2 + 10.1)	12 (3.2 + 9.2)
12	21 (1.3 + 10.2)	19 (9.1 + 11.2)	12 (7.2 + 5.4)	9 (5.3 + 4.3)
13	20 (10.3 + 11.4)	13 (3.2 + 12.4)	11 (6.1 + 5.1)	17 (12.3 + 7.3)
14	8 (3.4 + 5.3)	12 (11.3 + 2.4)	16 (11.2 + 6.2)	20 (10.1 + 10.3)
15	16 (6.4 + 11.3)	5 (2.4 + 3.3)	17 (1.2 + 6.3)	20 (13.3 + 10.3)
16	10 (2.3 + 8.2)	11 (8.4 + 3.1)	14 (11.1 + 4.4)	9 (3.3 + 6.2)
17	12 (7.3 + 5.2)	15 (5.3 + 11.2)	14 (11.4 + 4.1)	12 (13.4 + 2.2)
18	18 (8.3 + 13.2)	14 (5.1 + 9.3)	13 (12.4 + 3.2)	21 (11.2 + 1.2)
19	14 (11.2 + 4.2)	16 (5.1 + 1.2)	9 (3.1 + 6.3)	15 (9.1 + 6.1)

(Continued on page 59)

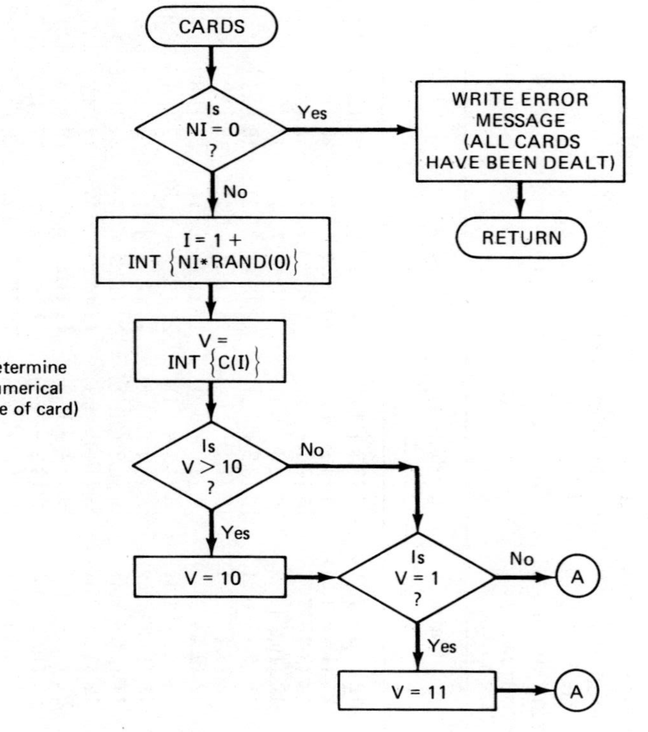
CARDS
Is
NI = 0
?
Yes
WRITE ERROR
MESSAGE
(ALL CARDS
HAVE BEEN DEALT)
RETURN
No
I = 1 +
INT {NI*RAND(0)}
V =
INT {C(I)}
(Determine
numerical
value of card)
Is
V > 10
?
No
Yes
V = 10
Is
V = 1
?
No
A
Yes
V = 11
A

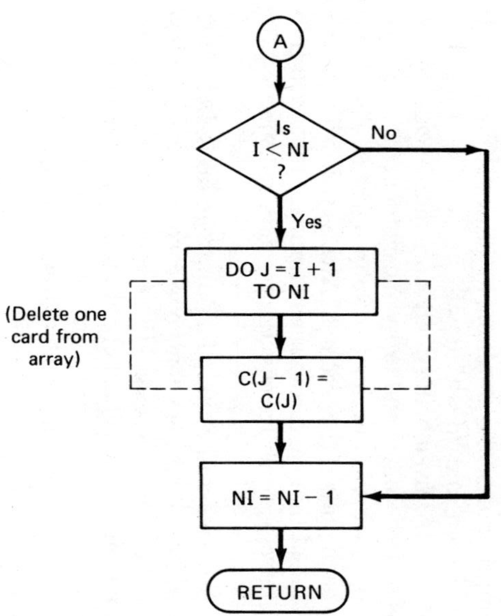
A
Is
I < NI
?
No
Yes
DO J = I + 1
TO NI
(Delete one
card from
array)
C(J − 1) =
C(J)
NI = NI − 1
RETURN

Figure 3.6

Game	Player 1	Player 2	Player 3	Player 4
20	12 (12.4 + 2.1)	20 (12.1 + 11.2)	18 (13.1 + 8.4)	19 (9.3 + 10.1)
21	13 (2.1 + 1.4)	19 (11.4 + 9.3)	13 (3.4 + 11.1)	17 (13.4 + 7.3)
22	8 (6.1 + 2.2)	15 (8.1 + 7.1)	20 (12.4 + 11.1)	19 (9.3 + 10.4)
23	16 (1.2 + 5.1)	8 (2.2 + 6.1)	18 (12.1 + 8.1)	11 (4.1 + 7.3)
24	12 (10.2 + 2.3)	17 (7.4 + 10.4)	11 (3.4 + 8.3)	12 (12.2 + 2.4)
25	17 (7.4 + 10.4)	13 (12.2 + 3.2)	8 (4.2 + 4.1)	12 (11.3 + 2.2)

(Notice that each game begins with a complete, randomly shuffled deck.)

The results of the above simulation can be summarized by grouping the scores:

Score	Number of hands	Score	Number of hands
21	6	12	14
20	8	11	8
19	5	10	3
18	7	9	4
17	7	8	6
16	6	7	2
15	6	6	1
14	6	5	1
13	9	4	1
			100

Since there is a total of 100 hands, we can conclude that there is approximately a 6 percent chance of obtaining an initial score of 21 (blackjack!), an 8 percent chance of obtaining a score of 20, and so on. The accuracy of the estimates can be expected to improve as the total number of hands being simulated increases.

The results presented in this example do not, of course, apply to a complete game of blackjack, although the computational strategy can easily be modified to accommodate the remaining rules of the game. This is particularly true if the simulation program is written in a conversational mode, so that each player receives additional cards only when they are requested.

3.2 RANDOM WALK PROCESSES

Random walk processes are concerned with the random migration of objects in space. Such processes are quite common in nature. Molecular diffusion and Brownian motion, for example, are both processes of this type.

The example below is concerned with a well-known, elementary random walk process, which is particularly easy to simulate.

Example 3.3—The Staggering Drunk

A drunk standing at a lamp post begins to stagger off into the dark, attempting to find his way home. Each time he moves he travels one unit from his prior location. However, the direction in which he travels is completely arbitrary.

In order to simulate the random movement of the drunk, we will select a coordinate system whose origin is the lamp post. After the ith move, the location of the drunk, relative to the lamp post, will be given by the coordinates (x_i, y_i), and the distance from the lamp post will be determined as

$$d_i = \sqrt{x_i^2 + y_i^2}$$

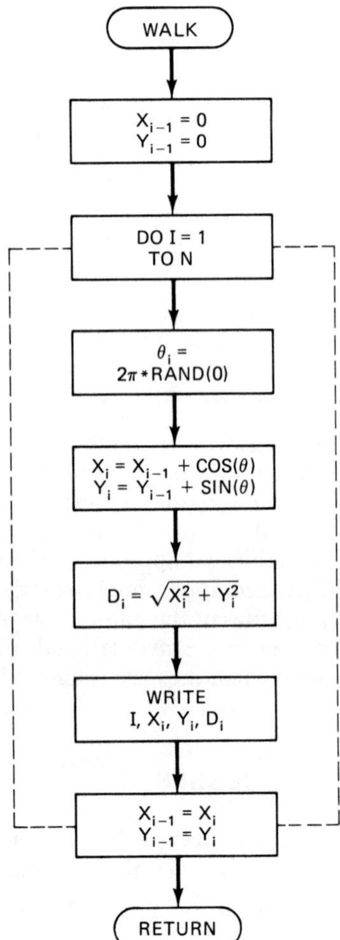

Figure 3.7

Suppose that θ_i represents the direction of the ith move (measured from the x-axis). Since the direction of each move is chosen at random, θ_i will be a uniformly distributed random variate within the interval $(0, 2\pi)$. (Note that θ_i is expressed in radians.) Once a value has been determined for θ_i, the new location can be expressed as

$$x_i = x_{i-1} + \cos \theta_i$$

$$y_i = y_{i-1} + \sin \theta_i$$

Figure 3.7 presents a flowchart of a subprogram (WALK) that will simulate N consecutive moves, starting at the lamp post. This subprogram, in conjunction with an appropriate main program, can be used to obtain a detailed tabulation of the drunk's movements as he staggers from one location to another. From this information we can obtain a distribution of locations and distances from the lamp post. Of particular interest is an assessment of the likelihood that the drunk will eventually return to (or very near to) the lamp post where he first started.

For example, suppose that the drunk is allowed to make 100 random moves of one unit each, beginning at the lamp post. At the end of 100 moves we record the drunk's final distance from the lamp post. If this situation is simulated repeatedly, we obtain the following distribution of final distances.

Distance from lamp post (units)	Relative frequency	Cumulative distribution
0– 1.99	0.02	0.02
2– 3.99	0.10	0.12
4– 5.99	0.12	0.24
6– 7.99	0.14	0.38
8– 9.99	0.18	0.56
10–11.99	0.21	0.77
12–13.99	0.09	0.86
14–15.99	0.05	0.91
16–17.99	0.03	0.94
18–19.99	0.03	0.97
20–21.99	0.02	0.99
22–23.99	0.01	1.00
	1.00	

The cumulative distribution data are plotted in Fig. 3.8.

We can easily obtain the expected value of the final distance from the lamp post as

$$\bar{d} = (0.02)(1) + (0.10)(3) + \ldots + (0.01)(23) = 9.48$$

and

$$s = \sqrt{(0.02)(1 - 9.48)^2 + (0.10)(3 - 9.48)^2 + \ldots + (0.01)(23 - 9.48)^2} = 4.58$$

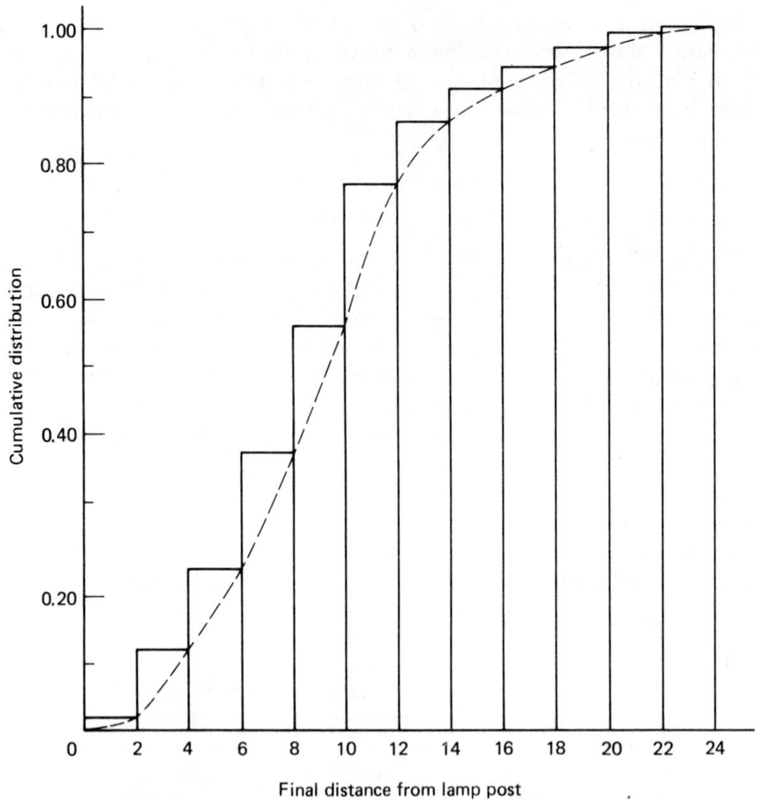

Figure 3.8

Moreover, from Fig. 3.8 we can see that there is a 10 percent chance that the final distance from the lamp post will be less than 3.5 units; we also see that there is a 10 percent chance of obtaining a final distance greater than 15.5 units; and so on.

3.3 ELEMENTARY BUSINESS APPLICATIONS

Stochastic process simulation is very useful for analyzing business situations involving profitability, return on investment, etc. We will discuss applications of this type in Chap. 5. For now, however, let us introduce a very simple business application in order to illustrate the overall approach.

Example 3.4—Estimating Profitability

A company that has just developed a new product wishes to assess its profitability. The company estimates that there is a 35 percent chance of selling between 40,000 and 60,000 units a year, a 40 percent chance of selling between 60,000 and 80,000 units a year, and a 25 percent chance of selling between 80,000 and 100,000 units a year. Based upon current market conditions, it appears unlikely that the company will sell fewer than 40,000 or more than 100,000 units a year.

The cost of manufacturing and distributing this product also appears somewhat uncertain. There is a 20 percent chance that the cost will be between $60 and $70 per unit, a 35 percent chance that it will be between $70 and $80 per unit, a 30 percent chance that it will be between $80 and $90 per unit, and a 15 percent chance that it will be between $90 and $100 per unit. It is likely that the cost will be less than $60 or greater than $100 per unit.

The company wishes to determine the expected yearly profit corresponding to some specified selling price. (Most likely, the profitability associated with several different selling prices will eventually be investigated.) In addition, the company would like to estimate the chances that the yearly profit will be much higher or much lower than the expected value. This latter information will provide some indication of the degree of risk associated with the proposed venture.

Let us begin by defining the following variables.

S = price per unit
C = total cost per unit
V = number of units sold per year
P = yearly profit

The yearly profit can now be expressed as

$$P = (S - C)\ V$$

where C and V (and hence P) are random variates and S is a specified input quantity (that is, a decision variable).

In order to simulate this situation, we must be able to evaluate the yearly sales volume (V) and the cost per unit (C). One way to do so is to generate a uniformly distributed random number, u_1, within the interval (0, 1). If u_1 does not exceed 0.35, we assign a value of 50,000 units to V (note that this is the midpoint of the first category given above). If u_1 is greater than 0.35 but does not exceed 0.75, then we assign a value of 70,000 units to V; and if u_1 exceeds 0.75, we set V = 90,000 units. A corresponding subprogram (SALES) is outlined in Fig. 3.9.

Similarly, we can assign a value to C by generating another uniformly distributed (0, 1) random number, u_2. If u_2 does not exceed 0.20, we set C = $65 per unit. If u_2 is greater than 0.20 but does not exceed 0.55, then C = $75. Likewise, C = $85 if $0.55 < u_2 \leq 0.85$, and C = $95 if $u_2 > 0.85$. Figure 3.10 shows a flowchart of the corresponding subprogram (COST).

Figure 3.11 shows the overall computational strategy. The computation begins by reading in values for S and N (where N is the desired number of simulations). This is followed by a loop that repeatedly generates random values for the sales volume and the cost, calculates the corresponding profit, and then prints the results obtained during each pass. After the loop has been completed the profit data are grouped to form a cumulative distribution, and values are

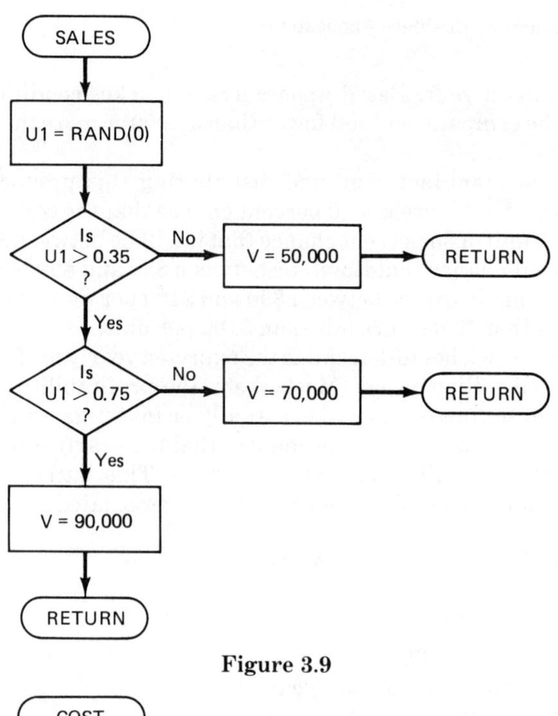

Figure 3.9

Figure 3.10

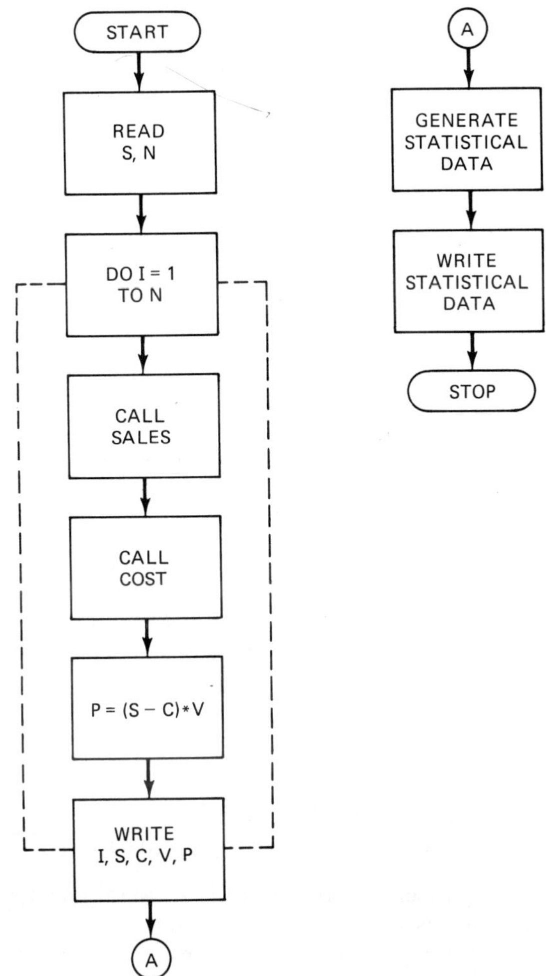

Figure 3.11

determined for the expected profit and the standard deviation. This statistical information is then printed out, and the computation is terminated.

Now suppose that the proposed venture has been simulated, using a selling price of $100 per unit. The following distribution of yearly profits is representative of the results that might be obtained.

Yearly profit (millions of dollars)	Relative frequency	Cumulative distribution
0.25	0.03	0.03
0.35	0.05	0.08
0.45	0.04	0.12
0.75	0.14	0.26
1.05	0.15	0.41
1.25	0.19	0.60
1.35	0.06	0.66
1.75	0.18	0.84
2.25	0.08	0.92
2.45	0.04	0.96
3.15	0.04	1.00
	1.00	

(In this example the yearly profit can take on only the discrete values listed above, because of the manner in which numerical values are assigned to the sales volume (V) and the cost per unit (C). In the next chapter we will consider a more sophisticated method for assigning values in such situations. The use of this method would result in a continuous range of values for the yearly profit.)

We can easily obtain a mean value and a standard deviation from the above distribution. Thus, we can write

$$\overline{P} = (0.03)(0.25) + (0.05)(0.35) + \ldots + (0.04)(3.15) = 1.343$$

and

$$s = \sqrt{(0.03)(0.25 - 1.343)^2 + (0.05)(0.35 - 1.343)^2 + \ldots + (0.04)(3.15 - 1.343)^2} = 0.6753$$

We see than that the expected yearly profit is 1.343 million dollars. Moreover, the standard deviation is 0.6753 million dollars. The relatively large value obtained for s indicates that the yearly profit may differ significantly from the expected value, which suggests a relatively high degree of risk for this particular venture.

Figure 3.12 shows a plot of the above cumulative distribution. This figure can be used to determine (or at least estimate) the chances of obtaining various values for the yearly profit. For example, there is a 20 percent chance that the yearly profit will be less than 0.63 million dollars. Similarly, there appears to be a 20 percent chance of obtaining a yearly profit in excess of 1.65 million dollars, and so on.

3.4 EVALUATION OF INTEGRALS

Before concluding our discussion of elementary applications, let us briefly consider a purely technical problem, namely, the use of stochastic process simulation to evaluate integrals. This is an interesting applica-

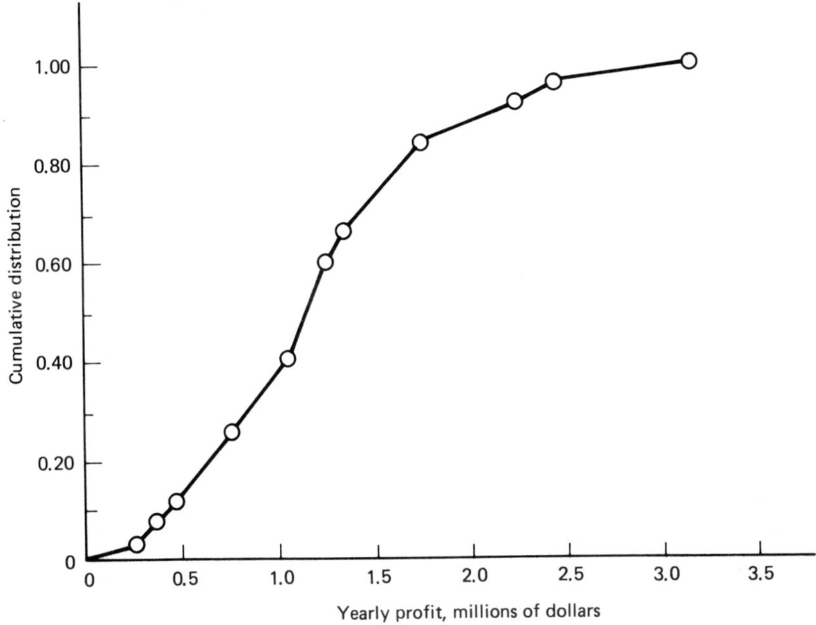

Figure 3.12

tion in the sense that a probabilistic method will be used to solve a purely deterministic problem. The procedure, which is frequently referred to as the *Monte-Carlo method*, requires nothing more than a uniformly distributed random number generator.

Suppose that we wish to evaluate the integral

$$I = \int_a^b f(x)\ dx$$

where the function $f(x)$ represents a continuous curve within the interval $a \le x \le b$. We will require that $0 \le f(x) \le f_{max}$. The Monte-Carlo method then proceeds as follows.

1. Generate a uniformly distributed random number, u_x, whose value lies between a and b.
2. Evaluate $f(u_x)$.
3. Generate a second uniformly distributed random number, u_y, whose value lies between 0 and f_{max}. The two random numbers (u_x, u_y) will represent the coordinates of a point in space.
4. Compare u_y with $f(u_x)$. If u_y does not exceed $f(u_x)$, then the point (u_x, u_y) will fall on or below the given curve.

5. Steps 1 though 5 are repeated N times. Upon completion, the fraction of points falling on or under the curve (FRACT) is determined. The value of the integral is then obtained as

$$I = \text{FRACT} \times (b - a) \times f_{max}$$

The value for I that is obtained in this manner will be reasonably accurate provided N is large.

Figure 3.13 presents a flowchart of the above procedure. In this flowchart the parameter K is a counter that indicates the number of points falling on or below the given curve. The fraction of points falling

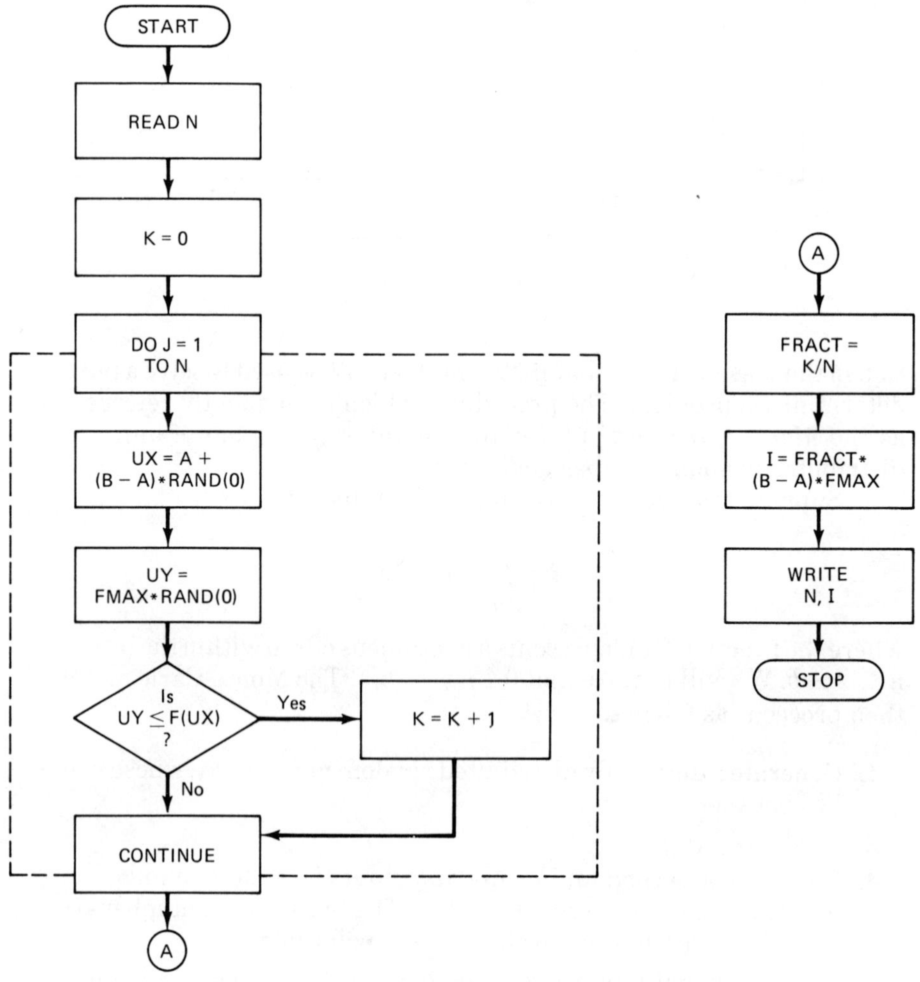

Figure 3.13

on or below the given curve is then determined as the final value for K divided by N.

Example 3.5—The Normal Distribution

Consider the integral

$$I = \frac{1}{\sqrt{2\pi}} \int_{-5}^{1} e^{-x^2/2} \, dx$$

(This is the *standard normal distribution*, which will be discussed in the next chapter.) We see that $a = -5$, $b = 1$, and $f_{max} = 1/\sqrt{2\pi} = 0.3989$.

Suppose that the first few calculations yielded the following results.

u_x	$f(u_x)$	u_y	Is $u_y \leq f(u_x)$?
0.8452	0.2791	0.1836	Yes
−2.6024	0.0135	0.3034	No
−0.9698	0.2493	0.2014	Yes
−2.7860	0.0082	0.3283	No
−1.4818	0.1331	0.3422	No
−2.4056	0.0221	0.1169	No
−0.7514	0.3008	0.0727	Yes

Thus we see that three of the calculations satisfy the condition $u_y \leq f(u_x)$.

Now suppose that we had carried out 1000 such calculations, and 344 of these calculations satisfied the condition that $u_y \leq f(u_x)$. Then FRACT would have a value of $344/1000 = 0.344$, and the integral would be assigned a value of

$$I = 0.344 \times [1 - (-5)] \times 0.3989 = 0.8233$$

(The correct value is 0.8413.)

It should be understood that the Monte-Carlo method is not an efficient way to solve elementary integration problems. There are, however, a few very difficult integration problems that cannot be solved by any other method.

PROBLEMS

3.1. Write a complete computer program that will simulate a succession of N games of craps, based upon the flowcharts presented in Figs. 3.1 to 3.3. Include a provision to print out the results of each game. Use the program to simulate 1000 games. Estimate the likelihood of winning, based upon the simulated results.

3.2. Write a conversational computer program, to be run from a time-sharing terminal, that will simulate a game of craps. Let each throw of the dice be initiated by depressing the SPACE bar on the terminal. Print out the result of each throw (that is, the value for each die and the total score), and print an appropriate message when each game has been completed. Write the program in such a manner that successive games can be simulated as long as desired.

3.3. Write a complete computer program that will simulate dealing the first two cards in a game of blackjack, based upon the flowcharts presented in Figs. 3.5 and 3.6. Allow for M different players per game (including the dealer), and N successive games. Print out each player's score after every game.

Use the program to simulate dealing the first two cards to five players in 100 successive games. Print each player's score after each game. From these results estimate the likelihood of a player obtaining every possible score (that is, 21, 20, 19, . . . , 4), and compare with the results presented in Ex. 3.2.

3.4. Extend the simulation program developed in Prob. 3.3 so that each player may be dealt one or more additional cards. Assume that these additional cards will not be dealt until all players have received their first two cards. Then let each player receive all of his or her additional cards before going on to the next player.

For simplicity, assume that the decision whether or not to receive an additional card is determined by the following probabilities.

Score	Probability of requesting an additional card
21	0
20	0
19	0.03
18	0.10
17	0.25
16	0.50
15	0.75
14	0.90
13	0.95
12	0.98
11 or less	1.00

Every time one such card is dealt, a new score must be determined and consideration given to requesting yet another card. A player will automatically be considered "out" if his or her score exceeds 21. (Remember, however, to change the value of an Ace from 11 to 1 whenever possible. Thus, it will be necessary to "tag" all Aces that are dealt to each player.)

Use the program to simulate 100 games with five players per game (hence, generate 500 complete scores). Begin each game with a complete, randomly shuffled deck of cards. Obtain and print out a distribution of final scores, with all scores that exceed 21 lumped into a single category.

3.5. Another way to simulate the dealing of playing cards from a deck is to use a one-dimensional array, as described in Ex. 3.2. Rather than delete a card from the array once it has been dealt, however, we can simply tag each such card by placing a minus sign in front of its value. We can then avoid dealing such cards more than once simply by testing for the minus sign whenever a card is selected at random. This method is less elegant but simpler than that outlined in Fig. 3.6.

Repeat Probs. 3.3 and 3.4 using this method. Compare the use of this method with the method presented in Ex. 3.2. Consider the following three factors in making your comparison.
(a) Logical simplicity
(b) Programming effort
(c) Computational efficiency (particularly execution times)

3.6. Write a conversational computer program, to be run from a time-sharing terminal, that will simulate a complete game of blackjack between one player and the dealer (hence a total of two players). Carry out the calculations in the following manner.
   (a) First deal one card to each player, followed by a second card. Print out both cards and the total score for the first player, but print only the value of the first card for the dealer. (Thus the first player will not know the dealer's total score.)
   (b) Allow the first player to request additional cards by depressing the space bar on the time-sharing terminal for each additional card. Use a carriage return to indicate that the player does not want any additional cards.
   (c) If the first player's final score does not exceed 21, then the dealer must draw additional cards. This will continue until the dealer's score exceeds 16. (The dealer will automatically lose if his score exceeds 21. Otherwise, the player with the highest score wins.)
   (d) Print each player's final score, followed by a message that indicates whether the first player has won or lost.

3.7. Baccarat is a simple card game that is played at many gambling casinos. Like blackjack, baccarat is played against the dealer. Initially, each player is dealt two cards. Picture cards and 10s count as zero; all other cards are assigned their face values. The object of the game is to obtain the highest possible score without exceeding 9. The following rules apply.
(a) An 8 or a 9 on the first two cards is a win (a 9 beats an 8).
(b) If the total score of the first two cards exceeds 10, then the score is adjusted to be the excess over 10 (for example, 18 becomes 8).
(c) If a player's adjusted score (based upon the first two cards) is less than 5, one additional card must be drawn.
(d) If a player's adjusted score (first two cards) is equal to 5, one additional card may be drawn, if desired.
Subsequent draws are not allowed.

Write a complete computer program that will simulate N successive games of baccarat. Allow for M different players per game, and begin each game with a complete, randomly shuffled deck of cards. (Be certain that no card is dealt more than once.) Assume a 40 percent probability that a player whose total score is 5 will request an additional card. Print out each player's score after every game.

Use the program to simulate 200 consecutive games between one player and the dealer (hence a total of two players). Assume that both participants play by the same rules. Determine and print out a distribution of final scores.

3.8. Write a conversational computer program, to be run from a time-sharing terminal, that will simulate a complete game of baccarat between one player and the dealer (hence a total of two players), using the rules given in Prob. 3.7. Identify (print out) the cards that are dealt to each player.

If the first player has an initial score of 5 (based on the first two cards), let the player request an additional card by depressing the space bar (depress the carriage return if the player does not want the additional card). Let the dealer's decision to request an additional card be determined automatically, using the 40 percent probability (see Prob. 3.7).

Print each player's final score, followed by a message that indicates whether the first player has won or lost.

3.9. Roulette is played with a wheel containing 38 different squares along its circumference. Two of these squares, numbered 0 and 00, are green; 18 squares are red, and 18 are black. The red and black squares alternate in color, and are randomly numbered 1 through 36. A small marble is spun within the wheel, which eventually comes to rest within one of the squares. The game is played by betting on the outcome of each spin, in any one of the following ways.

(a) By selecting a single red or black square, at 35-to-1 odds. Thus, if a player were to bet $1.00 and win, the player would receive $36.00 (the original $1.00 plus an additional $35.00).

(b) By selecting a color (either red or black) at 1-to-1 odds. Thus, if a player chose red on a $1.00 bet, the player would receive $2.00 if the marble came to rest in any red square.

(c) By selecting either the odd or the even numbers (excluding 0 and 00) at 1-to-1 odds.

(d) By selecting either the low 18 or the high 18 numbers at 1-to-1 odds. The player will automatically lose if the marble comes to rest in one of the green squares (0 or 00).

Write a conversational computer program, to be run from a time-sharing terminal, that will simulate a roulette game. Allow the players to select whatever type of play they may wish. Then print the outcome of each game, followed by an appropriate message indicating whether the players have won or lost.

3.10. Write a conversational computer program, to be run from a time-sharing terminal, that will simulate a game of solitaire.

3.11. Write a computer program that will randomly select birthdays for N different people. Consider only the month and day (for example, May 24), not the year of birth. Run the program several times, for

(a) $N = 10$ people
(b) $N = 25$ people
(c) $N = 50$ people
In each case, determine if two or more people have their birthdays on the same day of the year. How likely is this to happen for each of the above values of N?

3.12. Write a complete computer program that will simulate the random walk process described in Ex. 3.3. Allow the program to carry out N successive simulations, with M consecutive moves in each simulation. Each simulation must begin at the lamp post (that is, the origin). Print out the distance from the lamp post at the end of each simulation.

Use the program to carry out 100 different simulations, each consisting of 100 moves (hence $M = N = 100$). Generate a distribution of final distances from the lamp post, similar to that shown in Ex. 3.3. Use the distribution to determine the likelihood that the drunk will end up
(a) less than 2 units from lamp post
(b) less than 5 units from the lamp post
(c) more than 10 units from the lamp post
(d) more than 20 units from the lamp post

3.13. Repeat Prob. 3.12 for the case where $M = 200$ consecutive moves per simulation, and $N = 50$ successive simulations. Are the results significantly different from those obtained in Prob. 3.12, when $M = N = 100$?

3.14. Repeat Prob. 3.12, allowing the distance of each move, u_i, to increase progressively as follows.

$$u_i = 1 + (i/M)$$

where i is a running index, that is, $i = 1, 2, 3, \ldots, M$. (Note that the equations for x_i and y_i must now be modified to read

$$x_i = x_{i-1} + u_i \cos \theta_i$$

$$y_i = y_{i-1} + u_i \sin \theta_i$$

where θ_i is determined randomly, as in Ex. 3.3.) Run the program for the case where $M = N = 100$. Are the results significantly different from those obtained in Prob. 3.12?

3.15. Write a complete computer program that will solve the profitability problem described in Ex. 3.4. Include a provision for reading in S, the selling price per unit, and N, the number of simulations to be carried out. Print out a cumulative distribution for the yearly profit, similar to that shown in Ex. 3.4.

Use the program to determine the profitability associated with a selling price of \$80 per unit, based upon 200 consecutive simulations. From the calculated distribution, estimate the likelihood of

(a) a loss

(b) a yearly profit in excess of $0.5 million

(c) a yearly profit in excess of $1 million

(d) a yearly profit between $0.25 and $0.75 million

3.16. For the proposed venture described in Ex. 3.4, determine the value for S (the selling price per unit) that will result in a mean yearly profit of $0.8 million.

3.17. Evaluate the profitability model presented in Ex. 3.4 using the following distributions for total cost per unit and yearly sales volume.

Total cost per unit ($)	Probability of occurrence
60	0.15
80	0.30
100	0.35
120	0.20

Yearly sales volume	Probability of occurrence
35,000	0.15
50,000	0.25
75,000	0.35
100,000	0.25

Obtain a cumulative distribution for the yearly profit, based upon a sales price of $100 per unit. Carry out 200 consecutive simulations to generate the distribution.

3.18. Suppose that the expression for yearly profit given in Ex. 3.4 is modified to read

$$P = (S - C)\,V - T$$

where T represents the yearly taxes. If the taxes are determined as

$$T = 0.48\,(S - C)\,V - \$250{,}000$$

then what will be the distribution of the net yearly profit after taxes corresponding to a selling price of $S = \$100$ per unit?

3.19. For the proposed venture described in Ex. 3.4, what selling price will result in a mean yearly profit of $0.8 million if tax payments are included in the model (see Prob. 3.18)? Compare your answer with that obtained in Prob. 3.16.

3.20. Write a complete computer program to evaluate an integral using the Monte-Carlo method presented in Sec. 3.4. Use the program to evaluate the integral

$$I = \frac{1}{\sqrt{2\pi}} \int_0^2 e^{-x^2/2}\, dx$$

based upon a value of $N = 1000$. (The correct answer is $I = 0.4773$.)

3.21. Use the computer program written for Prob. 3.20 to evaluate the integral

$$I = \int_0^3 x^2\, dx$$

based upon a value of $N = 1000$.

4

NONUNIFORM RANDOM VARIATES

We now turn our attention to the generation of random variates that are governed by various distribution functions other than the uniform. Such random variates are usually required when simulating a realistic problem situation. In fact, many realistic simulation problems require the generation of several different types of random variates in order to evaluate the governing mathematical model.

In this chapter we will see how uniformly distributed (0, 1) random numbers can be used to generate nonuniform random variates. Some general methods for generating nonuniform random variates will be presented, and then applied to several specific, commonly used distributions. These applications will include empirical, as well as theoretical, and discrete, as well as continuous, distributions. In addition, a method is presented for determining the manner in which the desired output parameter (that is, the measure of system performance) is distributed.

Flowcharts will again be used to illustrate how the various methods can be implemented on a computer.

4.1 THE INVERSE TRANSFORMATION METHOD

Suppose that we are given a probability density function $f(x)$, and we want to generate a random variate that is governed by this probability density function. The *inverse transformation method* offers a simple and straightforward approach to this problem.

We begin by obtaining the cumulative distribution, $F(x)$, corresponding to the given probability density function. Thus

$$F(x) = \int_{-\infty}^{x} f(x')dx' \tag{4.1}$$

where $0 \leq F(x) \leq 1$. Figure 4.1 shows a plot of a typical cumulative distribution function.

The cumulative distribution function is then solved for x; that is, if $y = F(x)$, then we can write

$$x = F^{-1}(y) \tag{4.2}$$

This expression allows us to determine the particular value of x that corresponds to a given value of y. Let us refer to these two values as x_0 and y_0, respectively. The relationship between x_0 and y_0 is illustrated in Fig. 4.1.

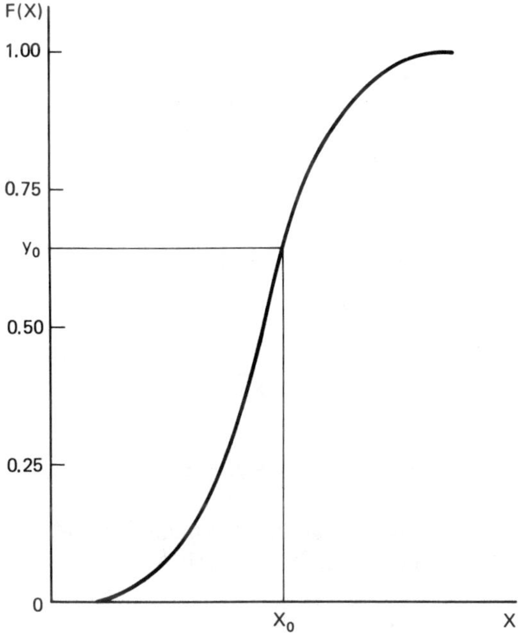

Figure 4.1

Now suppose that X is a random variate governed by the given probability density function, and that $Y = F(X)$ is a corresponding value of the cumulative distribution. We can write

$$\text{Prob}\left\{Y \le y_0\right\} = \text{Prob}\left\{X \le x_0\right\} = F(x_0)$$

But

$$F(x_0) = y_0$$

Hence

$$\text{Prob}\left\{Y \le y_0\right\} = y_0 \tag{4.3}$$

which is the expression for the cumulative uniform distribution within the interval (0, 1). This tells us that Y *is uniformly distributed within the interval (0, 1), regardless of the distribution of* X.

Therefore in order to generate a value for X using the inverse transformation method, we first represent Y by a uniformly distributed (0, 1) random number, U. We can then obtain the corresponding value for X by evaluating the expression

$$X = F^{-1}(U) \tag{4.4}$$

Example 4.1

Apply the inverse transformation method to the probability density function $f(x) = x/4$, $1 \le x \le 3$. Use the method to generate ten successive values for X, given the following uniformly distributed random variates: 0.35, 0.97, 0.22, 0.15, 0.60, 0.43, 0.79, 0.52, 0.81, 0.65

We first determine the cumulative distribution function. Thus

$$y = F(x) = \int_1^x (x/4)dx = (x^2 - 1)/8$$

for $1 < x \le 3$. Solving for x, we obtain

$$x = \sqrt{8y + 1}$$

which can be written as

$$X_i = \sqrt{8U_i + 1}$$

where U_i is a uniformly distributed (0, 1) random number.

Since $U_1 = 0.35$, we can obtain the corresponding value of X_1 as

$$X_1 = \sqrt{(8)(0.35) + 1} = 1.95$$

Similarly, since $U_2 = 0.97$, we obtain

$$X_2 = \sqrt{(8)(0.97) + 1} = 2.96$$

The values obtained for the ten required random variates are summarized below.

i	U_i	X_i
1	0.35	1.95
2	0.97	2.96
3	0.22	1.66
4	0.15	1.48
5	0.60	2.41
6	0.43	2.11
7	0.79	2.71
8	0.52	2.27
9	0.81	2.73
10	0.65	2.49

Notice that most of the calculated X's are located toward the right portion of the overall interval ($1 \leq x \leq 3$), even though the U's are uniformly distributed. This is caused by the shape of the given density function.

In the next few sections of this chapter we will consider some applications of the inverse transformation method to several well-known, commonly used distribution functions. The reader should bear in mind, however, that this method cannot be used with all distributions, since there are some probability densities (such as the normal) that cannot be integrated analytically. Moreover it may not be possible to obtain an explicit equation for x, even if an analytical expression for the cumulative distribution function can be obtained. We will consider some other techniques that can be used in such situations later in this chapter.

4.2 EMPIRICAL DISTRIBUTIONS

In many realistic problems the probability that an event will occur is expressed in terms of empirical, grouped data. Figure 4.2 illustrates a typical set of grouped data. We see that there are several adjacent subintervals, which are numbered consecutively (that is, $j = 1, 2, \ldots$ m). Each subinterval is represented as a rectangle whose upper and lower interval bounds are XU_j and XL_j, respectively. The height of each subinterval, f_j, represents the probability that the value of X for a random event will fall into the jth subinterval, that is, $XL_j \leq X \leq XU_j$. Since the f_j's represent probabilities, their sum must equal 1, that is,

$$f_1 + f_2 + \ldots + f_m = 1.$$

It is very easy to construct a cumulative distribution, as shown in Fig. 4.3, from such a set of grouped data. To do so, we simply calculate the cumulative sums of the previous f_j's, that is,

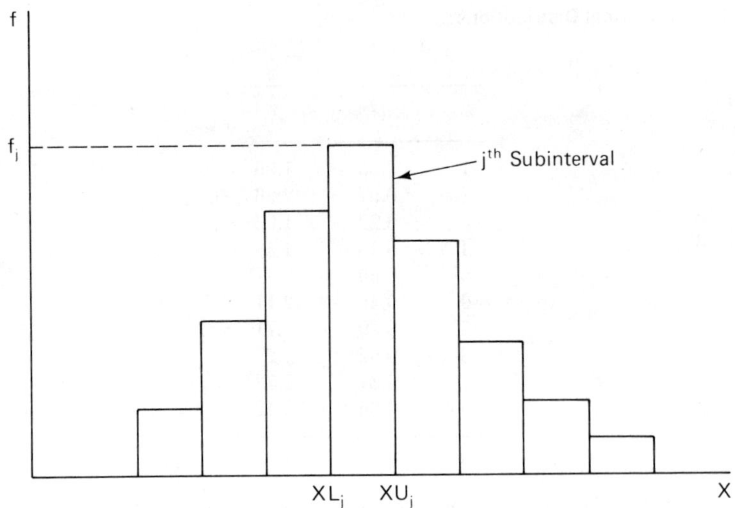

Figure 4.2

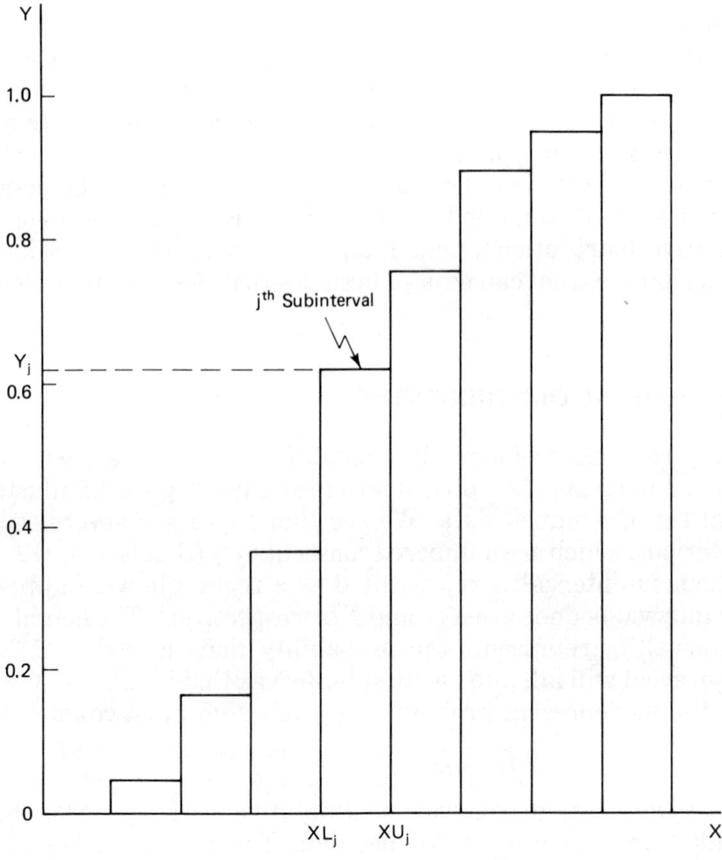

Figure 4.3

$$Y_1 = f_1$$

$$Y_2 = f_1 + f_2$$

$$.$$

$$.$$

$$.$$

$$Y_j = f_1 + f_2 + \ldots + f_j$$

$$.$$

$$.$$

$$.$$

$$Y_m = f_1 + f_2 + \cdots + f_m = 1$$

The height of each subinterval, Y_j, now represents the probability that the value of X for a random event does not exceed XU_j.

In order to make use of the inverse transformation method, we must pass a continuous curve through the cumulative distribution. The easiest way to do this is to use straight-line segments, as shown in Fig. 4.4. We can then generate a uniformly distributed $(0, 1)$ random number, U, and obtain the corresponding value for X by linear interpolation, that is,

$$X = XL_j + \left[\frac{U - Y_{j-1}}{Y_j - Y_{j-1}} \right] (XU_j - XL_j) \tag{4.5}$$

Figure 4.4 illustrates the relationship between U and X.

The only remaining problem is that of determining the appropriate subinterval. This matter can easily be resolved by starting with the left-most subinterval (that is, $j = 1$) and then successively comparing U with each of the Y_j's. The desired interval will be the first interval for which $U \leq Y_j$. The procedure is illustrated by the flowchart given in Fig. 4.5.

When implementing the method on a computer, the values of a and b and the Y_j's must be entered as input data. The interval boundaries, XL_j and XU_j, can either be read in or, for equally spaced intervals, be calculated using the expressions

$$XL_j = a + \left(\frac{b - a}{m} \right) (j - 1) \tag{4.6}$$

$$XU_j = a + \left(\frac{b - a}{m} \right) j \tag{4.7}$$

It is also quite easy to represent a portion of the cumulative distribution by a smooth curve rather than a straight-line segment. This is accomplished by passing an nth-degree polynomial through the data points within some local region of interest. (Typically, a quadratic or a cubic function is selected, that is, $n = 2$ or $n = 3$.) The reader is referred to a textbook on numerical analysis (specifically, polynomial interpolation) for more information on this topic.

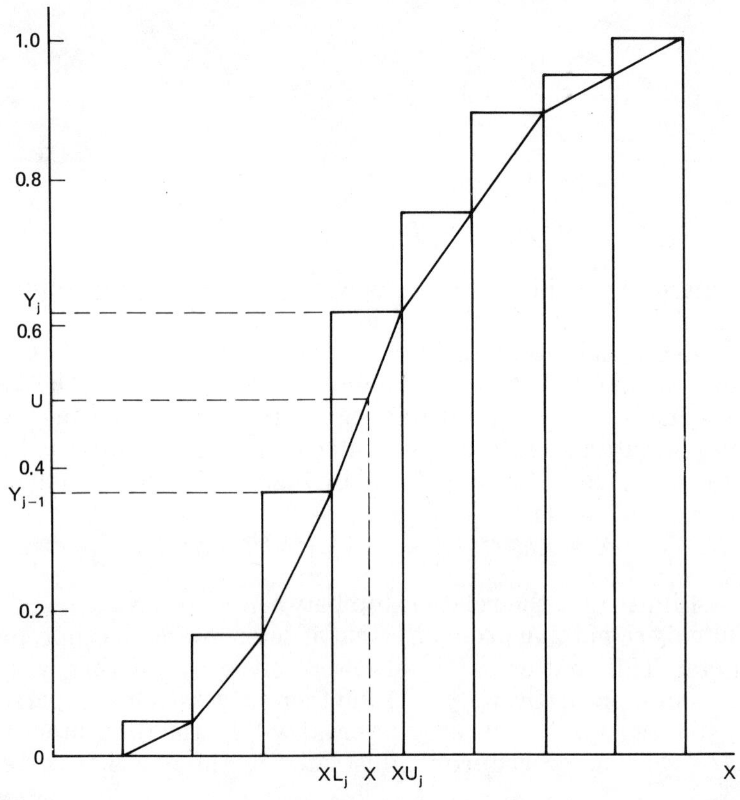

Figure 4.4

Example 4.2

The following empirical distribution describes the anticipated range of production costs for a newly developed product.

Production costs ($ per unit)	Probability of occurrence
60–70	0.20
70–80	0.35
80–90	0.30
90–100	0.15

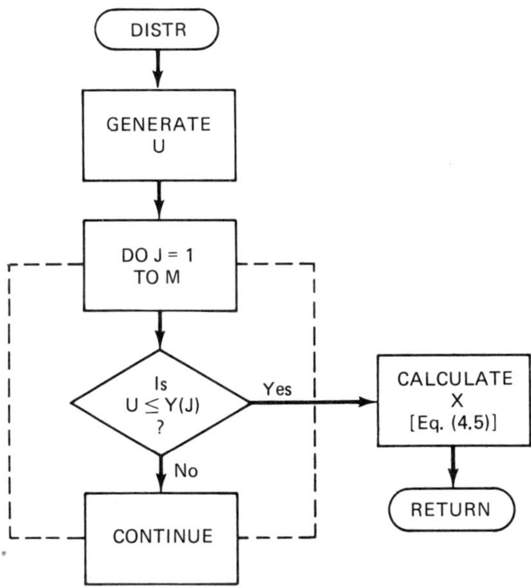

Figure 4.5

Generate ten successive random values for the production cost, using the following uniformly distributed random numbers: 0.35, 0.97, 0.22, 0.15, 0.60, 0.43, 0.79, 0.52, 0.81, 0.65.

We begin by constructing a cumulative distribution for the given data. This can be accomplished as follows:

j	XL_j	XU_j	Y_j
1	$60	$70	0.20
2	70	80	0.55
3	80	90	0.85
4	90	100	1.00

The distribution is shown in Fig. 4.6.

The first random number, 0.35, falls into the second category (because U_1 exceeds Y_1 but not Y_2). Therefore, $j = 2$, and

$$X_1 = 70 + \left[\frac{0.35 - 0.20}{0.55 - 0.20}\right](80 - 70) = 74.29$$

Similarly, the second random number, 0.97, falls into the fourth category. Hence

$$X_2 = 90 + \left[\frac{0.97 - 0.85}{1.00 - 0.85}\right](100 - 90) = 98.00$$

All ten values are summarized:

i	U_i	X_i
1	0.35	$74.29
2	0.97	98.00
3	0.22	70.57
4	0.15	67.50
5	0.60	81.67
6	0.43	76.57
7	0.79	88.00
8	0.52	79.14
9	0.81	88.67
10	0.65	83.33

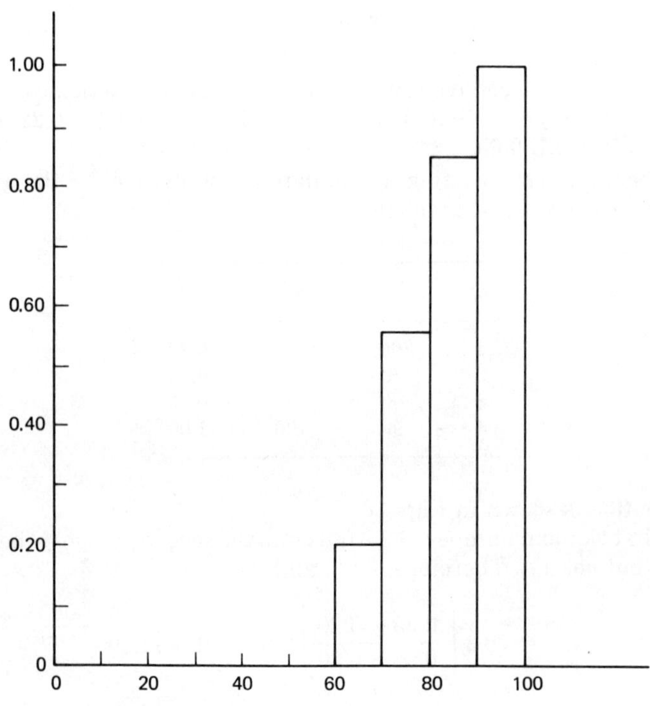

Figure 4.6

4.3 THE EXPONENTIAL DISTRIBUTION

Many simulation problems require the use of the exponential distribution. This is especially true of problems that involve a sequence of arrivals and departures, such as the simulation of a bank teller's window, a supermarket checkout counter, an airport, a job shop. We shall therefore give some consideration to the physical significance of the exponential distribution before discussing the generation of exponential random variates.

Suppose that x now represents time. We will assume that the probability of a random event occurring between times x and $(x + \Delta x)$ is $\alpha \Delta x$, where α is a known positive constant. Therefore the probability that the event will *not* occur within this time interval is $(1 - \alpha \Delta x)$.

Now consider a large time interval that ranges from 0 to x. Let us subdivide this interval into n equal subintervals of length Δx, so that $x = n \Delta x$. The probability that a random event does not occur within the given interval can now be expressed as

$$\lim_{\substack{\Delta x \to 0 \\ n \to \infty}} (1 - \alpha \Delta x)^n = \lim_{\Delta x \to 0} (1 - \alpha \Delta x)^{x/\Delta x}$$

$$= \lim_{\Delta x \to 0} \left[(1 - \alpha \Delta x)^{-1/\alpha \, \Delta x} \right]^{-\alpha x}$$

$$= e^{-\alpha x}$$

where e is the base of the *natural* (or *Naperian*) system of logarithms (e is an irrational number whose value is approximately $e = 2.7182818 \ldots$). Hence the probability that such an event *does* occur is

$$\text{Prob } (0 \leq X \leq x) = F(x) = 1 - e^{-\alpha x} \tag{4.8}$$

We can easily obtain the corresponding probability density as

$$f(x) = \alpha \, e^{-\alpha x} \tag{4.9}$$

Moreover, it is easy to show that the mean, μ, for the exponential distribution is simply $\mu = 1/\alpha$. (DeGroot 1975).

In order to make use of the inverse transformation method, we first solve Eq. (4.8) for x. Thus

$$x = -(1/\alpha) \; ln \; [1 - F(x)] \tag{4.10}$$

Since $F(x)$ is uniformly distributed, the quantity $1 - F(x)$ will also be uniformly distributed. Therefore we can write

$$X = -(1/\alpha) \; ln \; U \tag{4.11}$$

where X is the desired exponentially distributed random variate, and U is a uniformly distributed $(0, 1)$ random number, as before.

Now suppose that x is required to be greater than or equal to some specified positive value, x_0, that is, $0 < x_0 \leq x$. Equation (4.11) must be modified to read

$$X = x_0 - (1/\alpha) \; ln \; U \tag{4.12}$$

Also, the relationship between α and μ now becomes

$$\alpha = 1/(\mu - x_0) \tag{4.13}$$

(Notice that these relationships reduce to those presented earlier when $x_0 = 0$.)

In many waiting-line problems the probability of a random arrival between times x and $x + \Delta x$ is indeed proportional to the time interval Δx, as implied by the exponential distribution. Such arrivals can therefore be described by this distribution, where the random variable X represents the time between two successive arrivals.

Figure 4.7 shows a flowchart of the computational procedure.

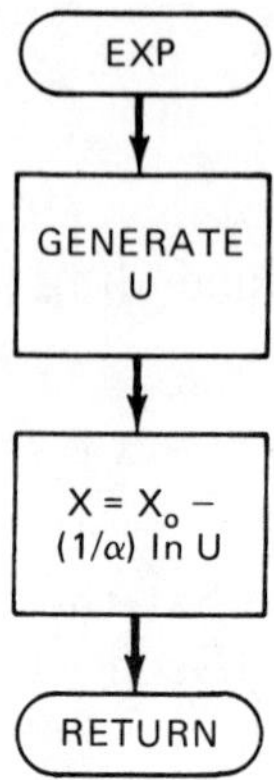

Figure 4.7

Example 4.3

Generate ten exponentially distributed random variates, using the following uniformly distributed random numbers: 0.35, 0.97, 0.22, 0.15, 0.60, 0.43, 0.79, 0.52, 0.81, 0.65. Assume that $x_0 = 2$ and $\mu = 6$ for the exponential distribution.

We begin by calculating a value for α, using Eq. (4.13). Thus,

$$\alpha = 1/(6 - 2) = 0.25$$

Equation (4.12) can now be used to calculate the desired random variates. Thus we obtain

$$X_1 = 2 - (1/0.25)\ ln\ (0.35) = 6.20$$
$$X_2 = 2 - (1/0.25)\ ln\ (0.97) = 2.12$$

and so on. All ten values are summarized:

i	U_i	X_i
1	0.35	6.20
2	0.97	2.12
3	0.22	8.06
4	0.15	9.59
5	0.60	4.04
6	0.43	5.38
7	0.79	2.94
8	0.52	4.62
9	0.81	2.84
10	0.65	3.72

4.4 THE GEOMETRIC DISTRIBUTION

Let us now consider a sequence of independent experiments, where the outcome of each experiment is either a success or a failure. Suppose that p is the probability of success for each experiment (hence $0 < p < 1$), and $q = 1 - p$ is the probability of failure. The probability of experiencing x consecutive failures followed by a success is therefore

$$f(x) = p\ q^x \tag{4.14}$$

where x must be a nonnegative integer, that is $x = 0, 1, 2, \ldots$, etc.

Equation (4.14) represents the probability density for the *geometric distribution*, whose mean is given by

$$\mu = q/p \tag{4.15}$$

and whose variance is

$$\sigma^2 = q/p^2 = \mu/p \tag{4.16}$$

(DeGroot 1975).

The corresponding cumulative distribution can be expressed as

$$F(x) = f(0) + f(1) + f(2) + \ldots + f(x)$$
$$= p + p\,q + p\,q^2 + \ldots + p\,q^x$$
$$= \sum_{k=0}^{x} p\,q^k \tag{4.17}$$

Observe that $F(0) = p$, and $p \le F(x) \le 1$.

In order to make use of the inverse transformation method, we must express the cumulative distribution in a somewhat different form. To do so, we note that

$$\text{Prob}\left\{X > 0\right\} = 1 - F(0) = 1 - p = q$$

$$\text{Prob}\left\{X > 1\right\} = 1 - F(1) = 1 - p - pq = q^2$$

$$\text{Prob}\left\{X > 2\right\} = 1 - F(2) = 1 - p - pq - pq^2 = q^3$$

$$\vdots$$

$$\text{Prob}\left\{X > x\right\} = 1 - F(x) = q^{x+1}$$

where X is a geometrically distributed random variate. Thus we can write

$$\frac{1 - F(x)}{q} = q^x \tag{4.18}$$

Now we know that $F(x)$ is uniformly distributed within the interval $(p, 1)$. Therefore the function $[1 - F(x)]/q$ will be uniformly distributed over the interval $(0, 1)$. Moreover from Eq. (4.18), we conclude that

$$0 \le q^x \le 1 \tag{4.19}$$

Therefore in order to use the inverse transformation method, we can write

$$U = q^X \tag{4.20}$$

where U is a uniformly distributed $(0, 1)$ random number. Solving for X, we obtain

$$X = \text{INT}\left\{ ln\ U / ln\ q \right\} \tag{4.21}$$

where INT denotes truncation. (The truncation is required so that X takes on only integer values.)

A flowchart of the procedure is shown in Fig. 4.8.

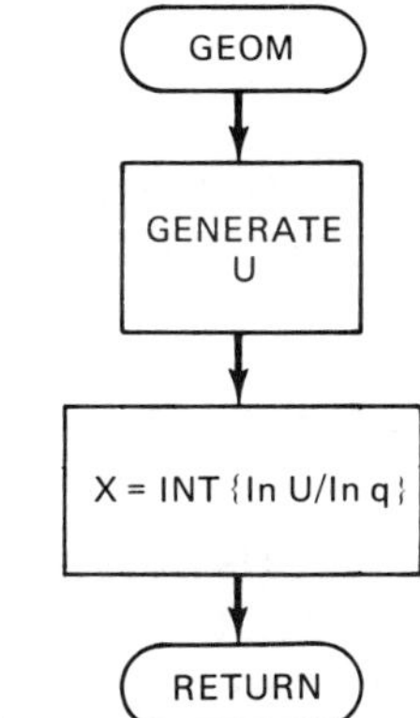

Figure 4.8

Example 4.4

Generate ten geometrically distributed random variates for a process whose probability of success is $p = 0.3$. Use the following set of uniformly distributed random numbers as a basis for the computation: 0.35, 0.97, 0.22, 0.15, 0.60, 0.43, 0.79, 0.52, 0.81, 0.65.

Since $p = 0.3$, we know that $q = 1 - p = 0.7$. Equation (4.21) can therefore be used to obtain

$$X_1 = \text{INT} \left\{ ln\ 0.35/ln\ 0.70 \right\} = 2$$

$$X_2 = \text{INT} \left\{ ln\ 0.97/ln\ 0.70 \right\} = 0$$

and so on. The results of the computation are summarized:

i	U_i	X_i
1	0.35	2
2	0.97	0
3	0.22	4
4	0.15	5
5	0.60	1
6	0.43	2
7	0.79	0
8	0.52	1
9	0.81	0
10	0.65	1

4.5 DIRECT SIMULATION—THE GAMMA DISTRIBUTION

Recall that the inverse transformation method can be used only if an analytical expression for the cumulative distribution function can be obtained and solved explicitly for x. There are many probability density functions for which this is not possible. An alternative technique must be used in such situations.

One such method involves direct simulation of the process under consideration. The details of the procedure are illustrated below for the *gamma distribution*, whose probability density is given by

$$f(x) = \frac{\alpha^\beta x^{(\beta - 1)} e^{-\alpha x}}{(\beta - 1)!} \tag{4.22}$$

where α is a positive constant and β is a positive, integer-valued constant. It can be shown that the mean for this distribution is $\mu = \beta/\alpha$ and the variance is $\sigma^2 = \beta/\alpha^2 = \mu/\alpha$ (DeGroot 1975). Moreover, it is not difficult to show that the variable x can be interpreted as the sum of β exponentially distributed random variables, each having an expected value of $1/\alpha$. Thus

$$x = x_1 + x_2 + \ldots + x_\beta \tag{4.23}$$

where

$$f(x_i) = \alpha e^{-\alpha x_i} \tag{4.24}$$

The probability density function for the gamma distribution [that is, Eq. (4.22)] cannot be integrated analytically; hence the inverse transformation method cannot be used to generate gamma random variates. We can, however, simulate the gamma process directly, by summing β exponential random variates. Thus, if we substitute Eq. (4.11) into Eq. (4.23), we obtain

$$X = -(1/\alpha) \sum_{i=1}^{\beta} ln \, U_i \tag{4.25}$$

where U_i is a uniformly distributed (0, 1) random number, as before. This expression can be written in a more convenient form as

$$X = -(1/\alpha) \, ln \prod_{i=1}^{\beta} U_i \tag{4.26}$$

since the logarithm of a product is equal to the sum of the logarithms of the individual factors. Equation (4.26) now provides a basis for a simple computational strategy, as outlined in Fig. 4.9.

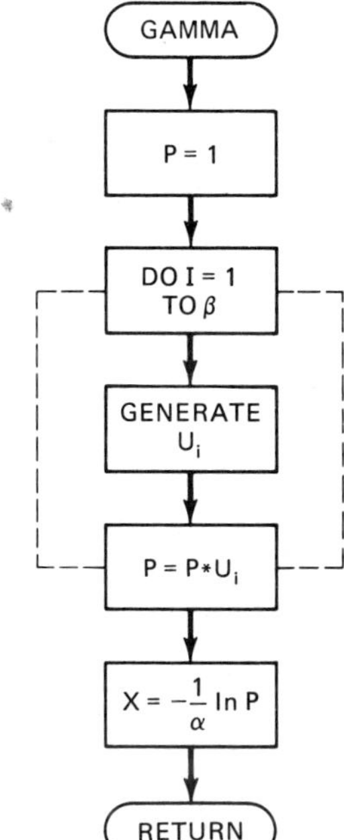

Figure 4.9

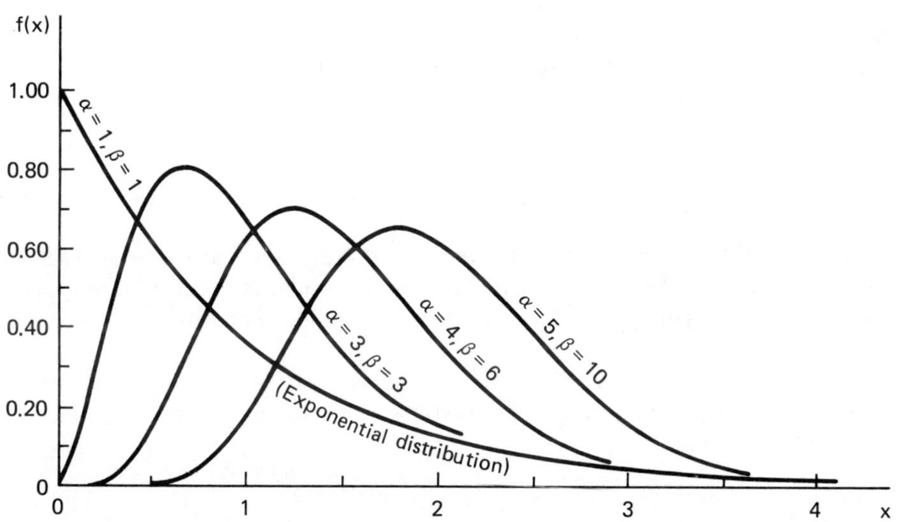

Figure 4.10

The form of the gamma distribution given above, in which β is restricted to integer values, is sometimes referred to as the *Erlang distribution*. This distribution is often used to represent empirical data because it can take on a variety of shapes, depending upon the values that are assigned to α and β. Figure 4.10 shows some representative curves. (Note that the gamma distribution reduces to the exponential distribution when $\alpha = \beta = 1$.)

The gamma distribution can also be defined for noninteger values of β, although physical applications of this nature are less common. The generation of gamma variates is considerably more difficult in this situation (Fishman 1973).

Example 4.5

Generate five random variates that are governed by the gamma distribution with $\alpha = 1$ and $\beta = 2$. Carry out the calculations using the following set of uniformly distributed random numbers: 0.35, 0.97, 0.22, 0.15, 0.60, 0.43, 0.79, 0.52, 0.81, 0.65.

The desired random variates can be obtained from Eq. (4.26). Thus,

$$X_1 = -(1/1) \, ln \, [(0.35)(0.97)] = 1.08$$
$$X_2 = -(1/1) \, ln \, [(0.22)(0.15)] = 3.41$$
$$X_3 = -(1/1) \, ln \, [(0.60)(0.43)] = 1.35$$
$$X_4 = -(1/1) \, ln \, [(0.79)(0.52)] = 0.89$$
$$X_5 = -(1/1) \, ln \, [(0.81)(0.65)] = 0.64$$

4.6 THE POISSON DISTRIBUTION

The Poisson distribution is closely related to the exponential distribution, and is therefore used in many simulation problems that involve a sequence of arrivals and departures. In particular, if the time between successive events is distributed exponentially, then the *number* of events that occur in some finite time interval will be distributed in accordance with the Poisson distribution, whose probability density is given by

$$f(x) = \frac{(\lambda t)^x}{x!} \, e^{-\lambda t} \tag{4.27}$$

where λ and t are positive constants. It can be shown that $\mu = \sigma^2 = \lambda t$ for this distribution (DeGroot 1975). The variable x must be a nonnegative integer, since it represents the number of individual events occurring in time t.

Poisson variates can easily be generated by direct simulation. To do so, note that the total amount of time between x successive events cannot exceed t, that is,

$$\sum_{i=1}^{x} t_i \leq t < \sum_{i=1}^{x+1} t_i \tag{4.28}$$

where t is specified and the t_i are exponential random variates, which can be expressed as

$$t_i = -(1/\lambda) \; ln \; U_i \qquad (4.29)$$

Hence, the smallest value of k that satisfies the inequality

$$\sum_{i=1}^{k+1} -(1/\lambda) \; ln \; U_i > t \qquad (4.30)$$

will be the desired Poisson variate.

In order to obtain an efficient computational procedure, let us rewrite Eq. (4.30) as

$$\sum_{i=1}^{k+1} ln \; U_i < -\lambda t$$

or

$$ln \prod_{i=1}^{k+1} U_i < -\lambda t$$

Now let us exponentiate both sides of this last expression. If we arbitrarily set $t = 1$, we obtain

$$\prod_{i=1}^{k+1} U_i < e^{-\lambda} \qquad (4.31)$$

The procedure, then, will be to form the product of successive uniformly distributed $(0, 1)$ random numbers, until Eq. (4.31) is satisfied. The desired random variate will be one less than the required number of U_i's. Figure 4.11 shows a flowchart of the computational procedure.

Example 4.6

Generate five random variates that are governed by the Poisson distribution with $\lambda t = 1.5$. Carry out the calculations using the following set of uniformly distributed random numbers: 0.35, 0.97, 0.22, 0.15, 0.60, 0.43, 0.79, 0.52, 0.81, 0.65, 0.20, 0.57, 0.10.

Since we have arbitrarily set $t = 1$, we therefore have a value of $\lambda = 1.5$. Hence, $e^{-\lambda} = e^{-1.5} = 0.223$. The desired random variates can now be obtained using Eq. (4.31). In order to obtain the first random variate, three uniformly distributed random numbers are required. Thus,

$$(0.35)(0.97)(0.22) = 0.075 < 0.223$$

Hence $X_1 = (3 - 1) = 2$. [Notice that Eq. (4.31) cannot be satisfied using only two uniformly distributed random numbers, that is, $(0.35)(0.97) = 0.340 > 0.223$.] Similarly,

$$X_2 = 0 \text{ since } 0.15 < 0.223$$
$$X_3 = 2 \text{ since } (0.60)(0.43)(0.79) = 0.204 < 0.223$$
$$X_4 = 3 \text{ since } (0.52)(0.81)(0.65)(0.20) = 0.055 < 0.223$$
$$X_5 = 1 \text{ since } (0.57)(0.10) = 0.057 < 0.223$$

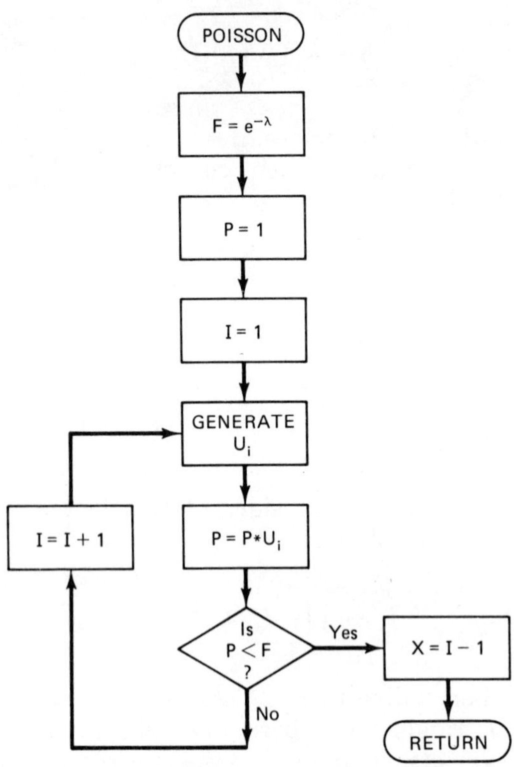

Figure 4.11

4.7 THE NORMAL DISTRIBUTION

In various realistic physical situations, there are many types of random events that are governed by the normal distribution. This distribution is characterized by a symmetric, bell-shaped probability density, given by

$$f(x) = \frac{1}{\sigma\sqrt{2\pi}}\ \exp\left\{-\frac{1}{2}\left[\frac{x-\mu}{\sigma}\right]^2\right\} \tag{4.32}$$

where μ is the mean and σ is the standard deviation. Like the gamma and the Poisson densities, the normal density function cannot be integrated analytically; hence the inverse transformation method cannot be used to generate normally distributed random variates. We can, however, once again generate the desired random variates by direct simulation. To do so we will consider a special case of Eq. (4.32), obtained by setting $\sigma = 1$ and

$$z = (x - \mu)/\sigma \tag{4.33}$$

Thus Eq. (4.32) becomes

$$f(z) = \frac{1}{\sqrt{2\pi}}\ e^{-z^2/2} \tag{4.34}$$

This is the probability density function for the *standard normal distribution*.

Now it is well known that a sequence of sample means obtained from *any* statistical distribution function will tend to be normally distributed about the true theoretical mean provided the sample size is sufficiently large (this is a consequence of the *central limit theorem*). In particular, if each sample mean is obtained from a set of N uniformly distributed (0, 1) random numbers, then the quantity

$$Z = \frac{(1/N)\ \sum\limits_{i=1}^{N} U_i - (1/2)}{\sqrt{1/(12N)}} \tag{4.35}$$

will be a standard normal variate, as governed by Eq. (4.34), provided N exceeds 10.

Let us now multiply the numerator and denominator of Eq. (4.35) by N, resulting in

$$Z = \frac{\sum\limits_{i=1}^{N} U_i - (N/2)}{\sqrt{N/12}} \tag{4.36}$$

For computational convenience, let us arbitrarily set $N = 12$. Equation (4.36) then simplifies to

$$Z = \sum\limits_{i=1}^{12} U_i - 6 \tag{4.37}$$

Equation (4.37) now provides a simple computational technique for obtaining a standard normal random variate. We simply sum twelve independent, uniform (0, 1) random numbers and then subtract six, resulting in a value for Z. Moreover if we wish to generate a normal random variate with mean μ and standard deviation σ, we first generate a standard normal, Z, using Eq. (4.37), and then calculate the desired value, X, as

$$X = \mu + \sigma Z \tag{4.38}$$

Figure 4.12 shows a flowchart of the procedure.

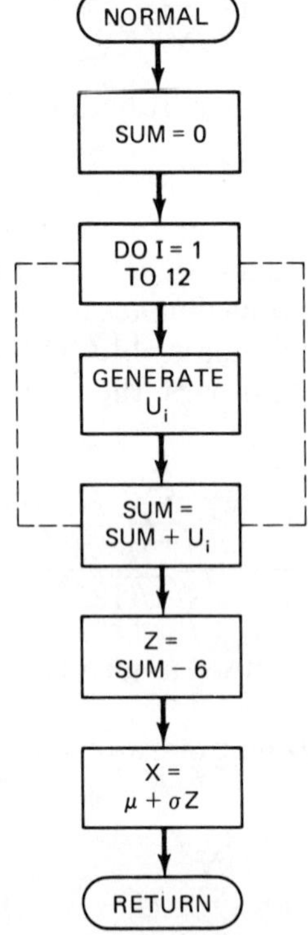

Figure 4.12

Example 4.7

Generate a random variate that is governed by a normal distribution whose mean is 5 and whose standard deviation is 2. Use Eq. (4.37) and the following set of uniformly distributed random numbers: 0.35, 0.97, 0.22, 0.15, 0.60, 0.43, 0.79, 0.52, 0.81, 0.65, 0.20, 0.57.

   If we determine the sum of the twelve given random numbers we obtain

$$\sum_{i=1}^{12} U_i = 0.35 + 0.97 + \ldots + 0.57 = 6.26$$

If we substitute this result into Eq. (4.37), we obtain the following standard normal random variate:

$$Z = (6.26 - 6) = 0.26$$

The desired normal random variate can now be obtained from Eq. (4.38). Thus

$$X = 5 + (2)(0.26) = 5.52$$

An alternative method for generating normally distributed random variates is to make use of either of the following two expressions:

$$Z = (-2 \ln U_1)^{1/2} \sin (2\pi U_2) \tag{4.39}$$

or

$$Z = (-2 \ln U_1)^{1/2} \cos (2\pi U_2) \tag{4.40}$$

Both expressions are known to result in standard normal random variates. The desired values can then be obtained from Eq. (4.38), as before.

Figure 4.13 presents a flowchart of the procedure, based upon the use of Eq. (4.39).

Notice that the use of Eq. (4.37) requires twelve different values of U_i for each Z, whereas Eqs. (4.39) and (4.40) each require only two U_i's. Thus this latter method appears to be computationally more efficient. On the other hand, the evaluations of a logarithm, a square root, and a sine or cosine function require more computational effort than the determination of a simple sum. Therefore it is not clear which of the methods is really more desirable.

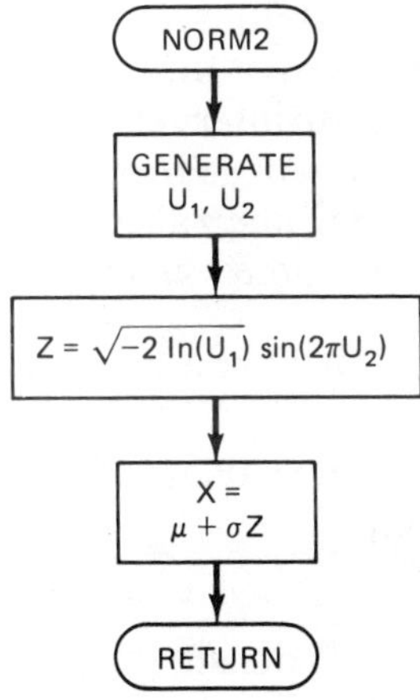

Figure 4.13

Example 4.8

Generate six random variates that are governed by a normal distribution whose mean is 5 and whose standard deviation is 2. Use Eq. (4.39) and the following set of uniformly distributed random numbers: 0.35, 0.97, 0.22, 0.15, 0.60, 0.43, 0.79, 0.52, 0.81, 0.65, 0.20. 0.57.

Substituting the first two random numbers into Eq. (4.39) results in

$$Z_1 = \sqrt{(-2)\ ln\ (0.35)}\ \ sin\ [(2\pi)(0.97)] = -0.27$$

and therefore

$$X_1 = 5 + (2)(-0.27) = 4.46$$

Similarly

$$Z_2 = \sqrt{(-2)\ ln\ (0.22)}\ \ sin\ [(2\pi)(0.15)] = 1.41$$

$$X_2 = 5 + (2)(1.41) = 7.82$$

$$Z_3 = \sqrt{(-2)\ ln\ (0.60)}\ \ sin\ [(2\pi)(0.43)] = 0.43$$

$$X_3 = 5 + (2)(0.43) = 5.86$$

$$Z_4 = \sqrt{(-2)\ ln\ (0.79)}\ \ sin\ [(2\pi)(0.52)] = -0.09$$

$$X_4 = 5 + (2)(-0.09) = 4.82$$

$$Z_5 = \sqrt{(-2)\ ln\ (0.81)}\ \ sin\ [(2\pi)(0.65)] = -0.53$$

$$X_5 = 5 + (2)(-0.53) = 3.94$$

$$Z_6 = \sqrt{(-2)\ ln\ (0.20)}\ \ sin\ [(2\pi)(0.57)] = -0.76$$

$$X_6 = 5 + (2)(-0.76) = 3.48$$

4.8 THE REJECTION METHOD

The *rejection method* provides a general procedure for generating random variates with *any* distribution whose probability density $f(x)$ is continuous and bounded within a finite region, that is, we require that $0 \leq f(x) \leq f_{max}$ within the finite interval $a \leq x \leq b$. The method is similar to that presented in Sec. 3.4, where we considered the use of the Monte-Carlo method for evaluating integrals.

In order to obtain a random variate, X, we proceed as follows:

1. Generate a pair of uniformly distributed (0, 1) random numbers, U_1 and U_2.

2. Obtain a uniform random variate, Z, within the interval $a \leq Z \leq b$, using the relationship

$$Z = a + (b - a)\ U_1$$

3. Evaluate the probability density at point Z, that is, determine $f(Z)$.

4. Obtain a uniform random value, Y, within the interval $0 \leq Y \leq f_{max}$, using the relationship

$$Y = f_{max}\ U_2$$

(Note that Y and Z will represent the coordinates of some point in space, as illustrated in Figs. 4.14 and 4.15.)

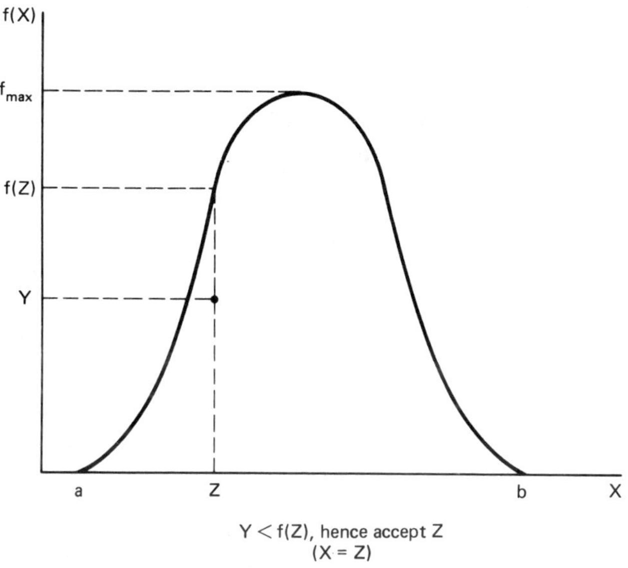

Figure 4.14

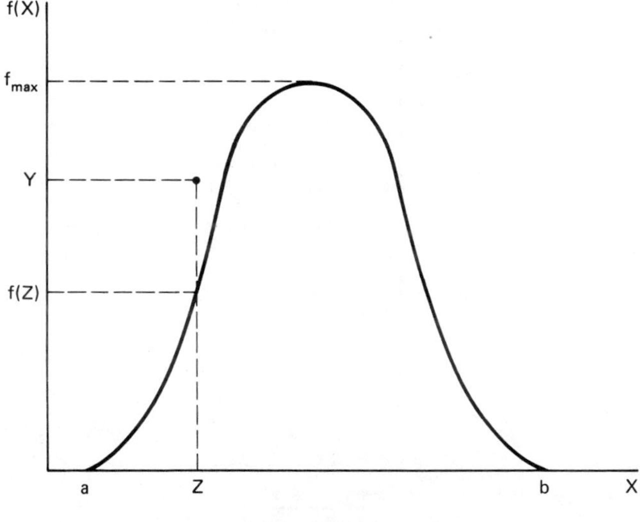

Figure 4.15

5. Compare Y with $f(Z)$.
   (a) If Y does not exceed $f(Z)$, then the point (Y, Z) will fall on or below the probability density curve, as indicated in Fig. 4.14. We therefore accept Z as the desired random variate, that is, we set, we set $X = Z$.
   (b) If Y is greater than $f(Z)$, then the point (Y, Z) will lie above the probability density curve, as illustrated in Fig. 4.15. We therefore reject this point and try again (that is, return to step 1).

6. The entire procedure (steps 1 to 5) is repeated successively until an acceptance is encountered in step 5, resulting in a value for the desired random variate.

The procedure is illustrated schematically in Fig. 4.16.

Although the rejection method can be used with many different probability density functions, it is relatively inefficient. In fact it can be shown that the expected number of trials (that is, the expected number of

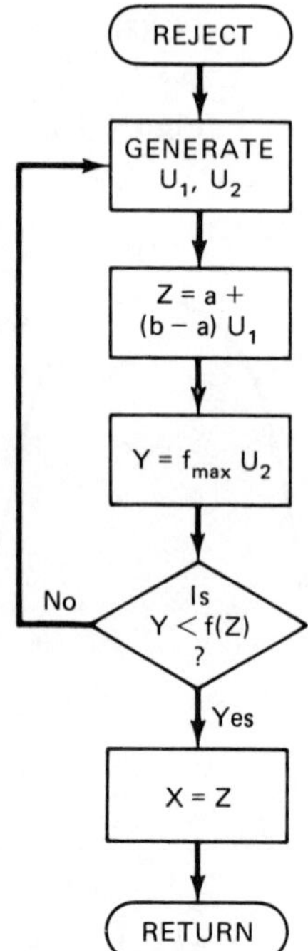

Figure 4.16

pairs of uniformly distributed random variates) required to obtain one value of X is equal to $(b - a)f_{max}$. Thus the rejection method is generally used only if there is no other alternative.

4.9 THE BETA DISTRIBUTION

In order to illustrate the use of the rejection method, let us consider the *beta distribution*. This distribution has a probability density given by

$$f(x) = \frac{(\beta_1 + \beta_2 - 1)! \; x^{(\beta_1 - 1)} \; (1 - x)^{(\beta_2 - 1)}}{(\beta_1 - 1)! \; (\beta_2 - 1)!} \tag{4.41}$$

where β_1 and β_2 are positive integers and $0 \le x \le 1$. It can be shown that the mean for this distribution is $\mu = \beta_1/(\beta_1 + \beta_2)$, and the variance is $\sigma^2 = \mu\beta_2/(\beta_1 + \beta_2)(\beta_1 + \beta_2 + 1)$. Furthermore it can be established that the variable x can be interpreted as

$$x = x_1/(x_1 + x_2) \tag{4.42}$$

where x_1 is a gamma variable with parameters (α, β_1) and x_2 is a gamma variable with parameters (α, β_2).

Example 4.9
The easiest way to generate beta-distributed random variates is to use direct simulation, based upon Eq. (4.42). This technique is discussed in Prob. 4.12 at the end of this chapter. For now, however, we will make use of the (less efficient) rejection method. In particular let us generate several beta variates with $\beta_1 = 2$ and $\beta_2 = 3$, based upon the following sequence of consecutive, uniformly distributed random numbers: 0.35, 0.97, 0.22, 0.15, 0.60, 0.43, 0.79, 0.52, 0.81, 0.65, 0.20, 0.57.

We first note that $Z = U_1$ in this example, since $a = 0$ and $b = 1$ (that is, $0 \le x \le 1$). Furthermore the probability density can be written

$$f(x) = \frac{(4!)}{(1!)(2!)} \, x(1 - x)^2 = 12x(1 - x)^2$$

This function takes on its maximum value at $x = 1/3$. Hence

$$f_{max} = 12(1/3)(2/3)^2 = 16/9 = 1.78$$

so that $Y = 1.78 \, U_2$.

The results obtained, using the method described previously, are summarized:

U_1	U_2	Z	$f(Z)$	Y	$Y \leq f(Z)$?	X
0.35	0.97	0.35	1.77	1.73	Yes	0.35
0.22	0.15	0.22	1.61	0.27	Yes	0.22
0.60	0.43	0.60	1.15	0.77	Yes	0.60
0.79	0.52	0.79	0.42	0.93	No	
0.81	0.65	0.81	0.35	1.16	No	
0.20	0.57	0.20	1.54	1.01	Yes	0.20

Thus we have generated four beta variates, whose values are $X_1 = 0.35$, $X_2 = 0.22$, $X_3 = 0.60$, and $X_4 = 0.20$.

4.10 OTHER DISTRIBUTION FUNCTIONS

In this chapter we have considered only the more common distribution functions. There are many other statistical distributions, such as the binomial, negative binomial, hypergeometric, log normal, nonintegral gamma, and nonintegral beta, as well as various multivariate distributions that may be required in certain types of simulation problems. Detailed information on the generation of random variates for such distributions is given by Fishman (1973) and by Naylor, Balintfy, Burdick and Chu, (1966).

4.11 CHOOSING A DISTRIBUTION FUNCTION

The choice of a distribution function is sometimes troublesome to the beginning practitioner. We therefore present some guidelines that should be helpful.

There are four considerations in the selection of a distribution function. They are

1. Special characteristics of a particular distribution function
2. Accuracy with which a distribution function can represent a given set of empirical data
3. Ease with which a distribution function can be fitted to a given set of empirical data
4. Computational efficiency when generating random variates

The exponential distribution (or its counterpart, the Poisson distribution) is frequently chosen to represent random arrivals to a system because of its special applicability to such situations (as discussed in Sec. 4.3). Moreover exponentially distributed random variates can be generated efficiently, and if the mean has not been determined, the distribution can easily be fitted to a set of empirical data simply by drawing a straight line through the data, when plotted in the proper manner using semilogarithmic coordinates (Gottfried 1979). The use of this distribution function is therefore recommended without reservations for those situations to which it applies.

The normal distribution is also used extensively, not necessarily because of any special intrinsic property, but because so many naturally occurring phenomena seem to be governed by this distribution. Unfortunately it is less efficient to work with than the exponential distribution (because the inverse transformation method does not apply). It can, however, be fitted to a set of empirical data quite easily, simply by specifying the mean and the standard deviation, or by plotting the data in the proper manner, using special probability coordinates, and then passing a straight line through the data (Gottfried 1979). The accuracy of the fit can also be determined with this latter method.

The gamma distribution is often used to represent a skewed set of empirical data. This is not a convenient function to work with, however, since it cannot easily be fitted to empirical data (nonlinear regression is required), and its use is relatively inefficient from a computational standpoint. The Weibull function may be a better choice if a mathematical distribution function is really required (see Prob. 4.11). In many practical applications, however, an empirical distribution will be entirely adequate; such distributions can easily be fitted to empirical data, and are computationally efficient.

Some applications require certain specialized distribution functions that cannot easily be fitted to the available data. Computerized curve-fitting techniques are required in such situations. The reader is referred to a textbook on regression analysis for more information on this important topic.

4.12 BUILDING AN EMPIRICAL DISTRIBUTION

So far this chapter has been concerned with techniques for generating random variates from known distribution functions. Let us now consider the reverse problem; namely, the construction of an empirical distribution from a given set of random variates. This problem arises when processing the output data (that is, the measure of system performance) that is generated during a simulation study.

Suppose that $X_1, X_2, \ldots, X_n$ represent random variates that were obtained by repeatedly evaluating a mathematical model. We wish to *group* the data into several successive subintervals and then determine the corresponding cumulative distribution. Let us begin by determining how many random variates fall into each subinterval. To do so we compare each X_i with the successively increasing interval bounds XU_1, $XU_2, XU_3, \ldots, XU_j, \ldots$, until we find an interval bound that satisfies the condition

$$X_i < XU_j \tag{4.43}$$

The given random variate will then fall within the jth subinterval. (Note that we are using the same notation as in Sec. 4.2. In particular, the XU_j's represent the upper interval bounds; that is, $(a < XU_1 < XU_2 < \ldots < XU_m = b)$. These values may be specified or, in the case of equally spaced intervals, they may be calculated using Eq. (4.7) provided, of course, that a, b, and m are known.)

Figure 4.17 shows a flowchart of a procedure for grouping the data, based upon the use of Eq. (4.43). The first loop sets $K(J)$, the counter for each subinterval, equal to zero. (If the XU_j's are to be computed, that calculation can also be included within this loop.) The double loop then proceeds to group the data. The inner loop determines the proper subinterval for a given X_i, and the outer loop causes the procedure to be repeated for each X_i. Once all of the random variates have been grouped, the relative frequencies $F(1), F(2), \ldots, F(M)$ can be determined as $F(J) = K(J)/N$, and the cumulative distribution can be evaluated as $Y(1) = F(1)$ and $Y(J) = Y(J-1) + F(J)$.

Example 4.10

A proposed business venture has been simulated, resulting in 1000 different values for the yearly profit. All of the values were greater than zero but less than \$10,000. It was therefore decided to group the data into ten equally spaced intervals, with $a = 0$ and $b = 10$. Hence $XU_1 = 1$, $XU_2 = 2, \ldots, XU_{10} = 10$.

Suppose that the first five values for the yearly profit, in thousands of dollars, are 2.56, 7.73, 2.41, 5.00, and 8.32. Let us group the data using the method presented in Fig. 4.17.

(a) $X_1 = 2.56$

$$X_1 > XU_1 \text{ and } XU_2$$

However

$$X_1 < XU_3.$$

Therefore

$$J = 3 \text{ so that } K(3) = 1.$$

(b) $X_2 = 7.73$

$$X_2 > XU_1, XU_2, \ldots, XU_6$$

but

$$X_2 < XU_8.$$

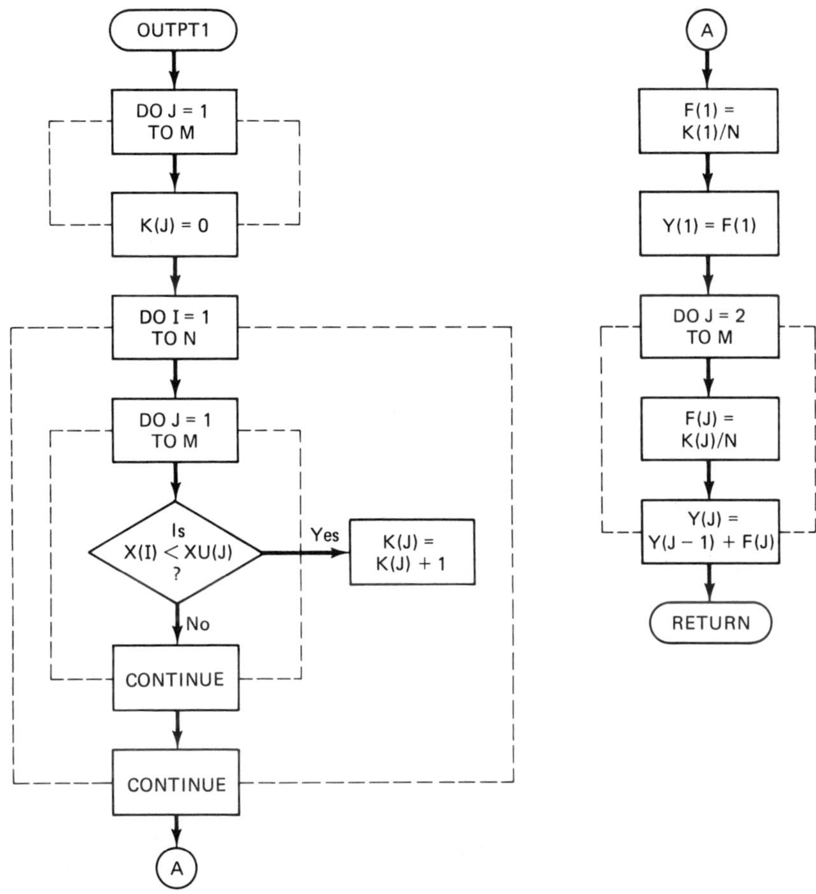

Figure 4.17

Therefore

$$J = 8 \text{ and } K(8) = 1.$$

(c) $X_3 = 2.41$

$$X_3 > XU_1 \text{ and } XU_2$$

but

$$X_3 < XU_3.$$

Therefore

$$J = 3 \text{ and } K(3) = 2.$$

(d) $X_4 = 5.00$

$$X_4 > XU_1, XU_2, XU_3, \text{ and } XU_4$$

but

$$X_4 = XU_5.$$

Hence

$$X_4 < XU_6, \text{ so that } J = 6 \text{ and } K(6) = 1.$$

$$\text{(e) } X_5 = 8.32$$

$$X_5 > XU_1, XU_2, \ldots, XU_8$$

but

$$X_5 < XU_9.$$

Hence

$$J = 9 \text{ and } K(9) = 1.$$

Now suppose that all of the 1000 random variates have been placed in their respective intervals, resulting in the following grouping of the data.

J	$K(J)$
1	70
2	97
3	149
4	137
5	118
6	104
7	96
8	82
9	78
10	69
	1000

The relative frequencies and the cumulative distribution can then be obtained:

J	$F(J)$	$Y(J)$
1	$70/1000 = 0.070$	$0.070 \qquad\quad = 0.070$
2	$97/1000 = 0.097$	$0.070 + 0.097 = 0.167$
3	$149/1000 = 0.149$	$0.167 + 0.149 = 0.316$
4	$137/1000 = 0.137$	$0.316 + 0.137 = 0.453$
5	$118/1000 = 0.118$	$0.453 + 0.118 = 0.571$
6	$104/1000 = 0.104$	$0.571 + 0.104 = 0.675$
7	$96/1000 = 0.096$	$0.675 + 0.096 = 0.771$
8	$82/1000 = 0.082$	$0.771 + 0.082 = 0.853$
9	$78/1000 = 0.078$	$0.853 + 0.078 = 0.931$
10	$69/1000 = 0.069$	$0.931 + 0.069 = 1.000$

These values, together with the corresponding mean and the standard deviation, would probably be printed out at the end of the simulation.

Figure 4.18 shows an alternative procedure for obtaining the cumulative distribution without first determining the relative frequencies.

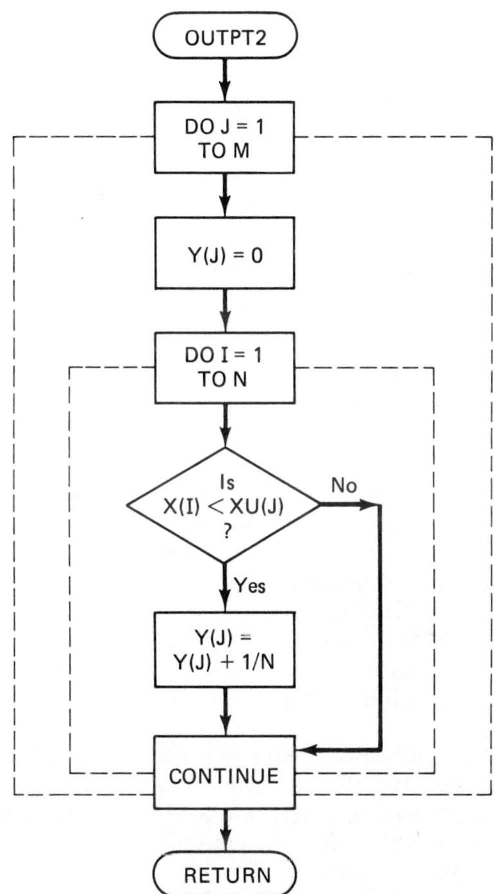

Figure 4.18

PROBLEMS

4.1. Use the inverse transformation method to obtain an explicit formula for X as a function of U for each of the following probability density functions.

(a) $f(x) = \begin{cases} 1/4, & 1 \le x \le 5 \\ 0 & \text{elsewhere} \end{cases}$

(b) $f(x) = \begin{cases} x^2/21, & 1 \le x \le 4 \\ 0 & \text{elsewhere} \end{cases}$

(c) $f(x) = \begin{cases} (x-1)/3, & 1 \le x \le 3 \\ (8-2x)/3, & 3 \le x \le 4 \\ 0 & \text{elsewhere} \end{cases}$

(d) $f(x) = \begin{cases} 1/6, & 0 \le x \le 2 \\ 1/3, & 2 \le x \le 4 \\ 0 & \text{elsewhere} \end{cases}$

4.2. Use the rejection method to generate several random variates governed by the following probability density function.

$$f(x) = \begin{cases} (x-1)/3, & 1 \le x \le 3 \\ (8-2x)/3, & 3 \le x \le 4 \\ 0 & \text{elsewhere} \end{cases}$$

The calculations should be based upon the following set of uniformly distributed (0, 1) and random numbers: 0.18, 0.64, 0.27, 0.92, 0.39, 0.81, 0.58, 0.73, 0.10, 0.41, 0.54, 0.07, 0.83, 0.31.

4.3. Write a self-contained computer subprogram to generate random variates for each of the distributions given below. Use the method described in the text for each distribution. Write each subprogram in such a manner that any required constants are entered as arguments.
  (a) An empirical distribution (see Sec. 4.2)
  (b) The exponential distribution (see Sec. 4.3)
  (c) The geometric distribution (see Sec. 4.4)
  (d) The gamma distribution (see Sec. 4.5)
  (e) The Poisson distribution (see Sec. 4.6)
  (f) The normal distribution using Eqs. (4.37) and (4.38)
  (g) The normal distribution, using Eqs. (4.39) and 4.38)
  (h) The beta distribution (see Sec. 4.9)

4.4. Write a self-contained computer subprogram that will group a set of N random variates and then determine the relative frequencies and the cumulative distribution, using the procedure outlined in Fig. 4.17. Include a provision to print out the following information.
  (a) Interval number [that is, J]
  (b) Interval size [that is, $XL(J)$ and $XU(J)$]
  (c) Relative frequency [that is, $F(J)$]
  (d) Cumulative distribution [that is, $Y(J)$]
  In addition, calculate and print out the expected value (that is, the mean) and the standard deviation for the given set of random variates.

4.5. Repeat Prob. 4.4 using the method outlined in Fig. 4.18. (Note that the calculation of the relative frequencies must be deleted when using this method.)

4.6. Write a self-contained computer subprogram that will generate a bar
graph of an empirical cumulative distribution, such as that obtained in
Prob. 4.4 or 4.5 above. Represent each bar with four adjacent lines of
asterisks. Include a provision for labeling the ordinate (from 0 to 1.000) and
the abscissa (with the proper values for the interval bounds). Note that the
bar graph will have to be generated sideways on a line printer, as
illustrated in Fig. 4.19 (Gottfried 1972).

4.7. Write a complete computer program that will read and store a set of N
random variates, and then process these data by accessing the subprograms that were written for Probs. 4.4 and 4.6, or Probs. 4.5 and 4.6,
above. (The program should generate and print out an empirical

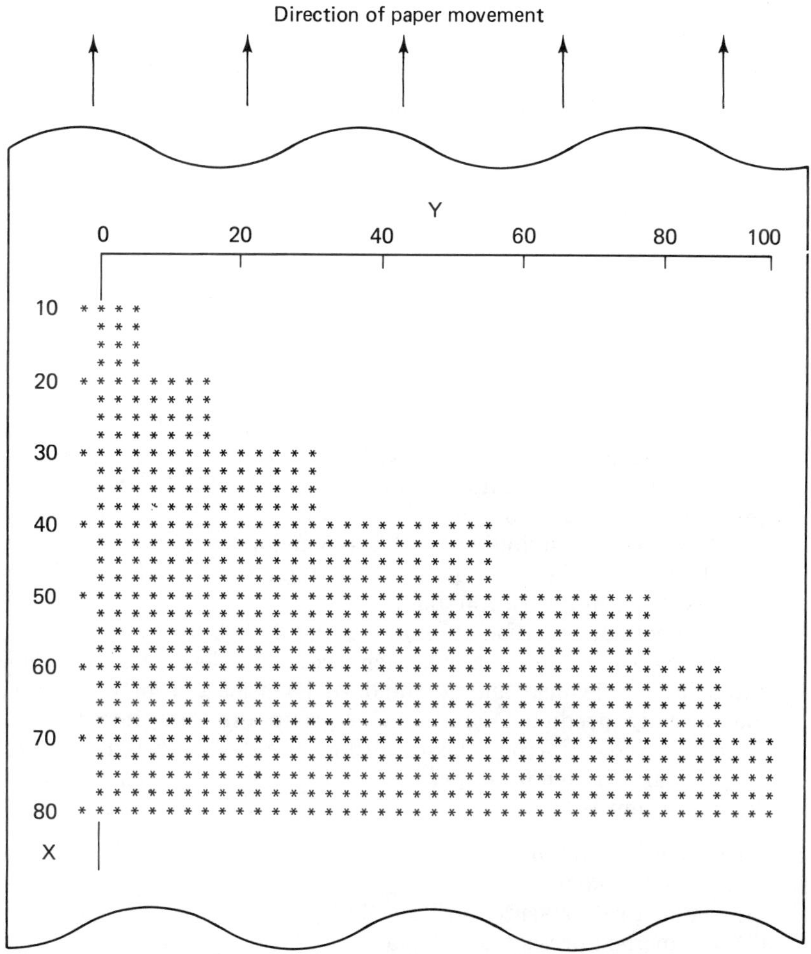

Figure 4.19

distribution, a corresponding bar graph, and a value for the mean and the standard deviation.)

Test the program with the following set of data:

−0.29399	2.29765	0.53428	4.11226
1.92471	2.52207	2.57476	2.77242
2.16467	0.41535	5.06051	0.52010
3.93196	1.25962	4.0995	−0.44103
7.37279	6.76416	2.34776	7.97699
4.89658	3.15635	5.01888	3.34819
−2.07186	4.06033	5.82695	3.93691
3.11837	1.12052	2.44334	0.80254
2.34600	7.09554	1.76621	2.90435
−7.88595	1.50665	2.02803	1.82497

4.8. Write a complete computer program that can be used to test the subprograms written for Prob. 4.3 by generating and storing N random variates for each of the distributions outlined below.

(a) An empirical distribution consisting of four subintervals, as follows:

J	$XL(J)$	$XU(J)$	$Y(J)$
1	60	70	0.20
2	70	80	0.55
3	80	90	0.85
4	90	100	1.00

(b) The exponential distribution, with $x_0 = 2$ and $\mu = 6$.
(c) The geometric distribution, with $p = 0.3$.
(d) The gamma distribution, with $\alpha = 1$ and $\beta = 2$.
(e) The Poisson distribution, with $\lambda = 1.5$.
(f) The normal distribution, with $\mu = 5$ and $\sigma = 2$.
[Test each of the routines for generating normal random variates, as described in Probs. 4.3f and g.]
(g) The beta distribution, with $\beta_1 = 2$ and $\beta_2 = 3$.
Generate 100 random variates in each case (that is, let $N = 100$). Calculate and print out the mean and the standard deviation and compare with the theoretically predicted values. Also group the data into ten equal intervals, and print out the resulting cumulative distribution. (The subprogram written for either Prob. 4.4 or 4.5 can be used to obtain the desired output.)

4.9. Write a complete computer program that will generate N normally distributed random variates, using
(a) the method presented in Prob. 4.3f
(b) the method presented in Prob. 4.3g
Calculate and print out the mean and the standard deviation, but do not print the individual random variates.

Use the program to generate 200 normally distributed random variates, with $\mu = 3$ and $\sigma = 1.5$. Run the program using each of the above methods (two separate runs). Determine which method requires less computer time.

4.10. Write a complete computer program to solve Prob. 3.17. Use the subprogram written for Prob. 4.3a to generate random values for the cost per unit and the yearly sales volume. Also use the subprogram written for Prob. 4.4 to process the output data (distribution of yearly profits). Carry out 200 consecutive simulations and compare with the results obtained in Prob. 3.17.

4.11. The probability density for the *Weibull distribution* is given by

$$f(x) = \frac{\alpha(x - x_0)^{\alpha-1}}{\beta^\alpha} e^{-[(x - x_0)/\beta]^\alpha}$$

where α, β, and x_1 are positive constants and $x \geq x_0$.

Write a self-contained computer subprogram for generating Weibull-distributed random variates using the inverse transformation method.

4.12. Write a self-contained computer subprogram that will generate beta-distributed random variates by direct simulation. To do so, first generate two gamma variates, X_1 and X_2, using the expressions

$$X_1 = -\ln \prod_{i=1}^{\beta_1} U_i$$

$$X_2 = -\ln \prod_{i=\beta_1+1}^{\beta_1+\beta_2} U_i$$

[See Eq. (4.26).] Then use Eq. (4.42) to obtain the desired beta variate. (Note that this method requires that β_1 and β_2 be positive integers.)

4.13. Write a self-contained computer subprogram that will generate normally distributed random variates using the rejection method. To do so, assume that $f(x) = 0$ for $x > \mu + 3\sigma$ and $x < \mu - 3\sigma$.

4.14. Write a complete computer program that will generate N beta-distributed random variates, using
(a) the rejection method (see Ex. 4.9)
(b) direct simulation, as described in Prob. 4.12.
Calculate and print out the mean and the standard deviation, but do not print the individual random variates.

Use the program to generate 200 random variates, with $\beta_1 = 3$ and $\beta_2 = 4$. Run the program using each of the above methods (two separate runs). Determine which method requires less computer time.

4.15. Repeat Prob. 4.9, comparing
(a) the method based upon Eqs. (4.37) and (4.38)
(b) use of the rejection method, as described in Prob. 4.13.
Based upon these results, does the rejection method appear to be an efficient way to generate normally distributed random variates?

5

INDUSTRIAL AND BUSINESS
APPLICATIONS

Let us now consider a variety of realistic problem situations that typically arise in business and industry. In particular we will consider some problems involving facility utilization, equipment maintenance, inventory control, networks (including PERT), and financial risk analysis. Included are several examples which illustrate, in simple terms, the general characteristics of these various problem areas. Once the underlying ideas are understood, the reader should have little difficulty in altering these models to describe related problem situations.

All of the problems presented in this chapter share a common characteristic, namely, that the consecutive random events are processed essentially independently of one another. This permits us to approach these problems from a relatively common point of view. A more difficult class of problems, to which the processing of one random event is dependent upon the completion of its predecessor, will be considered elsewhere in this text.

Most of the models in this chapter are *event based*, which means that the analysis is based upon the occurrence of various events (arrivals, departures, etc.) within the system. Some *time-based* models, based upon a sequence of consecutive time intervals, will also be considered.

5.1 GENERAL COMPUTATIONAL PROCEDURE

Let us begin by considering the procedure that is used to evaluate a simulation model once that model has been properly formulated and the required data have been obtained. (Figure 5.1 shows a schematic diagram of the overall solution procedure.)

1. The input data (that is, the values for the decision variables and the system parameters) are read into the computer.
2. The following sequence of steps is repeated N times:
    a. A value is generated for each of the required random variates $(X_1, X_2, \ldots, X_l)$.
    b. The model is then solved, resulting in a numerical value for each of the state variables.
    c. The system performance criterion (Y) is then evaluated, using the random variates generated in step 2a and the values for the state variables obtained in step 2b.

    (Note that the repeated execution of this step will result in N different values for Y.)
3. Compute and print the following statistical information for the system performance criterion:
    a. The expected value (that is, the mean), $\overline{Y}$.
    b. The standard deviation, s.

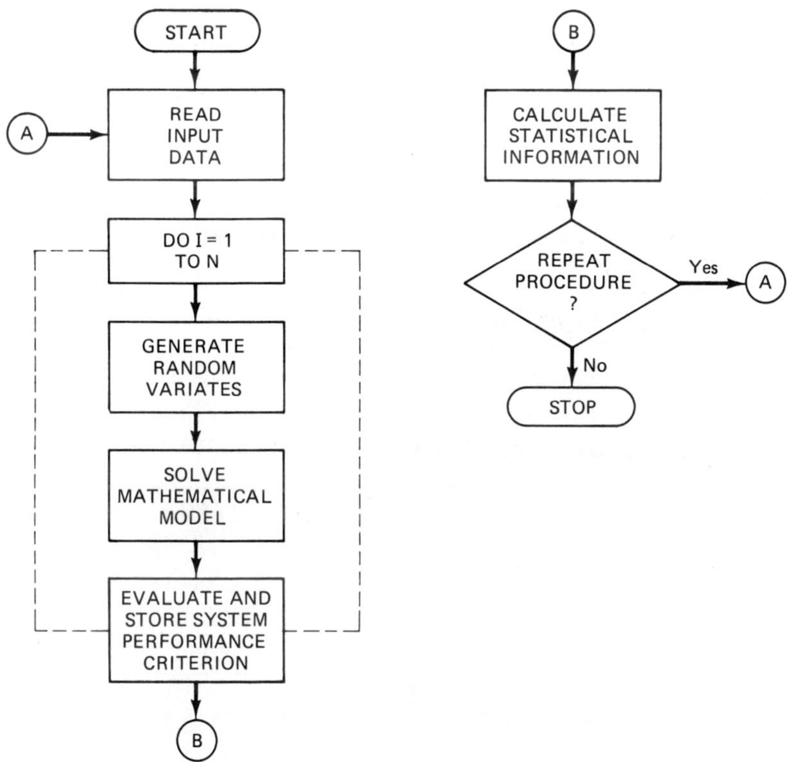

Figure 5.1

   c. The relative frequencies and the cumulative distribution. It is also a good idea to print some of the input data (in order to identify the problem), and perhaps certain of the state variables.

4. In many situations it will be desirable to change some particular values of the input data and then repeat the entire simulation (that is, steps 1 to 3 above).

The remainder of this chapter will be concerned with the development of specific mathematical models, and their corresponding computational strategies, for a variety of representative problem situations.

5.2 FACILITY UTILIZATION

Many different kinds of problem situations are concerned with the most efficient use of a facility (that is, a machine, a service area, etc.). In problems of this type we normally seek a balance between under-utilization and overutilization of the facility. The former is undesirable because it results in excessive idle time, whereas the latter causes either excessive delays or lost business. We wish to determine the degree of utilization associated with specified demand and service distributions.

This discussion will be confined to situations in which there is no backlog of demand. Thus an incoming order will be lost if a facility is not immediately available to process that order. (In Chap. 7 we will consider a more complicated class of problems, where incoming orders wait in line to be processed.)

Let us begin by defining the following symbols:

A_i = arrival time of the ith order (that is, the time when the ith order enters the facility)

D_i = departure time of the ith order (that is, the time when the ith order leaves the facility)

AT_i = time interval between the arrival of the $(i-1)$st and the ith orders (a random variate)

PT_i = time required for the facility to process the ith order (a random variate)

IT_i = idle time (that is, time interval when the facility is not being used) prior to processing the ith order

$TOTIT$ = cumulative idle time

F = fraction of time the facility is in use (the system performance criterion)

(The subscript i refers only to those orders that are *actually processed;* that is, lost orders are not included.)

The mathematical model consists of simply three equations, describing consecutive arrivals, consecutive departures, and an order "history." Thus we can describe the consecutive arrivals as

$$A_i = A_{i-1} + AT_i \tag{5.1}$$

Similarly the consecutive departures can be expressed as

$$D_i = D_{i-1} + PT_i + IT_i \tag{5.2}$$

Finally the arrival and departure of each order can be related by writing

$$D_i = A_i + PT_i \tag{5.3}$$

We must also develop a mathematical expression for the condition that any order that arrives while the facility is busy will be lost. This condition can be written as

$$A_i \geq D_{i-1} \tag{5.4}$$

Hence if a calculated value of A_i does not satisfy this condition, that particular order is considered lost. A new value for AT_i is then generated, and A_i is increased by this amount; that is,

$$A_i = A_i + AT_i \tag{5.5}$$

It is easy to see that $IT_i \geq 0$ provided Eq. (5.4) is satisfied. Thus from Eqs. (5.2) and (5.3),

$$
\begin{aligned}
IT_i &= D_i - (D_{i-1} + PT_i) \\
     &= (A_i + PT_i) - (D_{i-1} + PT_i) \\
     &= A_i - D_{i-1} \geq 0
\end{aligned}
\tag{5.6}
$$

as required.

In order to initiate the computation, let us arbitrarily begin with the arrival of the first order. We therefore set $A_1 = TOTIT = 0$ and $D_1 = PT_1$, where PT_1 is randomly generated. For each successive order, we then generate values for AT_i and PT_i, and proceed to calculate A_i using Eq. (5.1) and (if necessary) Eq. (5.5), D_i using Eq. (5.3), and IT_i from Eq. (5.2). The cumulative idle time can then be updated as

$$TOTIT = TOTIT + IT_i \tag{5.7}$$

This procedure is repeated consecutively, until either a desired number of orders or a specified time period has been simulated. The system performance criterion (that is, the fraction of time the facility is in use) can then be obtained as

$$F = 1 - [TOTIT/(D_n - A_1)] \tag{5.8}$$

where D_n represents the departure time of the last (that is, the nth) order processed. (Note that the quantity $(D_n - A_1)$ represents the total simulated time.)

Figure 5.2 shows a flowchart of the entire computational procedure. This figure includes a provision for carrying out the entire simulation N

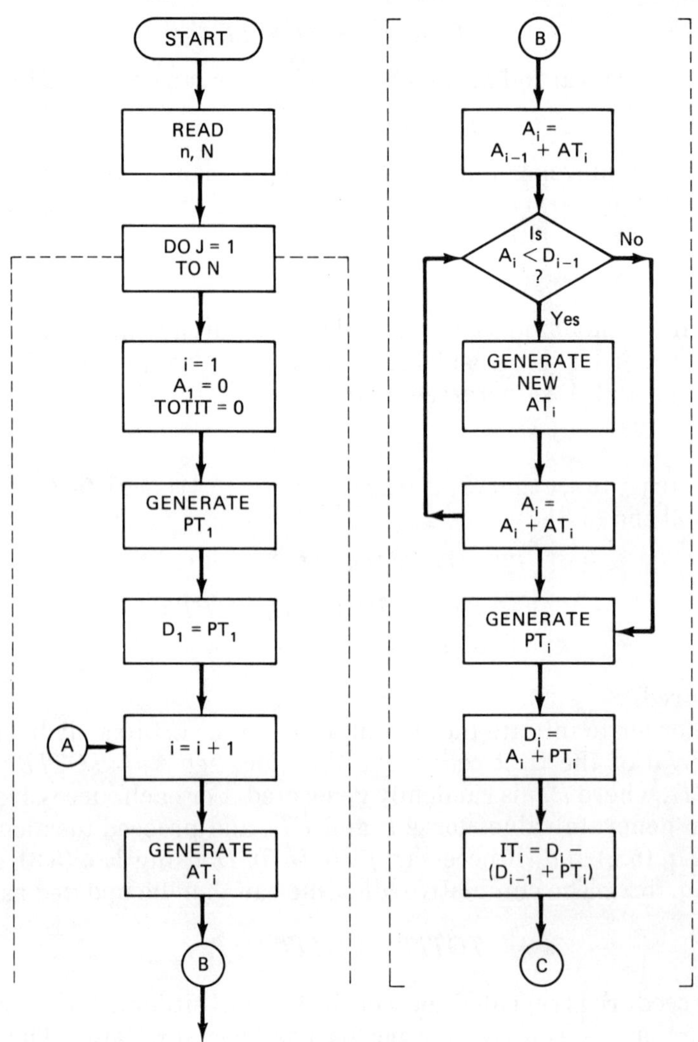

Figure 5.2

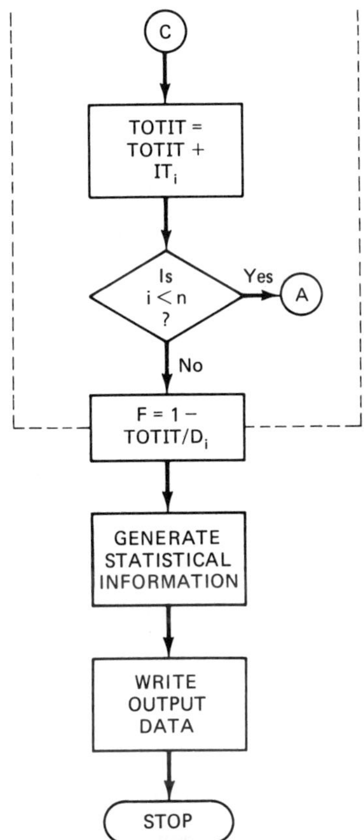

Figure 5.2 *(continued)*

times, as indicated in Sec. 5.1. A distribution for the fraction of time that the facility is in use can thus be obtained.

Example 5.1

A hospital wishes to study the utilization of a specially equipped emergency care facility. The interarrival times of patients requiring the use of the facility are known to be exponentially distributed, with a mean of 1.4 hours. The treatment times are believed to be normally distributed, with a mean of 0.8 hour and a standard deviation of 0.2 hour. Each patient must have immediate access to the facility when it is needed; otherwise the patient will be treated elsewhere within the hospital (though in a less effective manner).

The utilization of the facility can be simulated using the procedure shown in Fig. 5.2. In order to demonstrate the manner in which the calculations are carried out, however, let us consider the arrivals and departures of the first few patients.

i	AT_i^*	A_i	Is $(A_i \geq D_{i-1})$?	PT_i^*	D_i	IT_i	TOTIT
1		0		0.746	0.746	0	0
2	1.470	1.470	Yes ——	1.082	2.552	0.724	0.724
3	0.043	1.513	No				
	2.120	3.633	Yes ⟶	0.886	4.519	1.081	1.805
4	2.656	6.289	Yes ⟶	0.782	7.071	1.770	3.575
5	0.715	7.004	No				
	1.182	8.186	Yes ⟶	0.694	8.880	1.115	4.690
6	0.330	8.516	No				
	0.915	9.431	Yes ⟶	0.648	10.079	0.551	5.241
7	0.295	9.726	No				
	0.603	10.329	Yes ⟶	0.840	11.169	0.250	5.491
8	2.253	12.582	Yes ⟶	1.182	13.764	1.413	6.904

*Random variates.

Note that the second column (labeled AT_i) contains the randomly generated interarrival times, and the fifth column (labeled PT_i) contains the randomly generated treatment times. The table should be examined on a row-by-row basis, however, since each row corresponds to the arrival of a different patient. In some cases we see that the facility is not available when a patient arrives; hence that patient immediately proceeds elsewhere.

In the above example the total time elapsed is given by $(D_8 - A_1) = (13.764 - 0) = 13.764$ hours, and the cumulative idle time is 6.904 hours. Therefore for this simulated time period, the fraction of time that the facility is utilized is $1 - (6.904/13.764) = 0.502$ (or 50.2 percent). A much longer simulated time period would, of course, be required in order to obtain a more meaningful estimate of this parameter.

5.3 EQUIPMENT MAINTENANCE

In equipment maintenance problems, the objective is to determine a regular maintenance schedule that is reasonably inexpensive and not particularly disruptive. Thus we seek a schedule that is neither too frequent nor too infrequent, since both extremes may result in excessive downtime. The desired balance will be determined by such factors as the likelihood of failure, the corresponding repair time, the frequency of performing preventative maintenance, and the associated costs.

Let us consider a somewhat idealized situation in which a repair or maintenance team will always be immediately available whenever it is needed. Hence there will never be a backlog of demand. (Stated differently, any piece of equipment that requires service will never be kept waiting.) This assumption may be reasonably representative of some situations, and totally inappropriate for others.

Let us analyze each piece of equipment independently, based upon a viewpoint similar to that presented in the last section. Thus the equipment will be assumed to break down intermittently, with a random repair time for each breakdown. Now, however, we must take into consideration the fact that preventive maintenance will be performed at regularly scheduled intervals.

We now define the following symbols.

A_i = time when the ith breakdown occurs (equivalent to the time the repair team arrives)

D_i = time when the ith breakdown has been repaired (equivalent to the time the repair team departs)

DT_i = time interval between completion of the $(i-1)$st repair and the ith breakdown (equivalent to the length of time the equipment will function before breaking down—a random variate)

RT_i = time required to repair the ith breakdown (a random variate)

MT = time required to perform regularly scheduled preventative maintenance (assumed constant—a specified input parameter)

CT = cycle time (that is, the time between regularly scheduled maintenance services—a specified input parameter)

RC = repair cost, per unit of repair time (a specified input parameter)

MC = maintenance cost, per unit of regularly scheduled maintenance time (a specified input parameter)

$TOTC$ = cumulative total cost per maintenance cycle (the system performance criterion)

NC = number of maintenance cycles per simulated time period (a specified input parameter)

The mathematical model is based upon the use of the following two equations, which describe the occurrence of successive breakdowns for a given piece of equipment, and the time required to repair each breakdown.

$$A_i = D_{i-1} + DT_i \tag{5.9}$$

$$D_i = A_i + RT_i \tag{5.10}$$

In addition the cumulative total cost must be updated whenever the equipment is serviced. Thus

$$TOTC = TOTC + RC * RT_i \tag{5.11}$$

for each repair, and

$$TOTC = TOTC + MC * MT \tag{5.12}$$

whenever routine maintenance is performed.

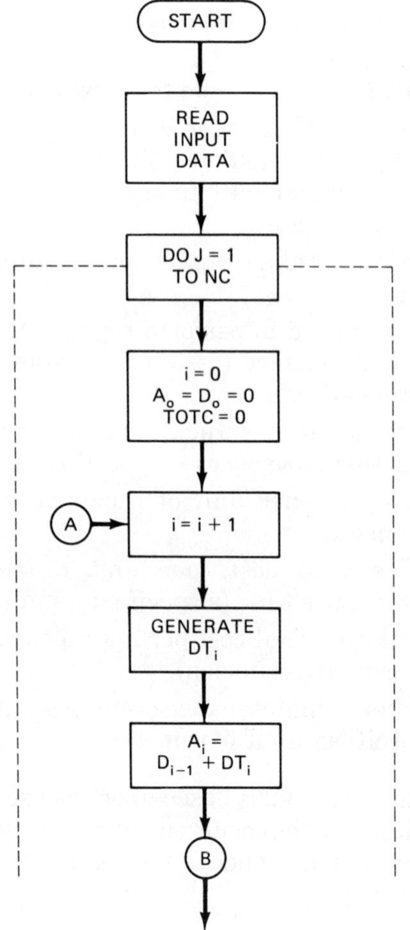

Figure 5.3

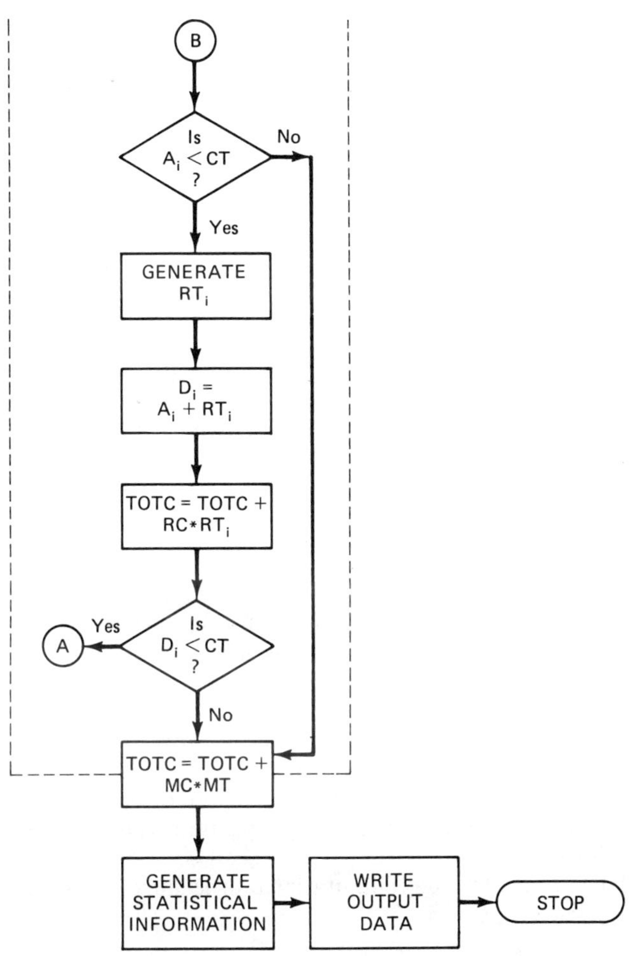

Figure 5.3 *(continued)*

The computation can best be initiated by assuming that the equipment has just been serviced and is ready for operation. Therefore we begin by setting $A_0 = D_0 = TOTC = 0$. We then generate successive values for DT_i and RT_i, and compute the corresponding values for A_i and D_i using Eqs. (5.9) and (5.10). The cumulative cost is then obtained using Eq. (5.11). This procedure is repeated until the simulated time becomes equal to or greater than the specified cycle time (CT). Once this occurs the cumulative cost is updated using Eq. (5.12), and another cycle is initiated. This procedure is repeated until the desired number of cycles (NC) has been simulated. Figure 5.3 shows a flowchart of the entire procedure.

Example 5.2

A machine shop owns twelve drill presses, which are used on an essentially continuous basis. Each machine is subject to periodic breakdowns, in accordance with the following distribution:

Time between breakdowns (Days)	Relative frequency	Time between breakdowns (Days)	Relative frequency
0- 1.99	0.021	16-17.99	0.092
2- 3.99	0.044	18-19.99	0.067
4- 5.99	0.079	20-21.99	0.047
6- 7.99	0.106	22-23.99	0.032
8- 9.99	0.119	24-25.99	0.018
10-11.99	0.128	26-27.99	0.008
12-13.99	0.123	28-29.99	0.003
14-15.99	0.113		1.000

The repair times are governed by a gamma distribution with $\alpha = 3$ and $\beta = 2$ (hence the expected value of the repair time is 2/3 day). The corresponding cost is $100 per day of downtime. For simplicity, assume that a repairman will always be available to service a machine that has broken down, that is, a machine will not have to wait to be serviced.

A periodic preventative maintenance program, requiring 6 hours of time per machine, could be initiated at a cost of $50 per machine. If this program were adopted, all of the machines would be serviced after 60 days of continuous operation. (Assume that all of the machines would receive service at the same time, during a weekend.)

In order to illustrate the computational procedure, the simulation of one maintenance cycle is outlined:

i	DT_i^*	A_i	Is $(A_i < 60)$?	RT_i^*	D_i	Is $(D_i < 60)$?	Cost	$TOTC$
1	9.64	9.64	Yes ⟶	0.36	10.00	Yes	$36	$36
2	24.05	34.05	Yes ⟶	1.14	35.19	Yes	114	150
3	7.51	42.70	Yes ⟶	0.45	43.15	Yes	45	195
4	6.10	49.25	Yes ⟶	0.80	50.05	Yes	80	275
5	13.77	63.82	No (stop)				50	325

*Random variates.

Observe that the second column (labeled DT_i) contains the randomly generated times between repair and subsequent breakdown, and the fifth column (labeled RT_i) contains the randomly generated repair times. However, the table should be read on a row-by-row basis, since each row corresponds to a different breakdown.

Notice that the fifth breakdown occurs after 63.82 days, which exceeds the 60-day cycle time. The computation therefore ceases at this time.

The total cost for this example would be equal to the cumulative repair costs ($275) plus the $50 maintenance cost—hence a total of $325. Normally, a distribution would be developed for this quantity.

Most equipment maintenance problems involve several pieces of equipment, as in the last example. If the pieces of equipment are independent of each other, as is usually the case, then their performance can be simulated on a one-at-a-time basis. This can be accomplished by means of a loop, which will be executed N times (where N is the number of pieces of equipment). Each pass through the loop will simulate the required number of maintenance cycles for a different piece of equipment. Thus we will have a loop within a loop. (The special case of service upon failure only, that is, no preventative maintenance, can be accommodated by setting $NC = 1$, $MC = 0$, and $CT = $ some large value.) Figure 5.4 illustrates the general procedure that is used for multiple pieces of equipment.

In Chap. 7 we will extend the analysis to problems in which a piece of equipment that is in need of repair may have to wait for service. Such problems are considerably more complicated than those discussed in this section.

5.4 INVENTORY CONTROL

We now consider the operation of a warehouse, which must maintain an inventory of many different items. Each item will be depleted on a more or less continuous basis, as a result of random customer demand. Additional merchandise is ordered from a supplier when the inventory reaches or drops below a certain level (the *reorder point*). The inventory level will continue to drop during the time required for the order to be processed (the *lead time*). Once the order has been filled, however, the inventory will be replenished to some relatively high level. Thus the inventory level will fluctuate periodically, as shown in Fig. 5.5.

In inventory control the objective is to maintain an adequate inventory level as inexpensively as possible. This requires the determination of suitable values for the recorder point and the order size. If either of these factors is too large the inventory level will tend to be too high, thus tying up capital and possibly resulting in spoilage. On the other hand, shortages may be experienced if the reorder point is too low

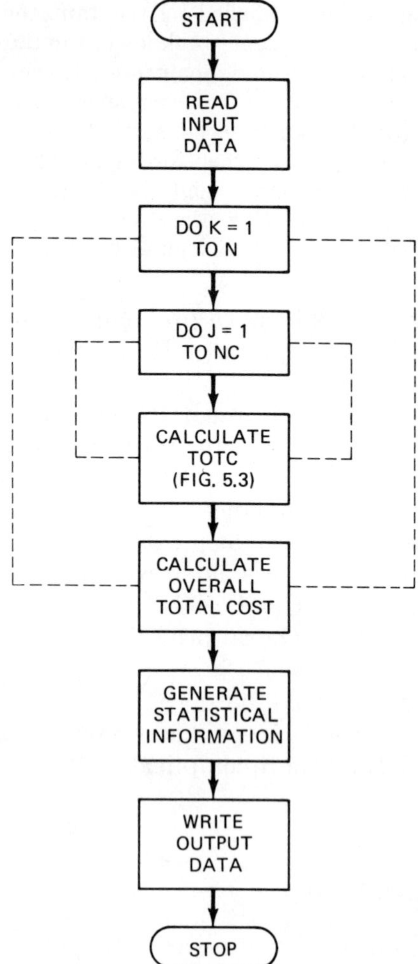

Figure 5.4

or the order size is too small. Thus we once again seek a balance between factors that can be excessively large on the one hand, or excessively small on the other.

This type of problem can best be analyzed by considering a time-based model, which utilizes a sequence of equal time intervals—typically, 1 day in duration. We therefore define the following symbols, based upon such a constant time interval.

D = customer demand, number of units per day (a random variate)

IL = inventory level, number of units

S = inventory shortage, number of units per day

ROP = reorder point, number of units

Q = order size, number of units

LT = lead time, days

t = cumulative time, days

t_f = final time, days (a stopping criterion)

T = time when current order will be filled

NB = number of units back-ordered at time t (to be delivered to customers once the inventory level has been replenished)

CI = carrying cost, per unit per day

CIT = cumulative carrying cost

CO = order cost, per order

COT = cumulative order cost

CS = shortage cost, per unit per day

CST = cumulative shortage cost

$TOTC$ = cumulative total cost (system performance criterion)

The mathematical model is essentially a scheme for making daily adjustments to the inventory level and the number of back-ordered

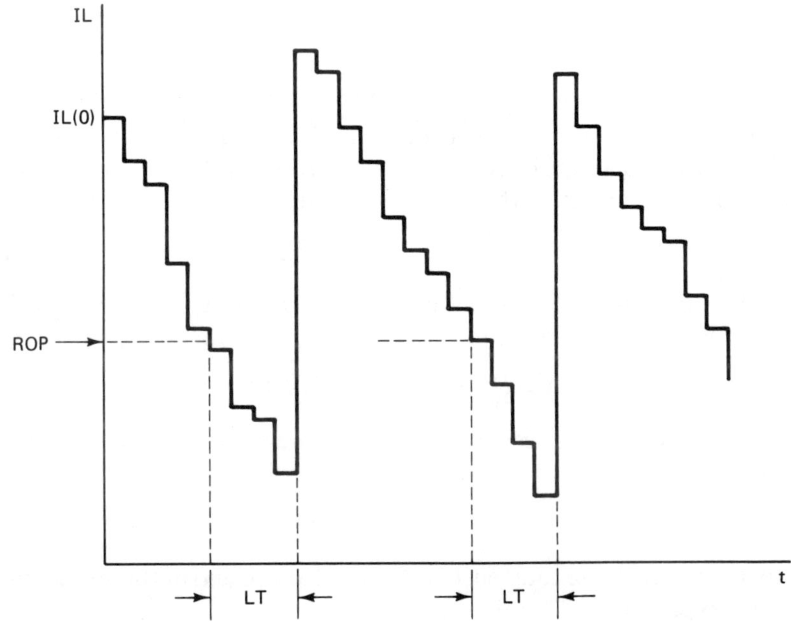

Figure 5.5

units. Thus the inventory level is adjusted upward as new orders are received, that is,

$$IL = IL + Q - NB \tag{5.13}$$

and downward in response to customer demand, that is,

$$IL = IL - D \tag{5.14}$$

with the provision that $IL \geq 0$. Similarly, the number of back-ordered units is adjusted upward as shortages occur, that is,

$$NB = NB + S \tag{5.15}$$

and downward as new orders are received, that is,

$$NB = NB - Q \tag{5.16}$$

with the restriction that $NB \geq 0$. The shortage, S, appearing in Eq. (5.15) is a positive quantity, which is equal in magnitude to any negative value for IL obtained from Eq. (5.14). That is, if the calculated value of IL is negative, we set $S = -IL$ and then set $IL = 0$.

After each day the cumulative time is adjusted as

$$t = t + 1 \tag{5.17}$$

Also whenever a new order is placed, the delivery time for that order is determined as

$$T = t + LT \tag{5.18}$$

To complete the model we must add the equations required to update the various cumulative costs. Thus the cumulative carrying cost is updated each day, after the inventory level has been adjusted, as

$$CIT = CIT + CI * IL \tag{5.19}$$

Similarly the cumulative shortage cost is updated whenever NB is adjusted, using the expression

$$CST = CST + CS * NB \tag{5.20}$$

and the cumulative order cost is updated whenever a new order is placed, using the equation

$$COT = COT + CO \tag{5.21}$$

Finally the cumulative total cost is obtained at the end of the simulation, using the expression

$$TOTC = CIT + CST + COT \tag{5.22}$$

Figure 5.6 shows a detailed flowchart of the computational procedure. The result of the computation will be a single value for the cumulative cost corresponding to the given input parameters (especially the decision variables ROP and Q). In order to obtain a cost distribution it will therefore be necessary to consider several successive time periods, each of length t_f, carrying out the computation for each time period in the manner indicated above. The individual values for the cumulative total cost can thus be grouped to form the desired distribution (see Sec. 5.1).

Example 5.3

A department store maintains an inventory of television sets in a warehouse. The demand for these sets is believed to be Poisson distributed, with a mean of 8.2 sets per day. The lead time required to obtain new sets from a supplier is 5 days.

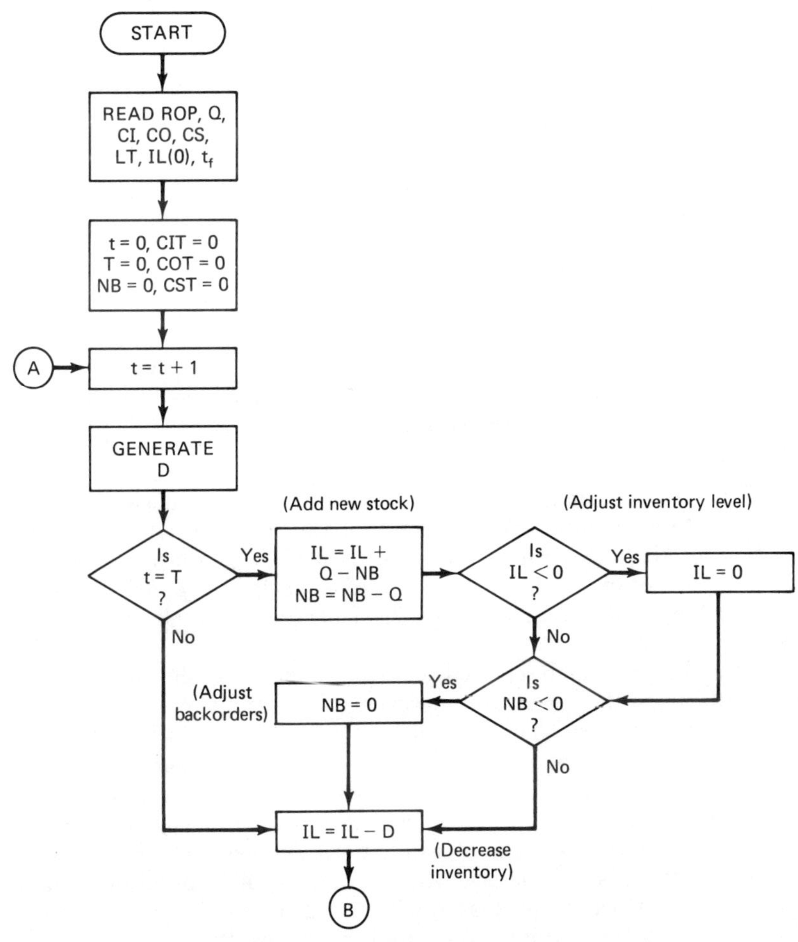

Figure 5.6

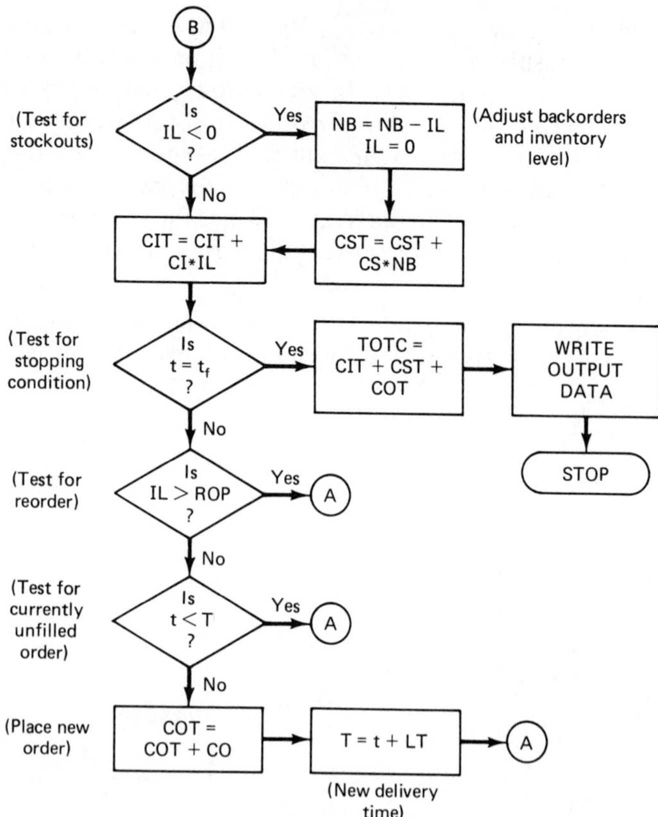

Figure 5.6 *(continued)*

We wish to determine by simulation an appropriate ordering policy (that is, appropriate values for the reorder point and the order size) if the carrying cost (inventory cost) is \$0.25 per set per day, the cost of placing an order is \$20 (plus, of course, the cost of each set), and the cost of a shortage is \$2 per set per day. Simulate a period of 1 year, assuming the department store is open 7 days per week.

The simulation can be carried out using the strategy shown in Fig. 5.6. In order to illustrate the procedure, however, a few sample calculations will be subsequently presented. These calculations are based on an initial inventory level of 60 sets, and values of $ROP = 40$ and $Q = 50$ as decision variables.

The second column (labeled D) contains the randomly generated daily demands. The table should be read on a row-by-row basis, however, since each row represents a separate day's transactions. Notice that an order for 50 additional units is placed whenever the inventory level drops below 40 units. In this example, new orders are placed when $t = 2$ and $t = 8$. Each order will be delivered 5 days after it is placed ($t = 7$ and $t = 13$), resulting in a replenishment of the inventory level.

t	D^*	IL	NB	Place new order?	CIT	CST	COT	$TOTC$
0		60	0		0	0	0	0
1	10	50	0	No	12.50	0	0	12.50
2	12	38 $(<ROP)$	0	Yes $(T = 7)$⌐	22.00	0	20	42.00
3	9	29	0	No	29.25	0	20	49.25
4	11	18	0	No	33.75	0	20	53.75
5	8	10	0	No	36.25	0	20	56.25
6	12	0 $(Q = 50)$	2	No	36.25	4	20	60.25
7	6	42	0	No	46.75	4	20	70.75
8	9	33 $(<ROP)$	0	Yes $(T = 13)$	55.00	4	40	99.00
9	11	22	0	No	60.50	4	40	104.50
10	5	17	0	No	64.75	4	40	108.75

etc.

*Random variate.

$$\left\{ \text{calculations based upon} \quad \begin{array}{l} ROP = 40 \text{ units} \\ Q = 50 \text{ units} \\ LT = 5 \text{ days} \end{array} \right\}$$

The last four columns show the cumulative carrying cost, cumulative shortage cost, cumulative order cost, and cumulative total cost, respectively. The last quantity is the system performance criterion; hence we would normally obtain a cumulative distribution of this parameter.

The reader should understand that the computational procedure presented here is representative of inventory control problems, but does not apply to all such problems. Many variations are encountered in actual practice; for example,

a. It may be possible to neglect back orders in some situations (the customer will go elsewhere if the item is not in stock).

b. Lead times may be negligible (all orders are delivered instantaneously).

c. Lead times may be random.

d. Orders may overlap (that is, a new order may be placed before a previous order has been delivered).

e. A complete inventory may (probably will) consist of many different items. Separate accounting will usually be required for each item. The placement of orders for multiple items from the same supplier will therefore have to be coordinated.

f. The merchandise may be perishable, thus restricting the length of time that an item can be stored in inventory.

Some of these variations are considered in the problems at the end of this chapter.

Before leaving this section let us consider some results from classical inventory theory. In particular, let us examine some analytical solutions for the reorder point and the order size which will minimize total inventory costs under certain idealized conditions. These results can be useful when debugging a simulation model.

(a) Fixed (deterministic) demand, constant lead time, no back orders

$$ROP = D*LT \tag{5.23}$$

$$Q \quad = \sqrt{2*CO*D/CI} \tag{5.24}$$

$$CT \quad = (CO*D/Q) + (CI*Q/2) \tag{5.25}$$

where CT represents total daily cost

(b) Random demand (Poisson distributed with parameter λ), zero lead time, no back orders

$$ROP = 0 \text{ (because } LT = 0) \tag{5.26}$$

$$Q \quad = \text{largest positive integer such that} \\ Q(Q-1) \le 2*CO*\lambda/CI \tag{5.27}$$

$$CT \quad = (CO*\lambda/Q) + [CI*(Q-1)/2] \tag{5.28}$$

Several other, more complicated solutions are also available (Hadley and Whitin 1963).

5.5 NETWORKS

A *network* consists of a number of *nodes* that are interconnected by *arcs*. In a *directed* network, each arc will be assigned an *origin*, a *destination*, and a *length*. The origin and destination will be nodes within the network, and the arc length will typically represent the distance between these nodes or the time required to travel from one node to the other.

In many networks the objective is to find the most likely path, and the corresponding expected path length, through the network (that is, from a given starting point to a specified final destination). The distribution of paths and path lengths is also of interest.

Suppose that the nodes are numbered consecutively, beginning with the starting point ($i = 1$) and ending with the final destination ($i = n$, where n is the total number of nodes). Let us define p_{ij} as the probability of direct traversal from node i to node j. Thus, if we are located at node i, p_{ij} is the probability of selecting the arc that leads to node j. Since every intermediate node must have some destination, we can write

$$\sum_{j} p_{ij} = 1 \qquad (5.29)$$

for $i = 1, 2, \ldots, (n-1)$. In addition, let the length of each arc be represented by t_{ij}. This may be a random variate, which is characterized by a mean, μ_{ij}, and a standard deviation, σ_{ij}. Finally let T represent the cumulative path length from the given starting point.

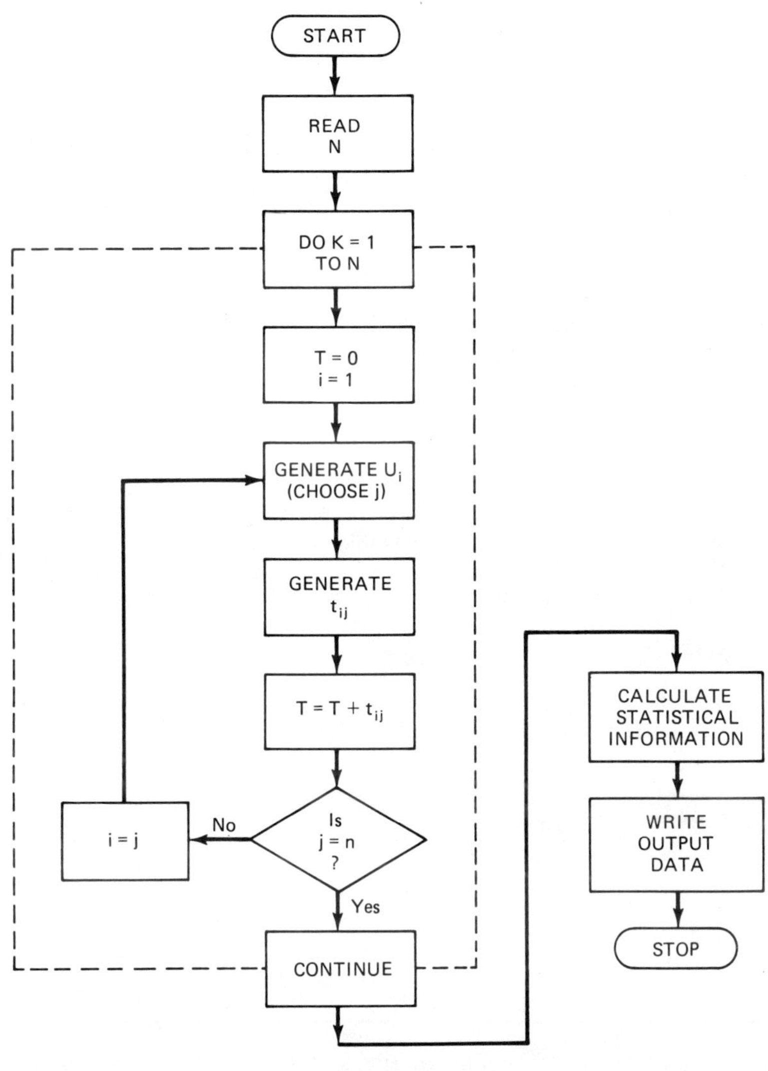

Figure 5.7

The procedure for determining arc length is based upon a randomly selected traversal through the network, beginning at the given starting point and ending at the final destination. At each node (i), a destination (j) is chosen by generating a uniformly distributed $(0, 1)$ random number and comparing this number with the specified probabilities p_{ij}. A corresponding arc length is then generated in accordance with a known distribution function. The cumulative path length is then updated, that is,

$$T = T + t_{ij} \tag{5.30}$$

and the destination becomes a new origin (that is, $i = j$), in preparation for the next arc. This procedure is continued until the entire network has been traversed, resulting in a unique path and an associated total path length, T. A distribution of paths, and a distribution of corresponding path lengths, can be obtained by repeating the entire procedure N times. Figure 5.7 shows a flowchart of the overall computational strategy.

Example 5.4

Consider the directed network shown in Fig. 5.8. We wish to obtain a distribution of paths from the starting point (node 1) to the final destination (node 8). In addition, we seek a distribution of path lengths through the network.

Suppose that each of the arc lengths is normally distributed. The numerical values of the means and standard deviations, and the associated arc selection probabilities, are tabulated:

Origin (i)	Destination (j)	Selection probability (p_{ij})	Mean arc length (μ_{ij})	Standard deviation (σ_{ij})
1	2	0.40	5	1.5
1	3	0.25	8	2.0
1	4	0.35	7	1.5
2	3	0.15	4	1.0
2	4	0.15	5	1.0
2	5	0.10	16	3.0
2	6	0.25	13	3.5
2	7	0.35	12	3.0
3	5	1.00	10	2.5
4	6	1.00	8	1.5
5	7	0.35	6	1.5
5	8	0.65	10	3.0
6	7	0.45	7	1.5
6	8	0.55	12	3.5
7	8	1.00	4	1.0

Notice that all of the required information is stored in a two-dimensional array, with one row for each arc within the network.

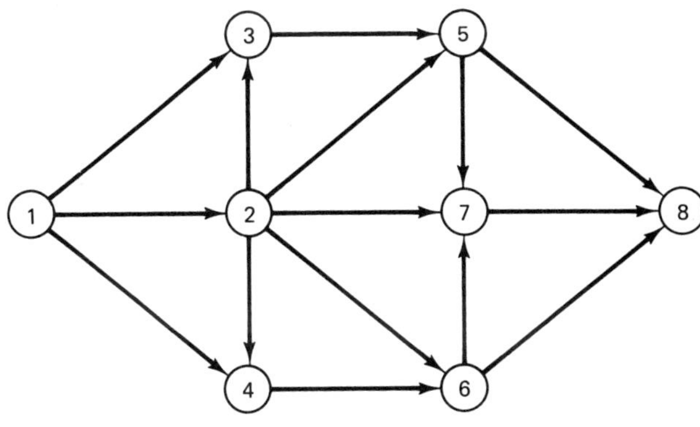

Figure 5.8

The desired distributions can now be obtained using the computational strategy presented in Fig. 5.7. It will also be necessary, however, to record all of the nodes that are traversed during each pass through the network in order to obtain a distribution of paths. This will require a small modification of the strategy shown in Fig. 5.7.

5.6 PERT

PERT (Project Evaluation and Review Technique) is a procedure that is used to assist in the scheduling of individual activities within a complex project (for example, the construction of a building). The entire project is represented by a directed network whose arcs represent *activities* and whose nodes represent *events* (that is, activity completions). The time required to complete each activity is represented by the length of the associated arc.

Unlike the networks described in the last section, PERT networks do not involve random path selection. Rather we seek the longest path through the network, from a given starting point to a specified final point. This will be the *critical path* within the network. It is important to identify the critical path because a delay in any activity along this path will result in a delay in the completion of the entire project. In addition we would like to know how much *slack* is associated with those activities that are *not* on the critical path. This represents the time that an activity can be delayed without affecting the completion of the overall project.

Let us again number the nodes (that is, the events) consecutively, beginning with the starting point (node 1) and ending with the final event (node n). The numbering system will thus correspond to the chronological sequencing of consecutive events within the project. In addition we will number the arcs (that is, the activities) consecutively, beginning with $k = 1$ and ending with $k = m$. We will designate the origin of a given arc as node i_k, and the destination as node j_k.

Now let T_k represent the time required to complete the kth activity. This may be a random variate, which is characterized by a mean, μ_k, and a standard deviation, σ_k. These values are usually obtained as

$$\mu_k = (a_k + 4b_k + c_k)/6 \tag{5.31}$$

and

$$\sigma_k = (c_k - a_k)/6 \tag{5.32}$$

where

a_k = most optimistic (smallest) time estimate to complete the kth activity

b_k = most likely time estimate to complete the kth activity

c_k = most pessimistic (largest) time estimate to complete the kth activity

Equations (5.31) and (5.32) are based upon the use of the beta distribution to represent the activity times.

In order to determine the critical path we first make a forward pass through the network, determining the *earliest start time* (ES_k) and the *earliest finish time* (EF_k) for each activity. For the initial activities (that is, those activities whose origin is node 1), we set

$$ES_k = 0 \tag{5.33}$$

and

$$EF_k = T_k \tag{5.34}$$

For each successive activity, the earliest start time is the *largest value* of the earliest finish times of the preceding activities. Thus

$$ES_k = \text{Max} \left\{ EF_l \right\} \tag{5.35}$$

where the subscript l represents those preceding activities whose destination becomes the origin of the current activity; that is, $j_l = i_k$. Moreover the earliest finish time is obtained as

$$EF_k = ES_k + T_k \tag{5.36}$$

After the earliest start time and the earliest finish time have been determined for all of the activities, we make a *backward* pass through the network to obtain the *latest finish time* (LF_k) and the *latest start time* (LS_k) for each activity. For the last activities (that is, those activities whose destination is node n), we set

$$LF_k = \text{Max} \left\{EF_k\right\} \tag{5.37}$$

and

$$LS_k = LF_k - T_k \tag{5.38}$$

For each successive activity, the latest finish time will be the *smallest value* of the latest start time of the succeeding activities. Thus

$$LF_k = \text{Min} \left\{LS_l\right\} \tag{5.39}$$

where the subscript l now represents those succeeding activities whose origin is the destination of the current activity; that is, $i_l = j_k$. Also, the latest start time is obtained as

$$LS_k = LF_k - T_k \tag{5.40}$$

Once ES_k, EF_k, LS_k, and LF_k have been obtained for all of the activities in the network, we can calculate the corresponding slack for each activity as

$$S_k = LS_k - ES_k \tag{5.41}$$

or, alternatively,

$$S_k = LF_k - EF_k \tag{5.42}$$

The critical path can then be identified as that path whose activities all have zero slack.

Figure 5.9 shows a flowchart of the entire procedure.

Example 5.5

Figure 5.10 illustrates a PERT network consisting of twelve activities (arcs) and eight events (nodes). The completion time for each activity is assumed to be normally distributed with a known mean and standard deviation, as indicated in the table.

Activity (k)	Origin $[I(k)]$	Destination $[J(k)]$	Mean $[\mu(k)]$	Standard deviation $[\sigma(k)]$
1	1	2	8	3
2	1	3	5	2
3	2	4	4	1
4	2	5	6	2
5	3	5	10	4
6	3	7	12	5
7	4	5	2	1
8	4	6	6	3
9	5	6	5	2
10	5	7	4	1
11	6	8	7	3
12	7	8	10	4

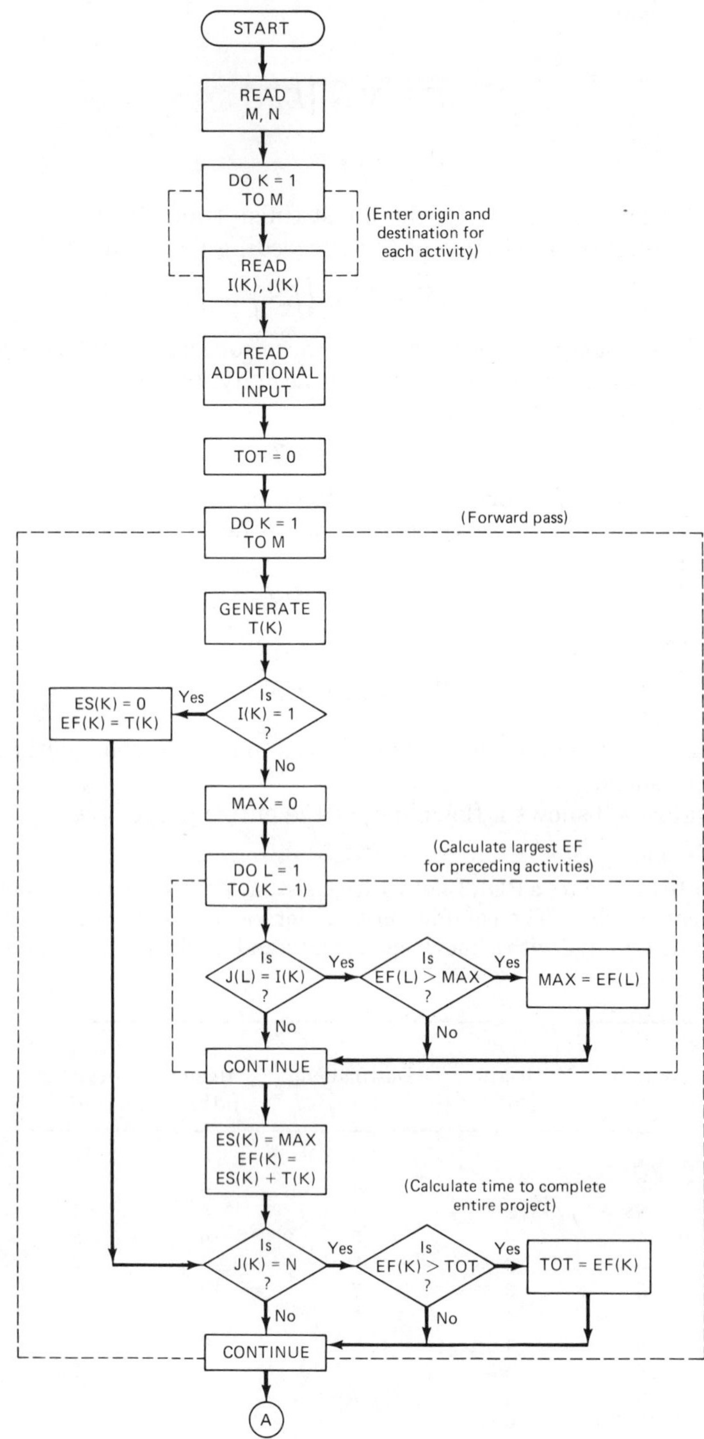

START
READ M, N
DO K = 1 TO M
READ I(K), J(K)
(Enter origin and destination for each activity)
READ ADDITIONAL INPUT
TOT = 0
DO K = 1 TO M
(Forward pass)
GENERATE T(K)
Is I(K) = 1 ?
Yes
No
ES(K) = 0
EF(K) = T(K)
MAX = 0
DO L = 1 TO (K − 1)
(Calculate largest EF for preceding activities)
Is J(L) = I(K) ?
Yes
No
Is EF(L) > MAX ?
Yes
No
MAX = EF(L)
CONTINUE
ES(K) = MAX
EF(K) = ES(K) + T(K)
(Calculate time to complete entire project)
Is J(K) = N ?
Yes
No
Is EF(K) > TOT ?
Yes
No
TOT = EF(K)
CONTINUE
A

Figure 5.9

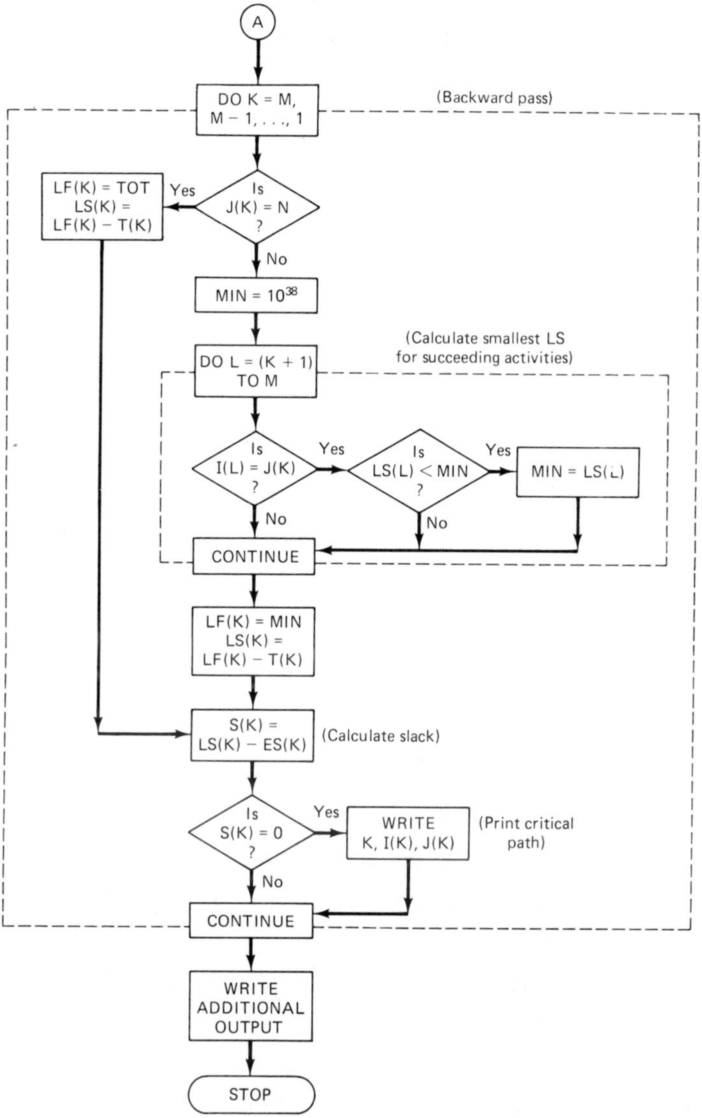

Figure 5.9 *(continued)*

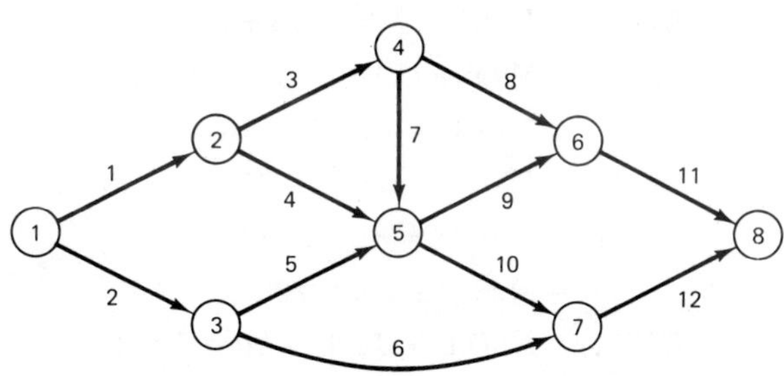

Figure 5.10

(Note that each randomly generated value for $T(k)$ must be greater than or equal to zero).

The critical path can be obtained using the procedure shown in Fig. 5.9. In order to illustrate the procedure, however, let us determine the critical path manually, using the mean completion times (rather than randomly generated values) for each activity. The forward pass begins at the given starting point. Thus for the first activity ($k = 1$),

$$ES(1) = 0$$
$$EF(1) = T(1) = 8$$

Similarly for the second activity ($k = 2$),

$$ES(2) = 0$$
$$EF(2) = T(2) = 5$$

Continuing in this manner we obtain

$$ES(3) = EF(1) = 8$$
$$EF(3) = ES(3) + T(3) = 8 + 4 = 12$$
$$ES(4) = EF(1) = 8$$
$$EF(4) = ES(4) + T(4) = 8 + 6 = 14$$
$$ES(5) = EF(2) = 5$$
$$EF(5) = ES(5) + T(5) = 5 + 10 = 15$$
$$ES(6) = EF(2) = 5$$
$$EF(6) = ES(6) + T(6) = 5 + 12 = 17$$
$$ES(7) = EF(3) = 12$$
$$EF(7) = ES(7) + T(7) = 12 + 2 = 14$$
$$ES(8) = EF(3) = 12$$
$$EF(8) = ES(8) + T(8) = 12 + 6 = 18$$
$$ES(9) = \text{Max } \{EF(4), EF(5), EF(7)\}$$
$$= \text{Max } \{14, 15, 14\} = 15$$
$$EF(9) = ES(9) + T(9) = 15 + 5 = 20$$
$$ES(10) = \text{Max } \{EF(4), EF(5), EF(7)\}$$
$$= \text{Max } \{14, 15, 14\} = 15$$
$$EF(10) = ES(10) + T(10) = 15 + 4 = 19$$
$$ES(11) = \text{Max } \{EF(8), EF(9)\}$$
$$= \text{Max } \{18, 20\} = 20$$
$$EF(11) = ES(11) + T(11) = 20 + 7 = 27$$

$$ES(12) = \text{Max } \{EF(6), EF(10)\}$$
$$= \text{Max } \{17, 19\} = 19$$
$$EF(12) = ES(12) + T(12) = 19 + 10 = 29$$

The backward pass begins at the final destination. Hence for the last two activities ($k = 12$ and $k = 11$).

$$LF(12) = \text{Max } \{EF(11), EF(12)\}$$
$$= \text{Max } \{27, 29\} = 29$$
$$LS(12) = LF(12) - T(12) = 29 - 10 = 19$$
$$LF(11) = \text{Max } \{EF(11), EF(12)\}$$
$$= \text{Max } \{27, 29\} = 29$$
$$LS(11) = LF(11) - T(11) = 29 - 7 = 22$$

Continuing in this manner,

$$LF(10) = LS(12) = 19$$
$$LS(10) = LF(10) - T(10) = 19 - 4 = 15$$
$$LF(9) = LS(11) = 22$$
$$LS(9) = LF(9) - T(9) = 22 - 5 = 17$$
$$LF(8) = LS(11) = 22$$
$$LS(8) = LF(8) - T(8) = 22 - 6 = 16$$
$$LF(7) = \text{Min } \{LS(9), LS(10)\}$$
$$= \text{Min } \{17, 15\} = 15$$
$$LS(7) = LF(7) - T(7) = 15 - 2 = 13$$
$$LF(6) = LS(12) = 19$$
$$LS(6) = LF(6) - T(6) = 19 - 12 = 7$$
$$LF(5) = \text{Min } \{LS(9), LS(10)\}$$
$$= \text{Min } \{17, 15\} = 15$$
$$LS(5) = LF(5) - T(5) = 15 - 10 = 5$$
$$LF(4) = \text{Min } \{LS(9), LS(10)\}$$
$$= \text{Min } \{17, 15\} = 15$$
$$LS(4) = LF(4) - T(4) = 15 - 6 = 9$$
$$LF(3) = \text{Min } \{LS(7), LS(8)\}$$
$$= \text{Min } \{13, 16\} = 13$$
$$LS(3) = LF(3) - T(3) = 13 - 4 = 9$$
$$LF(2) = \text{Min } \{LS(5), LS(6)\}$$
$$= \text{Min } \{5, 7\} = 5$$

$$LS(2) = LF(2) - T(2) = 5 - 5 = 0$$
$$LF(1) = \text{Min } \{LS(3), LS(4)\}$$
$$= \text{Min } \{9, 9\} = 9$$
$$LS(1) = LF(1) - T(1) = 9 - 8 = 1$$

The slacks can now be obtained using Eq. (5.41) [or Eq. (5.42)]. These calculations, together with the preceding calculations, are summarized:

k	$T(k)$	$ES(k)$	$EF(k)$	$LF(k)$	$LS(k)$	$S(k)$
1	8	0	8	9	1	1
2	5	0	5	5	0	0
3	4	8	12	13	9	1
4	6	8	14	15	9	1
5	10	5	15	15	5	0
6	12	5	17	19	7	2
7	2	12	14	15	13	1
8	6	12	18	22	16	4
9	5	15	20	22	17	2
10	4	15	19	19	15	0
11	7	20	27	29	22	2
12	10	19	29	29	19	0

We can now obtain the critical path, as those arcs (activities) having a zero slack. Thus for the given $T(k)$'s, the critical path is comprised of activities 2, 5, 10, and 12; connecting nodes (events) 1, 3, 5, 7, and 8.

PERT studies that are based upon mean activity times, such as in the preceding illustrative example, are generally unsatisfactory because they do not indicate the likelihood of various paths becoming critical. This deficiency can be removed however through the use of simulation. To do so, we repeatedly carry out the procedure presented in Fig. 5.9, using a different set of randomly generated activity times for each pass. At the end of each pass we identify and record the activities that lie along the critical path (which will not necessarily be unique, nor will it be the same from one pass to another), and the length of the critical path (that is, the overall project completion time). From this information we can obtain distributions of the activities along the critical path and the project completion times.

Example 5.6

Suppose that the PERT problem described in Ex. 5.5 has been solved 1000 times, using a different set of randomly generated activity times for each run. The individual activities fell on the critical path with the following relative frequencies:

Activity	Relative frequency
1	0.561
2	0.439
3	0.314
4	0.247
5	0.319
6	0.120
7	0.247
8	0.067
9	0.327
10	0.486
11	0.394
12	0.606

Also the overall project completion times were determined with the following frequencies:

Overall completion time	Relative frequency
20–21.99	0.007
22–23.99	0.022
24–25.99	0.056
26–27.99	0.187
28–29.99	0.221
30–31.99	0.194
32–33.99	0.157
34–35.99	0.103
36–37.99	0.044
38–39.99	0.009
	1.000

A cumulative distribution can easily be obtained from this last set of data, indicating the likelihood that the overall project completion time will be less than or greater than various specified values.

5.7 RISK ANALYSIS

Risk analysis is one of the most important and widely used applications of discrete-event simulation. The objective is to assess the desirability of a proposed investment, based upon some financial decision criterion such as present worth. (Other commonly used decision criteria, for example, future worth, uniform annual series, rate of return, can also be used.) Application of the method results in a cumulative distribution being generated for the decision criterion. Hence we can obtain not only the expected value of the decision criterion but also the likelihood of realizing a much higher or a much lower value.

Suppose we are considering an initial cash outlay that will generate a series of n yearly cash flows, as illustrated in Fig. 5.11. (Upward-pointing arrows indicate net cash *inflows:* downward-pointing arrows represent net cash *outflows*). The present value of each cash flow can be written as

$$PW_j = YCF_j/(1 + i)^j \tag{5.43}$$

where YCF_j represents the annual cash flow for the jth year in the future, and i is a specified annual interest rate, expressed as a decimal (this expression assumes *annual* compounding). Hence the present worth of the entire proposed investment can be expressed as

$$
\begin{aligned}
PW &= \left\{ \sum_{j=1}^{n} PW_j \right\} - I \\
&= \left\{ \sum_{j=1}^{n} YCF_j/(1 + i)^j \right\} - I
\end{aligned}
\tag{5.44}
$$

where I represents the initial cash outlay.

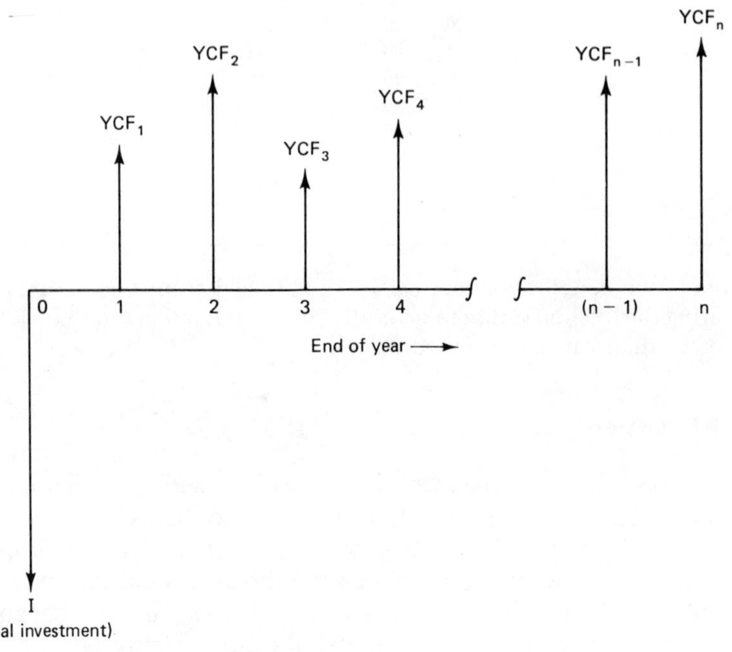

Figure 5.11

Each of the yearly cash flows is generally comprised of several components, such as yearly sales volume, production cost, taxes. These items are normally represented in terms of áppropriate distribution functions. Thus the computational procedure involves generating a random value for each cash flow component, resulting in a randomly generated value for PW_j. All of the yearly cash flows are evaluated in this manner. These values are then substituted into Eq. (5.44), resulting in a single value for PW. We then repeat the entire procedure N times, which allows us to obtain a distribution for PW. Figure 5.12 illustrates the general procedure.

Example 5.7

A large corporation is interested in diversifying its activities by purchasing a small manufacturer of car radios. An initial investment of $15 million will be required. The future sales will depend upon certain unknown factors, such as competitive position and market penetration. Therefore the sales price per unit and the yearly sales volume will have to be estimated. Let us assume that the sales price is normally distributed with a mean of $65.00 per unit and a standard deviation of $4.00 per unit. Also we will assume that the sales volume is governed by the following empirical distribution:

Sales units per year	Relative frequency
80,000–100,000	0.15
100,000–120,000	0.20
120,000–140,000	0.30
140,000–160,000	0.25
160,000–180,000	0.10

The production costs are assumed to be exponentially distributed with a mean of $32.00 per unit and a minimum of $20.00 per unit. We wish to obtain a cumulative distribution for the proposed investment's present worth, based upon a 12 percent interest rate, a corporate tax rate of 48 percent, and straight-line depreciation over a 10-year period.

The yearly cash flow can be expressed as

$$YCF = S - C - T$$

where S represents the annual sales revenue, C represents the total annual cost, and T represents the annual taxes. Now the annual sales revenue can be obtained as

$$S = P * V$$

where P represents the selling price per unit (which is normally distributed) and V represents the annual sales volume (empirically distributed). Similarly, the total cost can be determined approximately as

$$C = PC * V$$

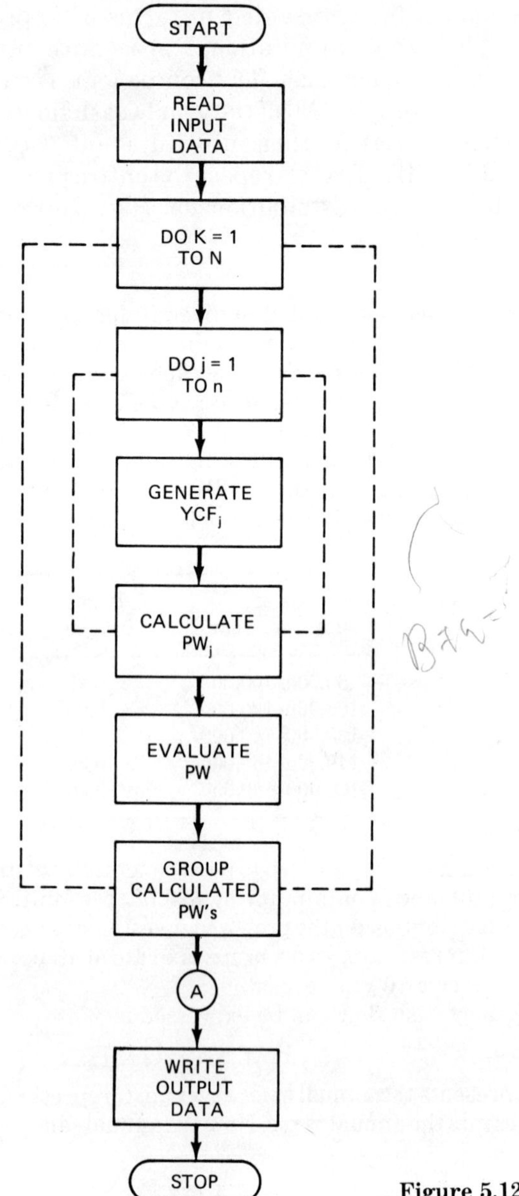

START
READ INPUT DATA
DO K = 1 TO N
DO j = 1 TO n
GENERATE YCF$_j$
CALCULATE PW$_j$
EVALUATE PW
GROUP CALCULATED PW's
A
WRITE OUTPUT DATA
STOP

Figure 5.12

where PC is the production cost per unit (exponentially distributed). Finally the annual taxes can be expressed as

$$T = R * (S - C - D)$$

provided $T \geq 0$ (T must be set equal to zero if its calculated value is negative). In this expression R is the tax rate (0.48) and D is the annual depreciation. For straight-line depreciation, the latter quantity becomes

$$D = I/n$$

where I is the initial investment ($15 million) and n is the project lifetime (10 years).

The expression for YCF then becomes

$$YCF = P * V - PC * V - T$$

$$= (P - PC) * V - T$$

where

$$T = \text{Max} \begin{cases} R * [(P - PC) * V - I/n] \\ 0 \end{cases}$$

By generating random variates for P, V, and PC, we can thus obtain a value for the yearly cash flow and its corresponding present worth. Moreover, the computation can be carried out separately for each year (that is, for $j = 1, 2, \ldots, n$) resulting in a calculated value for PW, the present worth of the entire proposed investment. Finally if the entire procedure is repeated, say, N times, as indicated in Fig. 5.12, we will obtain the desired distribution for PW.

When comparing several different investment opportunities it is customary to base the comparisons on the expected return for each proposal. Risk analysis can be used to obtain this information. The use of risk analysis in such situations provides important additional information, however, since it also allows us to assess the likelihood of a much higher or a much lower return. Comparisons can therefore be made on the basis of an acceptable level of risk as well as the expected value of the return.

Example 5.8

A cumulative distribution of present worth for each of two different proposed investments is shown in Figs. 5.13 a and b. The expected value (that is, the mean value) of the present worth is shown in each figure. Note that this value is higher for Proposal B than for Proposal A ($12.5 million versus $8.8 million). Moreover the shape of the distribution curves (and hence the corresponding standard deviations) is the same, which suggests that the degree of risk is the same in each case. Therefore Proposal B is more desirable than Proposal A, since it offers a higher expected return for the same degree of risk.

Proposal A
Cumulative distribution
1.00
0.80
0.60
0.40
0.20
0
0
5
10
15
20
$\overline{PW}_A$
Present worth (million $)
(a)

Proposal B
Cumulative distribution
1.00
0.80
0.60
0.40
0.20
0
0
5
10
15
20
25
$\overline{PW}_B$
Present worth (million $)
(b)

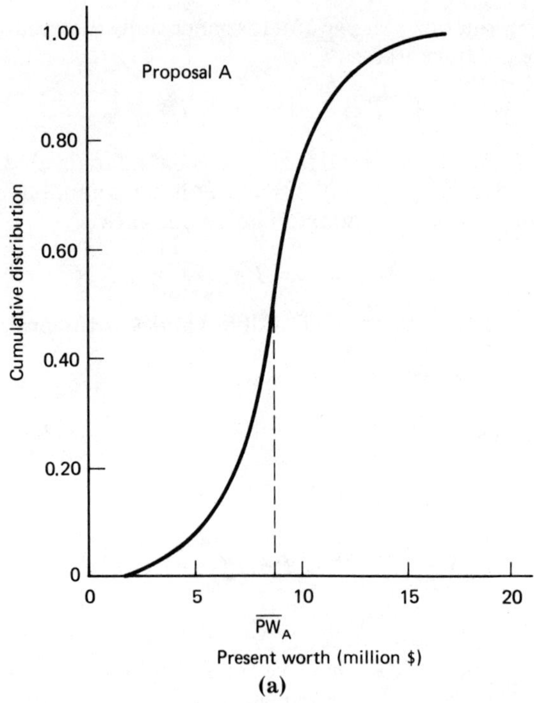

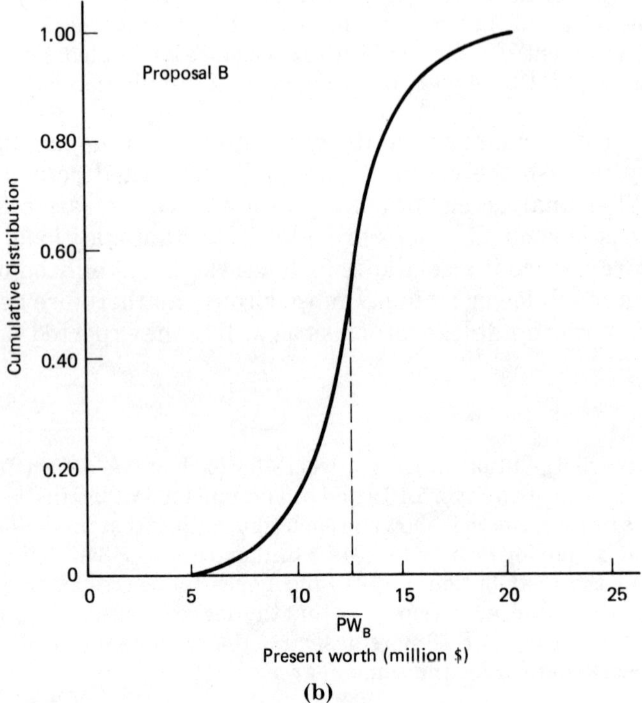

Figure 5.13

Now consider the situation shown in Figs. 5.14 a and b (see p. 148), where we again see a cumulative distribution of present worth for each of two different proposed investments. Both proposals have approximately the same expected present worth ($5 million). Proposal A represents a relatively conservative investment, since there is virtually no chance of a loss (that is, a negative present worth). There is also virtually no chance, however, that the present worth of this investment will exceed $10 million. On the other hand, Proposal B has an 8 percent chance of losing money, but it also has a 10 percent chance that the present worth will exceed $10 million. Proposal B is therefore riskier than Proposal A; it may also be more rewarding.

In this case it is not clear which proposal is better—the choice will depend upon the policies and investment goals of the parent company. A more conservative company will probably choose Proposal A. If the parent company wishes to invest some portion of its capital more aggressively, however, it will probably select Proposal B.

PROBLEMS

5.1. Solve the facility utilization problem described in Ex. 5.1 using simulation. Repeat the simulation 100 times in order to obtain a distribution for F (the fraction of time the facility is utilized). Determine
   (a) the expected value of F
   (b) the likelihood that F will be at least 10 percent greater than its expected value
   (c) the likelihood that F will be at least 5 percent less than its expected value

5.2. A printing company currently owns three presses, which are operated on a full-time basis. Overflow work is subcontracted to another printer, on a break-even basis, in order to maintain customer goodwill. The company is now considering the purchase of a fourth press, at a cost of $150,000. Management would like some estimate of the time required to recover this initial investment (neglecting the time-value of money). A simulation model has been proposed for this purpose.

The demand for the fourth press (that is, the time between arrivals of successive overflow orders) is assumed to be exponentially distributed, with a mean of 1.5 days. The service times (that is, the times required to process the orders) are believed to be normally distributed, with a mean of 1 day and a standard deviation of 0.4 days. In addition, there is a constant set-up time of 0.1 days required for each order. A net profit of $50 per hour ($400 per day) can be realized when the press is in operation.

For simplicity, assume that any new order that comes in while *all* presses are in operation (including the new press) is still subcontracted to another printer. Thus there will never be a backlog of orders waiting to be processed.

Simulate the operation of the fourth press for a period of time that is long enough that the cumulative net profit equals or exceeds the original

1.00
0.80
0.60
0.40
0.20
0
Cumulative distribution
Proposal A
0
5
10
15
20
PW_A
Present worth (million $)
(a)

1.00
0.80
0.60
0.40
0.20
0
Cumulative distribution
Proposal B
-5
0
5
10
15
20
PW_B
Present worth (million $)
(b)

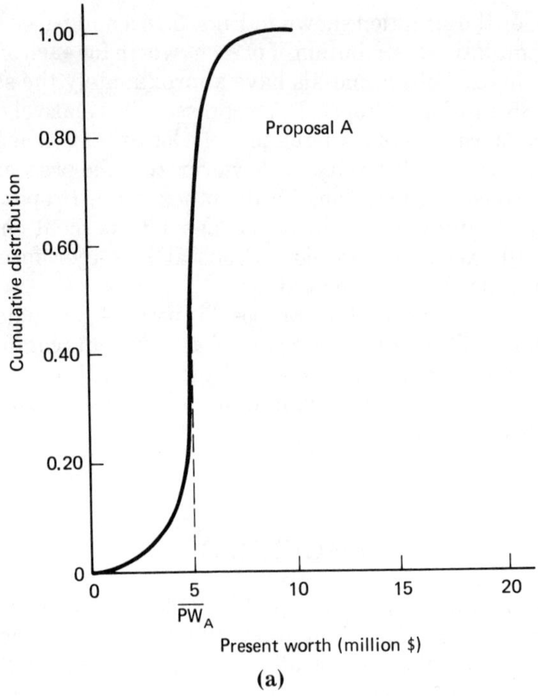

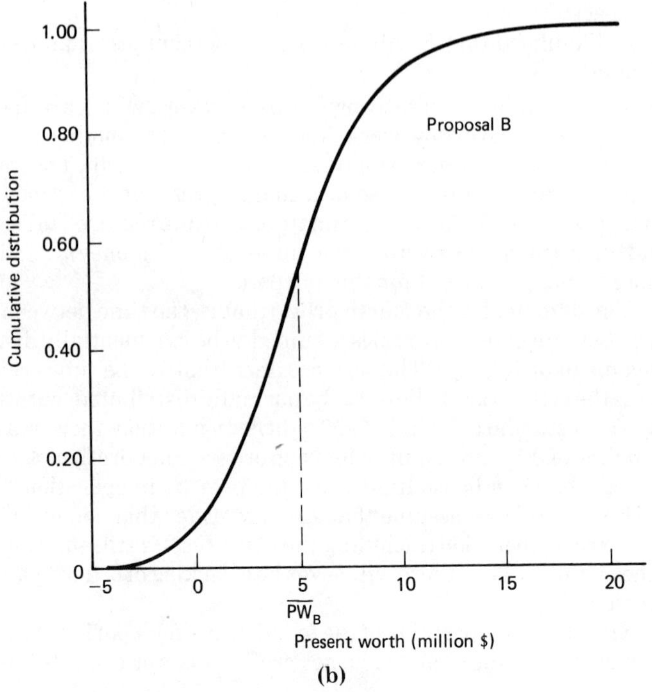

Figure 5.14

investment. Then calculate a value for the desired recovery time (in years) as

$$T = CT * (7/5.5) * (1/365) = 0.00349 * CT$$

where CT is the cumulative recovery time in working days (assuming $5\frac{1}{2}$ working days per week).

Also, determine the fraction of time that the press is in use.

5.3. Consider once again the situation described in Prob. 5.2 above. Simulate the operation of the fourth press 100 times in succession, in order to obtain a distribution of recovery times. Determine
  (a) the expected recovery time
  (b) the likelihood that the recovery time will be at least 5 percent greater than the expected value
  (c) the likelihood that the recovery time will be at least 10 percent less than the expected value

5.4. Determine by simulation how much money the printing company described in Prob. 5.2 can pay for a fourth press if the initial investment is to be recovered in 3 years. (Assume that all of the data given in Prob. 5.2 apply, except the initial cost of the press.)

5.5. Repeat Prob. 5.3 for the special case where a penalty of $65 is imposed for every incoming order that is lost (because all presses are in operation when the order is received).

5.6. Solve the equipment maintenance problem described in Ex. 5.2 by simulating each of the following situations and determining which is the least expensive.
  (a) Repair each machine only as the need arises (that is, on demand).
  (b) Perform preventative maintenance once a week, in addition to repairing machines on demand whenever necessary.
  (c) Preventative maintenance every 2 weeks, plus service on demand when necessary.
  (d) Preventative maintenance every 4 weeks, plus demand service when necessary.

Simulate fifty 1-year periods in each case, assuming that the machines are used 6 days per week.

5.7. Repeat Prob. 5.6 for the following set of conditions:
  (a) The time between breakdowns is normally distributed, with $\mu = 12.5$ days and $\sigma = 3.3$ days.
  (b) The repair times are gamma distributed, with $\alpha = 3$ and $\beta = 5$.
  (c) The repair cost is $25 per breakdown, plus $40 per hour of downtime.
  (d) The preventative maintenance cost is $65 per machine.

Compare the recommended policy with that obtained in Prob. 5.6.

5.8. Modify the flowchart shown in Fig. 5.3 so that the following items are calculated and printed for the simulated time period.
  (a) total downtime for each machine
  (b) number of failures for each machine

5.9. Solve the inventory control problem described in Ex. 5.3 by simulating fifty 1-year time periods for each of several inventory policies (that is, several sets of values for *ROP* and Q). Calculate the expected value of the total yearly inventory cost for each inventory policy, and determine which policy results in the lowest yearly cost. Assume an initial inventory level of 75 units when beginning each simulated time period.

5.10. Suppose that the inventory control problem described in Ex. 5.3 is now modified so that the lead time is random. In particular, assume that the lead time is exponentially distributed with a mean of 5 days and a minimum of 2 days (for simplicity, consider only integer-valued lead times).

Modify the flowchart given in Fig. 5.6 to accommodate this change. Then repeat Prob. 5.9, using these random lead times. Compare the recommended inventory policy with that obtained in Prob. 5.9.

5.11. Repeat the solution of the inventory control problem described in Ex. 5.3 for the case where back orders are not considered. Assume a fixed shortage cost of $3 per unit. Use the same solution procedure described in Prob. 5.9. Compare the recommended inventory policy with that obtained in Prob. 5.9.

5.12. Repeat the solution of the inventory control problem described in Ex. 5.3 for the case of instantaneous inventory replenishment (zero lead times). Use the same solution procedure described in Prob. 5.9. Compare the recommended inventory policy with that obtained in Prob. 5.9.

5.13. Suppose that a warehouse maintains an inventory of K different items, all of which are obtained from the same supplier. Each item is accounted for separately. When ordering new merchandise, however, only one order will be placed for all of the K different items. Therefore there may be a delay between the time the inventory level for a certain item drops below the reorder point and the time the order is actually placed.

Modify the flowchart shown in Fig. 5.6 to accommodate this situation.

5.14. Solve the PERT problem described in Ex. 5.5 using simulation, based upon the flowchart presented in Fig. 5.9. Repeat the simulation 100 times in order to obtain distributions of the activities along the critical path and the project completion times. Determine
   (a) the most likely critical path
   (b) the likelihood that activity 5 will fall on the critical path
   (c) the expected value of the overall project completion time
   (d) the likelihood that the overall project completion time will be less than 26 time units
   (e) the likelihood that the overall project completion time exceeds 33 time units.

5.15. Solve the risk analysis problem described in Ex. 5.7 using the flowchart presented in Fig. 5.12. Determine
   (a) the expected value of the present worth
   (b) the likelihood that the present worth will be less than $1 million
   (c) the likelihood that the present worth will exceed $2 million

5.16. A company is considering the desirability of the following proposed venture:

Initial investment: $15 million
Sales price: distributed as

Price per unit ($)	Relative frequency
20–25	0.08
25–30	0.17
30–35	0.22
35–40	0.27
40–45	0.13
45–50	0.09
50–55	0.04

Sales volume: normally distributed, with a mean of 200,000 units per year and a standard deviation of 50,000 units per year.
Annual costs: fixed cost of $100,000 per year, plus a unit cost which is distributed as

Cost per unit ($)	Relative frequency
10–12	0.12
12–14	0.22
14–16	0.36
16–18	0.20
18–20	0.10

Tax rate: 48%.
Depreciation: straight line, over a 10-year period.
Interest rate: 12 percent per year, compounded annually.
Determine
(a) the expected value of the present worth
(b) the likelihood that the present worth will be less than $1 million
(c) the likelihood that the present worth will exceed $2 million
Compare with the results obtained for Prob. 5.15. Which proposal is riskier? Which is more conservative?

What would happen if the interest rate were increased to 15 percent per year?

5.17. A hospital blood bank operates in the following manner: Whenever the inventory position of blood falls to r units, an order is placed for Q units. Lead time for such orders is exponentially distributed with a mean of 12 hours. The number of persons requiring blood is Poisson distributed with a mean of eight persons per day. The number of units of blood required per person is geometrically distributed with $p = 0.80$. If a person requires blood which cannot be supplied from the blood bank, an emergency order can be delivered within one-half hour. In addition to those units received through normal and emergency ordering procedures, the hospital receives blood through donations. Each blood donor contributes 1 unit. The number of blood donors per day is Poisson distributed with a mean of 5 per day.

The inventory system is to be maintained at such a level that the probability that ten or more emergency orders are placed per year is less than 0.05. Find values of r and Q such that the above criterion for operation is satisfied.

5.18. Metropolitan Gas serves the community of Potholeville—a large manufacturing center with a metropolitan population of about 2 million people.

As an assistant to the vice president in charge of gas procurement for Metropolitan, you must determine how to build up an adequate inventory of natural gas for the coming heating season. The following information is known:

(1) Metropolitan delivers about 3×10^{10} cu ft of gas to its industrial customers each month. This usage is independent of the weather.

(2) The demand for gas for heating purposes is about 1.14×10^{8} cu ft per degree day.*

(3) The cost of natural gas, delivered to Metropolitan in Potholeville, is $4.00 per 1000 cu ft (MCF) when purchased through ordinary, regulated, interstate sources. However, the delivery of interstate gas is governed by contractual agreements, which require that the *same* quantity of gas be purchased each month throughout the calendar year.

(4) If, on December 31, the supply of gas appears to be inadequate for the remainder of the heating season, there is an 80 percent chance that additional gas may be purchased for $7.50 per MCF. These purchases can be made on a month-to-month basis. However, there is a 1-month time lag between when the purchase is made and when the gas becomes available for delivery.

(5) The cost of storing gas underground is 15¢ per month per MCF.

(6) Metropolitan's executives estimate that a shortfall during the heating season costs the company the equivalent of $10.00 per MCF.

You are required to answer the following questions:

(a) What is a reasonable gas inventory policy, based upon a "typical" heating season?

(b) How much gas should be purchased from regular, interstate sources, and how much should be purchased at the higher price, if available?

---

*1 degree-day = temperature deficit of 1°F for 1 day where

$$\text{temperature deficit} = \begin{cases} 65° - T_{\text{av}}, & T_{\text{av}} < 65°\text{F} \\ 0, & T_{\text{av}} \geq 65°\text{F} \end{cases}$$

and

$$T_{\text{av}} = \text{average daily temperature, °F.}$$

(c) How likely is this policy, based upon a typical heating season, to be "undesirable" because of an abnormally cold or an abnormally warm winter?

(d) How will your results be altered if the cost of supplementary gas is increased to $8.50 per MCF and the availability reduced to 60 percent?

(e) How will your original results be altered if the cost of a shortfall is $12.00 per MCF?

Expected Daily Mean Temperature and Standard Deviation during the Heating Season in Potholeville, Pennsylvania

Month	Expected daily mean temperature (°F)	Standard deviation (°F)
October	53	7
November	41	8
December	31	7
January	29	6
February	29	6
March	37	7
April	49	8
May	60	7

The standard deviation is a measure of the "spread" of the daily mean temperatures. The variation of the daily mean temperature, about its expected value, can be assumed to be normally distributed.

6

CONDUCTING A COMPLETE SIMULATION STUDY

In the past several chapters our primary interest has been the generation of random variates, and the development of several different types of simulation models that make use of these random variates. We now turn our attention to a subject that is more general but equally important: namely, the overall development and implementation of a simulation model. For the most part, our discussions will apply to virtually all simulation models, irrespective of the particular application.

A complete simulation study generally involves several different, though complementary, activities. At the risk of oversimplifying, they are

1. Planning the study
2. Conceptualizing the model
3. Gathering the data
4. Implementing the model
5. Validating and refining the model
6. Using the model to draw practical conclusions

In practice however these activities are not carried out in the simple sequential manner outlined previously. Some of the steps (particularly steps 2 and 3) may be carried out in parallel. Moreover, the entire procedure is usually repeated several times, since the information obtained during the later activities often suggests refinements that should be incorporated into the earlier activities. Figure 6.1 illustrates the overall procedure.

In this chapter we will discuss each of the above steps in some detail. Our emphasis will be placed upon the implementation step, since this activity is generally less subjective than the others. This does not, however, suggest that the other activities are any less important; they are simply more difficult to describe adequately in an introductory-level textbook.

6.1 PLANNING THE STUDY

In the so-called "real world," most simulation studies begin with some relatively vague description of a general problem area. At this initial

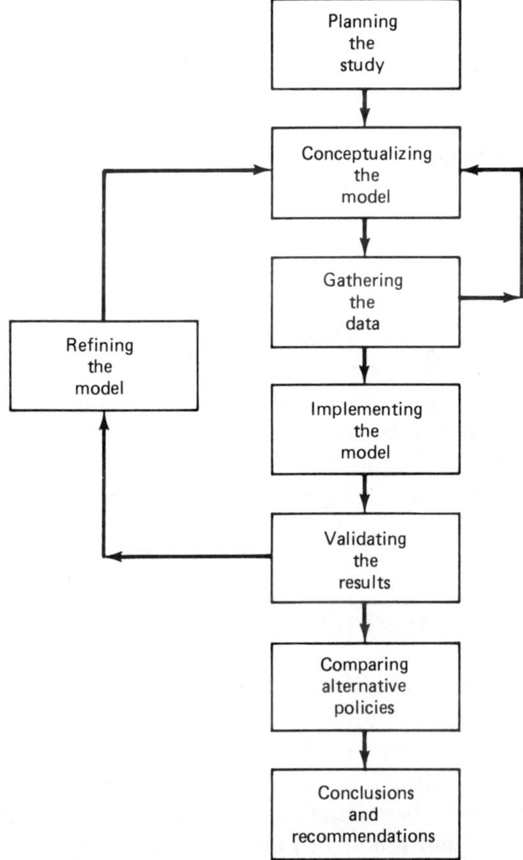

Figure 6.1

stage it is usually not known precisely what the problem consists of, much less how it should be solved. An active period of conceptualization is generally required before the problem can be perceived in more specific terms.

Perhaps the first matter to be considered when planning the study is the specific type of information being sought. This in turn raises questions about the overall characteristics of the problem, its principal parameters, the required data, and, finally, an appropriate solution procedure. Thus, we should ask whether the problem is primarily deterministic or stochastic, whether it can be represented by an established structured model (for example, linear programming), how much data are available, how reliable are the data, and so on. These matters are all interrelated and must therefore be considered at more or less the same time.

The primary factor in planning some overall solution strategy is its suitability for the particular problem being studied. There are, however, some cost–benefit factors that should also be considered. In particular, an estimate should be made of the time and effort required to do the following:

1. Construct an accurate model.
2. Gather the data.
3. Implement the model (presumably on a computer).
4. Validate the model.
5. Use the model to study a number of policy alternatives.
6. Obtain meaningful conclusions.

One must then attempt to assess the value of the information obtained, its cost, and whether or not the results are worth the effort. The answers to these questions may influence the choice of one approach over another, or they may suggest that a detailed problem analysis should not be carried out at all.

6.2 CONCEPTUALIZING THE MODEL

Let us now suppose that simulation has been selected as the solution procedure. An accurate mathematical description of the problem must then be created. This process of conceptualization is usually the most difficult and challenging part of the entire project, since the analyst must transform a loosely defined, broad-based problem statement into a detailed mathematical model. In order to accomplish this, the analyst must call upon insight, initiative, and resourcefulness, as well as technical knowledge.

On the other hand, the opportunity to develop a working model provides the analyst with a significant creative outlet for utilizing a variety of skills. An enormous amount of satisfaction can be derived from the successful development and implementation of a sophisticated simulation model.

In many respects the conceptualization process is as much an art as it is a science. Therefore it is not possible (or even desirable) to state a rigid set of rules which must always be followed. However the following questions are pertinent to the analysis of most problems, and they provide a set of guidelines that can assist the analyst in the development of a precise, quantitative model.

1. What are the specific objectives of the study?
2. Can these objectives be expressed quantitatively, in terms of system performance criteria?
3. What are the relevant problem parameters? (Specifically, what are the state variables, the decision variables, and the system parameters?)
4. What are the pertinent cause-and-effect relationships? How can they be expressed in mathematical terms?
5. What data are required? Are the data readily available? If not, how much effort is required to obtain meaningful data?

The development of a mathematical model can also be influenced by the availability of adequate computer facilities. This includes the computer's speed, the size of its primary and auxiliary memories, and the costs that are associated with its use. Of even greater significance are several programming considerations including the choice of a programming language (depending on availability), the design of an appropriate database, and the selection of an overall solution procedure. These factors can affect both the type of model to be constructed and the degree of detail to be included.

6.3 GATHERING THE DATA

We have already seen that the mathematical model has a major influence on the particular type of data that may be required. On the other hand, the availability of the data, or the quality of the data available, can influence the nature of the model. In particular the lack of a certain type of data might cause the model to be conceptualized in terms of a different set of parameters, so that the key variables are consistent with the available data. Thus the development of a mathematical model and the acquisition of supporting data are complementary activities that must be considered more or less simultaneously.

When gathering the data a distinction must be made between "hard" and "soft" data. *Hard data* generally refer to information that is readily available or easily obtained. Such data are usually known to a high degree of accuracy. *Soft data* however refer to information that is subject to considerable uncertainty. Data of this type are often based upon subjective estimates, particularly when future time periods are involved.

Soft data are required in most realistic simulation studies. It should be understood, however, that the use of such data will by their very nature, introduce some element of uncertainty into the simulated results. The extent to which this happens will depend upon the degree of influence that the soft data have upon the entire model.

Stochastic data can be represented most compactly in terms of mathematical distribution functions, though it is not essential to do so (empirical distributions are sometimes more convenient). We have already discussed this topic at some length in Chap. 4. The reader is again reminded, however, that there are numerous established graphical and analytical techniques (for example, regression analysis) for fitting mathematical distribution functions to tabular sets of data. Such techniques should be used whenever it is practical to do so.

Finally it should be recognized that the gathering of meaningful data is an activity that may involve a broad spectrum of people. In particular, the analyst may find it necessary to interact with management, production, and marketing personnel, as well as with various external sources (for example, consultants). This contrasts significantly with the planning phase, which is generally a management function, and the model-building phase, which involves principally the analyst. Thus resourcefulness and diplomacy may be as important as analytical ability in carrying out this phase of the study.

6.4 WRITING THE PROGRAM

Once a detailed model has been developed and the required data have been obtained, attention can be given to the development of a computer program to implement the model. The first matter to be considered is the choice of a programming language. In particular, a choice must be made between a general-purpose language such as FORTRAN or BASIC, and a special-purpose simulation language such as GPSS, SIMULA, GASP, Q-GERT, or SLAM (see Chap. 8). The general-purpose languages require more programming effort for most problems, though they are more universally available than the special-purpose languages. Moreover a general-purpose language allows the analyst to have more control over the flow of logic within the model (which is why the models presented in this textbook are oriented toward the use of a general-purpose language). Thus we have a trade-off between programming effort, logical control, and language availability (that is, program portability). Often, however, the decision will be made simply on the basis of whatever language happens to be available.

Program Organization

The special-purpose simulation languages all have their own unique programming characteristics (some of which will be described briefly in Chap. 8). We will not consider the characteristics or use of these languages at this time. On the other hand, some generalizations can be made about programs that are written in a general-purpose language such as FORTRAN. Such programs can often be enhanced considerably by careful, intelligent program organization. Usually, these enhancements can be realized with very little additional programming effort.

One desirable characteristic of a well-organized simulation program is *generality*. That is, the program should be written so that it can be run in a variety of different ways, within limits of practicality. This can usually be accomplished by allowing the values of key numerical parameters to be read into the computer at the start of execution rather than building them directly into the program. Also it may be desirable to include a few well-chosen branching operations (that is, IF statements) within the program, so that several different (though related) situations can each be simulated.

Modularity is another useful and desirable program characteristic. This involves the use of *subprograms* (that is, functions and subroutines) to represent major program elements that are relatively self-contained. Some typical program elements that lend themselves to modularization are random variate generators, output routines, and data collection routines (for example, routines used to build distribution functions for system performance criteria). Also it may be desirable for an entire program to be written in such a way that it can be converted into a subprogram. This allows major program elements to be imbedded within larger program elements. We will make use of this concept in the next chapter, when we discuss queuing problems.

There are three advantages to the use of a modularized program structure. First it becomes both possible and practical to build up a library of frequently used routines (modules). These routines can easily be interfaced with some calling program (that is, an executive routine) that has been tailored to a particular problem situation. Second the use of program modules contributes significantly to the development of orderly, error-free programs. Finally the level of programming effort can be reduced considerably through the use of standardized program modules.

Another desirable attribute of a well-written simulation program is *internal documentation*. This involves the judicious use of comments to identify key program parameters and to separate and identify major program segments. A block of comments can also be used to provide other kinds of documentation, such as a set of instructions in the use of a program and an explanation of program logic. The inclusion of such comments contributes enormously to an understanding of a program's underlying structure.

Finally every simulation program should be based upon a healthy balance between logical simplicity and computational efficiency. A program that utilizes a very simplistic logical structure may be easy to follow, but it may also be highly inefficient (that is, its running time and/or memory requirements may be excessive). Such programs should be avoided. On the other hand, a program whose logical structure is very complex may also be undesirable, even if it is highly efficient, since programs of this nature are often difficult to understand and to modify. Thus some trade-offs may be required between these different and sometimes conflicting program characteristics.

Example 6.1

Figures 6.2a to g show a complete FORTRAN program for the facility utilization problem described in Ex. 5.1. Notice that the program contains elements of all of the desirable program characteristics described above. In particular the program treats various system parameters (for example, mean values and standard deviation of random variates, seed for random number generator, number of orders processed per simulated time period) as input quantities rather than fixed numerical values. Thus the program is reasonably general. Moreover the program is modularized, since it consists of an executive routine and six independent subprograms. An easily identified comment block is included at the beginning of each module, thus providing substantial internal documentation. In addition, the executive routine contains a number of comments which indicate the logical structure of the program.

Output Data

We have already considered the statistical nature of the system performance criterion resulting from a simulation study. This suggests that the output data include a set of relative frequencies and a cumulative distribution for this parameter, in addition to the calculated mean and the standard deviation. Computer-generated graphs of the relative frequencies and the cumulative distribution may also be desirable, since they allow the analyst to visualize the statistical variation in the data.

```fortran
C*****************************************************************************
C       S A M P L E   F A C I L I T Y   U T I L I Z A T I O N   P R O B L E M   *
C                                                                            *
C                        ( E X A M P L E   5 . 1 )                           *
C                                                                            *
C IDENTIFICATION OF VARIABLES:                                              *
C       A       = ARRIVAL TIME OF THE ITH ORDER                             *
C       D       = DEPARTURE TIME OF THE ITH ORDER                           *
C       AT      = TIME INTERVAL BETWEEN ARRIVAL OF SUCCESSIVE ORDERS        *
C                     (EXPONENTIALLY DISTRIBUTED RANDOM VARIATE)            *
C       PT      = TIME REQUIRED TO PROCESS THE ITH ORDER                    *
C                     (NORMALLY DISTRIBUTED RANDOM VARIATE)                 *
C       IT      = TIME FACILITY IS IDLE BEFORE PROCESSING THE ITH ORDER     *
C       TOTIT   = CUMULATIVE IDLE TIME                                      *
C       FRACT   = FRACTION OF TIME THE FACILITY IS IN USE (OUTPUT ARRAY)    *
C       FAVG    = MEAN VALUE OF COMPUTED FRACT ARRAY (OUTPUT PARAMETER)     *
C       FSTDEV  = STANDARD DEVIATION OF COMPUTED FRACT ARRAY (OUTPUT PARAMETER)  *
C       XBAR    = MEAN VALUE OF TIME INTERVAL BETWEEN ORDERS (INPUT PARAMETER)   *
C       XAVG    = MEAN VALUE OF ORDER PROCESS TIME (INPUT PARAMETER)        *
C       STDEV   = STANDARD DEVIATION OF ORDER PROCESS TIME (INPUT PARAMETER)    *
C       NLOOP   = NUMBER OF CONSECUTIVE SIMULATIONS (INPUT PARAMETER)       *
C       KX      = SEED FOR RANDOM NUMBER GENERATOR (INPUT PARAMETER)        *
C       M       = NUMBER OF INTERVALS IN OUTPUT ARRAY (INPUT PARAMETER)     *
C       N       = NUMBER OF ORDERS PROCESSED PER SIMULATION                 *
C*****************************************************************************
      REAL IT
      DIMENSION FRACT(1000),BOUNDS(26),F(25),Y(25)
  100 FORMAT(3F4.1)
  200 FORMAT(I12,3I4)
  300 FORMAT('1',18X,'S A M P L E   F A C I L I T Y   U T I L I Z A T I
     1O N   P R O B L E M'///,11X,'EXPONENTIALLY DISTRIBUTED TIME INTERV
     2AL BETWEEN SUCCESSIVE ORDERS:  MEAN VALUE = ',F3.1,' HOURS'//,
     311X,'NORMALLY DISTRIBUTED PROCESS TIMES:  MEAN VALUE = ',F3.1,' HO
     4URS'/,48X,'STANDARD DEVIATION = ',F3.1,' HOURS'//)
  400 FORMAT('0',10X,'SEED (RANDOM NUMBER GENERATOR) = ',I12/,11X,'NUMBE
     1R OF CONSECUTIVE SIMULATIONS = ',I4/,11X,'NUMBER OF INTERVALS IN T
     2HE OUTPUT ARRAY = ',I4/,11X,'NUMBER OF ORDERS PROCESSED PER SIMULA
     3TION = ',I4//)
  500 FORMAT('0',10X,'SYSTEM PERFORMANCE CRITERION (FRACTION OF TIME THE
     1 FACILITY IS IN USE):'/)
C READ INPUT DATA
      READ (5,100) XBAR,XAVG,STDEV
      READ (5,200) KX,NLOOP,M,N
C ESTABLISH INTERVAL BOUNDS FOR OUTPUT ARRAY
      BOUNDS(1)=0.
      DO 1 J=1,M
    1 BOUNDS(J+1)=FLOAT(J)/M
C SET UP RANDOM NUMBER GENERATOR
      DUMMY=RAND(KX)
C CARRY OUT M SUCCESSIVE SIMULATIONS
      DO 4 K=1,NLOOP
C INITIALIZE VARIABLES FOR EACH PASS THROUGH THE LOOP
      I=1
      A=0.
      TOTIT=0.
C PROCESS FIRST ORDER (I=1)
      CALL NORMAL(XAVG,STDEV,PT)
      IF (PT.LT.0.) PT=0.
      D=PT
C PROCESS NEXT ORDER (I=2,3,...,N)
    2 I=I+1
    3 CALL EXPON(XBAR,0.,AT)
      A=A+AT
C HAS NEW ORDER ARRIVED BEFORE COMPLETION OF PREVIOUS ORDER?
      IF (A.LT.D) GO TO 3
      DOLD=D
      CALL NORMAL(XAVG,STDEV,PT)
      IF (PT.LT.0.) PT=0.
      D=A+PT
      IT=A-DOLD
```

(a)

Figure 6.2

```
C TEST FOR NEGATIVE IDLE TIME (NUMERICAL ERROR)
      IF (IT.LT.0.) IT=0.
      TOTIT=TOTIT+IT
C TEST FOR COMPLETION OF SIMULATION
      IF (I.LT.N) GO TO 2
      FRACT(K)=1.-TOTIT/D
    4 CONTINUE
C GENERATE STATISTICAL OUTPUT DATA
      CALL GROUP(FRACT,BOUNDS,F,Y,M,NLOOP)
      CALL PARAMS(FRACT,FAVG,FSTDEV,NLOOP)
C WRITE OUTPUT DATA
      WRITE (6,300) XBAR,XAVG,STDEV
      WRITE (6,400) KX,NLOOP,M,N
      WRITE (6,500)
      CALL OUTPUT(BOUNDS,F,Y,FAVG,FSTDEV,M,NLOOP)
      STOP
      END
```

(a)

```
      SUBROUTINE EXPON(XAVG,X0,X)
C****************************************************************************
C THIS SUBPROGRAM GENERATES AN EXPONENTIALLY DISTRIBUTED RANDOM VARIATE    *
C      WITH MEAN XAVG AND MINIMUM X0 USING THE INVERSE TRANSFORMATION METHOD *
C****************************************************************************
      ALPHA=1./(XAVG-X0)
      X=X0-ALOG(RAND(0))/ALPHA
      RETURN
      END
```

(b)

```
      SUBROUTINE GROUP(DATA,BOUNDS,F,Y,M,N)
C****************************************************************************
C THIS SUBPROGRAM GROUPS A SET OF DATA AND THEN CALCULATES THE RELATIVE    *
C      FREQUENCIES AND THE CORRESPONDING CUMULATIVE DISTRIBUTION           *
C                                                                          *
C IDENTIFICATION OF VARIABLES:                                             *
C      DATA   = INDIVIDUAL DATA POINTS (INPUT ARRAY)                       *
C      BOUNDS = INTERVAL BOUNDS (INPUT ARRAY)                              *
C      M      = NUMBER OF INTERVALS (INPUT SCALAR)                         *
C      N      = NUMBER OF DATA POINTS (INPUT SCALAR)                       *
C      K      = INTERVAL COUNTERS (INTERNAL ARRAY)                         *
C      F      = RELATIVE FREQUENCIES (OUTPUT ARRAY)                        *
C      Y      = CUMULATIVE DISTRIBUTION (OUTPUT ARRAY)                     *
C****************************************************************************
      DIMENSION DATA(1000),BOUNDS(26),F(25),Y(25),K(25)
C ZERO THE INTERVAL COUNTERS
      DO 1 J=1,M
    1 K(J)=0
C GROUP THE DATA
      DO 3 I=1,N
      DO 2 J=1,M
      IF (DATA(I).GE.BOUNDS(J+1)) GO TO 2
      K(J)=K(J)+1
      GO TO 3
    2 CONTINUE
    3 CONTINUE
C CALCULATE THE RELATIVE FREQUENCIES AND THE CUMULATIVE DISTRIBUTION
      F(1)=FLOAT(K(1))/N
      Y(1)=F(1)
      DO 4 J=2,M
      F(J)=FLOAT(K(J))/N
    4 Y(J)=Y(J-1)+F(J)
      RETURN
      END
```

(c)

Figure 6.2 *(continued)*

```fortran
      SUBROUTINE NORMAL(XAVG,STDEV,X)
C**********************************************************************
C THIS SUBPROGRAM GENERATES A NORMALLY DISTRIBUTED RANDOM VARIATE      *
C     WITH MEAN XAVG AND STANDARD DEVIATION STDEV USING A METHOD BASED  *
C     UPON THE CENTRAL LIMIT THEOREM                                    *
C**********************************************************************
      SUM=0.
      DO 1 I=1,12
    1 SUM=SUM+RAND(0)
      Z=SUM-6.
      X=XAVG+STDEV*Z
      RETURN
      END
```

(d)

```fortran
      SUBROUTINE OUTPUT(BOUNDS,F,Y,AVG,STDEV,M,N)
C**********************************************************************
C THIS SUBPROGRAM PRINTS THE MEAN, THE STANDARD DEVIATION, THE RELATIVE  *
C     FREQUENCIES AND THE CUMULATIVE DISTRIBUTION FOR A GIVEN DATA SET    *
C                                                                         *
C IDENTIFICATION OF VARIABLES:                                            *
C     BOUNDS  = INTERVAL BOUNDS                                           *
C     F       = RELATIVE FREQUENCIES                                      *
C     Y       = CUMULATIVE DISTRIBUTION                                   *
C     AVG     = MEAN                                                      *
C     STDEV   = STANDARD DEVIATION                                        *
C     M       = NUMBER OF INTERVALS                                       *
C     N       = NUMBER OF DATA POINTS                                     *
C**********************************************************************
      DIMENSION BOUNDS(26),F(25),Y(25)
  100 FORMAT('0',10X,'MEAN =',E12.5,10X,'STANDARD DEVIATION =',E12.5,
     1 10X,'NUMBER OF DATA POINTS =',I4//)
  200 FORMAT('0',14X,'INTERVAL',14X,'LOWER',14X,'UPPER',14X,'RELATIVE',
     1 14X,'CUMULATIVE'/,16X,'NUMBER',15X,'BOUND',14X,'BOUND',14X,
     2 'FREQUENCY',12X,'DISTRIBUTION'/)
  300 FORMAT(18X,I2,14X,F8.4,11X,F8.4,13X,F8.4,15X,F8.4)
      WRITE (6,100) AVG,STDEV,N
      WRITE (6,200)
      DO 1 J=1,M
      J1=J+1
    1 WRITE (6,300) J,BOUNDS(J),BOUNDS(J1),F(J),Y(J)
      RETURN
      END
```

(e)

```fortran
      SUBROUTINE PARAMS(DATA,AVG,STDEV,N)
C**********************************************************************
C THIS SUBPROGRAM CALCULATES A MEAN AND A STANDARD DEVIATION FOR A GIVEN  *
C     SET OF DATA                                                         *
C                                                                         *
C IDENTIFICATION OF VARIABLES:                                            *
C     DATA    = INDIVIDUAL DATA POINTS (INPUT ARRAY)                      *
C     N       = NUMBER OF DATA POINTS (INPUT SCALAR)                      *
C     AVG     = MEAN (OUTPUT SCALAR)                                      *
C     STDEV   = STANDARD DEVIATION (OUTPUT SCALAR)                        *
C**********************************************************************
      DIMENSION DATA(1000)
      SUM1=0.
      SUM2=0.
      DO 1 I=1,N
      SUM1=SUM1+DATA(I)
    1 SUM2=SUM2+DATA(I)**2
      AVG=SUM1/N
      STDEV=SQRT(SUM2/N-AVG**2)
      RETURN
      END
```

(f)

Figure 6.2 *(continued)*

```
      FUNCTION RAND(KX)
C***************************************************************************************
C THIS FUNCTION GENERATES A UNIFORMLY DISTRIBUTED RANDOM NUMBER                        *
C     BETWEEN ZERO AND ONE, USING THE POWER RESIDUE METHOD.                            *
C                                                                                      *
C KX SHOULD BE ASSIGNED A POSITIVE, ODD, INTEGER VALUE, NOT EXCEEDING                  *
C     34359738367, THE FIRST TIME THE FUNCTION IS CALLED.                              *
C     THEREAFTER, KX SHOULD BE ASSIGNED A VALUE OF ZERO.                               *
C                                                                                      *
C NOTE: THIS FUNCTION IS VALID ONLY FOR A COMPUTER HAVING A 36-BIT WORD.               *
C***************************************************************************************
      IF (KX.GT.0) IX=KX
      IY=262147*IX
      IF (IY.LT.0) IY=IY+34359738367+1
      RAND=FLOAT(IY)/34359738367
      IX=IY
      RETURN
      END
```

(g)

Figure 6.2 *(continued)*

Some problems involve more than one system performance criterion. In such cases a complete set of statistical data should be printed for each performance criterion.

In a simulation study, as in any other computer analysis, it is very important that all output be clearly labeled. This helps the decision maker to interpret the meaning and significance of the output. It is an unfortunate fact of life that the results of many computer studies are either misinterpreted or ignored because of obscure, poorly labeled output. This should be avoided at all costs.

Most realistic problem situations require that the computation be carried out several times, for different sets of input parameters (usually reflecting different physical conditions). Multiple sets of output will therefore be generated. In such situations it is important to print some of the input data along with each set of output, so that the various simulated results can be identified and distinguished from one other.

Example 6.2

Figure 6.3 shows the output that is generated by the simulation program presented in Ex. 6.1 for the following set of input.

Interarrival time of successive orders (exponentially distributed):

mean value = 1.4 hr

Process time for each order (normally distributed):

mean value = 0.8 hr

standard deviation = 0.2 hr

Seed for random number generator = 123456789

Number of consecutive simulations (where each simulation refers to a specified number of simulated orders) = 50

Number of orders simulated in each simulation = 100

```
S A M P L E   F A C I L I T Y   U T I L I Z A T I O N   P R O B L E M


EXPONENTIALLY DISTRIBUTED TIME INTERVAL BETWEEN SUCCESSIVE ORDERS:   MEAN VALUE = 1.4 HOURS

NORMALLY DISTRIBUTED PROCESS TIMES:   MEAN VALUE =  .8 HOURS
                                      STANDARD DEVIATION =  .2 HOURS



SEED (RANDOM NUMBER GENERATOR) =     123456789
NUMBER OF CONSECUTIVE SIMULATIONS =   50
NUMBER OF INTERVALS IN THE OUTPUT ARRAY =   10
NUMBER OF ORDERS PROCESSED PER SIMULATION =  100



SYSTEM PERFORMANCE CRITERION (FRACTION OF TIME THE FACILITY IS IN USE):

MEAN = 0.36611E+00          STANDARD DEVIATION = 0.20774E-01          NUMBER OF DATA POINTS =   50
```

INTERVAL NUMBER	LOWER BOUND	UPPER BOUND	RELATIVE FREQUENCY	CUMULATIVE DISTRIBUTION
1	0.0000	0.1000	0.0000	0.0000
2	0.1000	0.2000	0.0000	0.0000
3	0.2000	0.3000	0.0000	0.0000
4	0.3000	0.4000	0.9200	0.9200
5	0.4000	0.5000	0.0800	1.0000
6	0.5000	0.6000	0.0000	1.0000
7	0.6000	0.7000	0.0000	1.0000
8	0.7000	0.8000	0.0000	1.0000
9	0.8000	0.9000	0.0000	1.0000
10	0.9000	1.0000	0.0000	1.0000

Figure 6.3

Number of intervals in the output arrays (that is, the relative frequencies and the
cumulative distribution for the system performance criterion) = 10

Notice that all of the above input data are shown in the top portion of the
printout. This information is followed by the mean (0.366) and the standard
deviation (0.0208) for the system performance criterion, and the number of data
points (50) used to obtain these values. Finally tabulations are presented for the
relative frequencies and the cumulative distribution of the system performance
criterion. From these tabulations we see that 92 percent of the values for the
system performance criterion (that is, the fraction of time the facility is in use)
fall between 0.30 and 0.40; the remaining 8 percent of the values fall between 0.40
and 0.50.

Notice that the output data are clearly labeled and printed in a neat,
orderly, legible manner.

6.5 RUNNING THE PROGRAM

An important consideration in discrete-event simulation is the number
of random events to be included in a given run. Unfortunately it is not
possible to determine a unique value for this number. We can, however,
determine the number of random events required to establish a given
confidence interval for a calculated mean value of the system performance criterion. To do so, we must first define the following
symbols:

$\overline{Y}$ = the calculated mean value (that is, the sample mean) of the
system performance criterion

s = the calculated standard deviation of the system performance
criterion

n = the number of simulated values of the performance criterion
used to calculate $\overline{Y}$ (and s)

μ = the true mean value of the system performance criterion
(which is unknown)

Recall that the calculated quantities are determined as

$$\overline{Y} = \frac{1}{n} \sum_{i=1}^{n} Y_i \tag{6.1}$$

and

$$s = \left[\left\{ \frac{1}{n} \sum_{i=1}^{n} Y_i^2 \right\} - (\overline{Y})^2 \right]^{1/2} \tag{6.2}$$

as discussed in Sec. 5.1.

If the calculated Y_i's are normally distributed, then the statistic

$$\left(\frac{\overline{Y} - \mu}{s}\right)\sqrt{n-1} \tag{6.3}$$

is known to be distributed in accordance with the *t distribution*. Moreover this statistic is *approximately* t-distributed even if the Y_i's are not normally distributed, provided they are symmetrical about the mean. (This latter condition is satisfied in many simulation problems.) Thus we can write

$$-t_{n-1,\ 1-\alpha/2} < \left(\frac{\overline{Y} - \mu}{s}\right)\sqrt{n-1} < t_{n-1,\ 1-\alpha/2} \tag{6.4}$$

where $t_{n-1,\ 1-\alpha/2}$ represents a tabulated value of the t statistic having $(n-1)$ degrees of freedom (see Table 6.1), and $\alpha/2$ represents either of the shaded areas shown in Fig. 6.4. Note that the quantity $(1-\alpha)$ is the corresponding *confidence level*, that is, the likelihood that the value obtained from Eq. (6.3) will fall within the unshaded area in Fig. 6.4.

It is more convenient to rewrite Eq. (6.4) as

$$\left(\overline{Y} - t_{n-1,\ 1-\alpha/2}\ s/\sqrt{n-1}\right) < \mu < \left(\overline{Y} + t_{n-1,\ 1-\alpha/2}\ s/\sqrt{n-1}\right) \tag{6.5}$$

This equation tells us that the true mean (μ) falls within the interval

$$\overline{Y} \pm t_{n-1,\ 1-\alpha/2}\ s/\sqrt{n-1} \tag{6.6}$$

with $100*(1-\alpha)$ percent confidence. Thus, if the confidence level $(1-\alpha)$ is specified, an appropriate value of $t_{n-1,\ 1-\alpha/2}$ can be obtained from Table 6.1. The corresponding interval bounds can then be determined from Eq. (6.5) or Eq. (6.6).

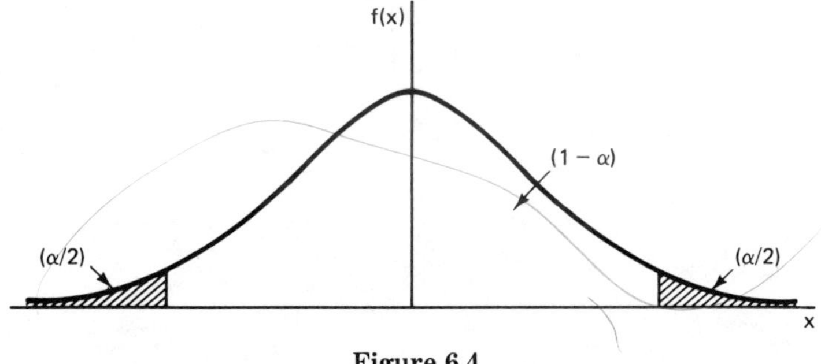

Figure 6.4

TABLE 6.1 Selected Values of the t Distribution

$n-1$	$t_{0.995}$	$t_{0.99}$	$t_{0.975}$	$t_{0.95}$	$t_{0.90}$	$t_{0.80}$	$t_{0.75}$	$t_{0.70}$	$t_{0.60}$	$t_{0.55}$
1	63.66	31.82	12.71	6.31	3.08	1.376	1.000	0.727	0.325	0.158
2	9.92	6.96	4.30	2.92	1.89	1.061	0.816	0.617	0.289	0.142
3	5.84	4.54	3.18	2.35	1.64	0.978	0.765	0.584	0.277	0.137
4	4.60	3.75	2.78	2.13	1.53	0.941	0.741	0.569	0.271	0.134
5	4.03	3.36	2.57	2.02	1.48	0.920	0.727	0.559	0.267	0.132
6	3.71	3.14	2.45	1.94	1.44	0.906	0.718	0.553	0.265	0.131
7	3.50	3.00	2.36	1.90	1.42	0.896	0.711	0.549	0.263	0.130
8	3.36	2.90	2.31	1.86	1.40	0.889	0.706	0.546	0.262	0.130
9	3.25	2.82	2.26	1.83	1.38	0.883	0.703	0.543	0.261	0.129
10	3.17	2.76	2.23	1.81	1.37	0.879	0.700	0.542	0.260	0.129
11	3.11	2.72	2.20	1.80	1.36	0.876	0.697	0.540	0.260	0.129
12	3.06	2.68	2.18	1.78	1.36	0.873	0.695	0.539	0.259	0.128
13	3.01	2.65	2.16	1.77	1.35	0.870	0.694	0.538	0.259	0.128
14	2.98	2.62	2.14	1.76	1.34	0.868	0.692	0.537	0.258	0.128
15	2.95	2.60	2.13	1.75	1.34	0.866	0.691	0.536	0.258	0.128

16	2.92	2.58	2.12	1.75	1.34	0.865	0.690	0.535	0.258	0.128
17	2.90	2.57	2.11	1.74	1.33	0.863	0.689	0.534	0.257	0.128
18	2.88	2.55	2.10	1.73	1.33	0.862	0.688	0.534	0.257	0.127
19	2.86	2.54	2.09	1.73	1.33	0.861	0.688	0.533	0.257	0.127
20	2.84	2.53	2.09	1.72	1.32	0.860	0.687	0.533	0.257	0.127
21	2.83	2.52	2.08	1.72	1.32	0.859	0.686	0.532	0.257	0.127
22	2.82	2.51	2.07	1.72	1.32	0.858	0.686	0.532	0.256	0.127
23	2.81	2.50	2.07	1.71	1.32	0.858	0.685	0.532	0.256	0.127
24	2.80	2.49	2.06	1.71	1.32	0.857	0.685	0.531	0.256	0.127
25	2.79	2.48	2.06	1.71	1.32	0.856	0.684	0.531	0.256	0.127
26	2.78	2.48	2.06	1.71	1.32	0.856	0.684	0.531	0.256	0.127
27	2.77	2.47	2.05	1.70	1.31	0.855	0.684	0.531	0.256	0.127
28	2.76	2.47	2.05	1.70	1.31	0.855	0.683	0.530	0.256	0.127
29	2.76	2.46	2.04	1.70	1.31	0.854	0.683	0.530	0.256	0.127
30	2.75	2.46	2.04	1.70	1.31	0.854	0.683	0.530	0.256	0.127
40	2.70	2.42	2.02	1.68	1.30	0.851	0.681	0.529	0.255	0.126
60	2.66	2.39	2.00	1.67	1.30	0.848	0.679	0.527	0.254	0.126
120	2.62	2.36	1.98	1.66	1.29	0.845	0.677	0.526	0.254	0.126
∞	2.58	2.33	1.96	1.645	1.28	0.842	0.674	0.524	0.253	0.126

Example 6.3

The utilization of a manufacturing facility has been simulated for 100 one-week periods, resulting in 100 different values for the fraction of time that the facility is in use (F). From these values we have determined that $\overline{F} = 0.626$ and $s = 0.108$. Determine the limits of μ that correspond to a 95 percent confidence level, assuming that the individual F's are normally distributed.

In this example we know that $\alpha = 0.05$ and $n = 100$. Hence we can obtain an appropriate value for t from Table 6.1, using linear interpolation between the tabulated values for $t_{60,\ 0.975} = 2.00$ and $t_{120,\ 0.975} = 1.98$. Thus

$$t_{99,\ 0.975} = 2.00 + \left[\frac{99 - 60}{120 - 60}\right] \ (1.98 - 2.00)$$

$$= 1.987$$

The limits of μ can now be determined from Eq. (6.6) as follows:

$$\overline{F} \pm t_{99,\ 0.975}\ s/\sqrt{n - 1} = 0.626 \pm (1.987)\ (0.108)/\sqrt{99}$$

$$= 0.626 \pm 0.022$$

Therefore

$$0.604 < \mu < 0.648$$

We conclude that the true mean value for F falls between 0.604 and 0.648 with 95 percent confidence.

If $n \geq 30$ (as is often the case in a simulation problem), then the standard normal distribution can be used in place of the t distribution. Equation (6.5) becomes

$$\left(\overline{Y} - Z_{0.5-\alpha/2}\ s/\sqrt{n - 1}\right) < \mu < \left(\overline{Y} + Z_{0.5-\alpha/2}\ s/\sqrt{n - 1}\right) \quad (6.7)$$

where $Z_{0.5-\alpha/2}$ is a tabulated value of the standard normal distribution, as given in Table 6.2. Thus the interval bounds for the true mean can be determined, with $100*(1 - \alpha)$ percent confidence, as

$$\overline{Y} \pm Z_{0.5-\alpha/2}\ s/\sqrt{n - 1} \quad (6.8)$$

Equations (6.7) and (6.8) are somewhat simpler to use than their counterparts—Eqs. (6.5) and (6.6)—since the number of degrees of freedom need not be considered.

Example 6.4

Repeat the calculations in Ex. 6.3 using the standard normal distribution rather than the t distribution. (We are able to do this because $n > 30$.)

Recall that $\overline{F} = 0.626$, $s = 0.108$, and $n = 100$. If we again choose a 95 percent confidence level, then $\alpha = 0.05$ and $Z_{0.475}$ can be obtained from Table 6.2 as $Z_{0.475} = 1.96$. Therefore from Eq. (6.8),

$$\overline{F} \pm Z_{0.475}\, s/\sqrt{n-1} = 0.626 \pm (1.96)\,(0.108)/\sqrt{99}$$

$$= 0.626 \pm 0.021$$

We therefore conclude that

$$0.605 < \mu < 0.647$$

which is essentially the same result obtained using the t distribution.

In most realistic situations, we wish to determine a value of n that will allow the true mean (μ) to fall within a desired interval at a specified confidence level. This is normally accomplished by trial and error. Thus a value for n is selected, and the corresponding interval bounds are calculated using the above procedures. If the resulting interval is too large, then n is increased and the computation is restarted. This continues until n has become large enough so that the calculated interval has become equal to or less than the desired interval.

In a practical sense, this procedure can be simplified by assuming that the calculated mean and standard deviation will not change appreciably as n is increased. This allows us to solve directly for n once $\overline{Y}$ and s have been determined. Thus if the desired interval is expressed as $\overline{Y} \pm \theta$, then, from Eq. (6.8),

$$Z_{0.5-\alpha/2}\, s/\sqrt{n-1} = \theta$$

Solving for n, we obtain

$$n = \left(Z_{0.5-\alpha/2}\, s/\theta \right)^2 + 1 \tag{6.9}$$

Example 6.5

Again consider the facility utilization problem described in Ex. 6.3. We wish to simulate a large enough number of 1-week periods so that the true mean value of F (that is, μ) falls within ± 1 percent of the calculated mean ($\overline{F}$), at a 95 percent confidence level. Determine an appropriate value for n.

We have already seen that n must exceed 100, because we obtained an interval size of 0.626 ± 0.022 when $n = 100$. This corresponds to a value of ± 3.5 percent of the calculated mean ($\overline{F} = 0.626$).

Let us assume that the calculated mean ($\overline{F}$) will remain approximately equal to 0.6, and the standard deviation (s) will remain approximately 0.1. The specified interval bound can then be expressed as

$$\theta = 0.01\,\overline{F} = (0.01)\,(0.6) = 0.006$$

Hence from Eq. (6.9),

$$n = [\,(1.96)\,(0.1)/(0.006)\,]^2 + 1$$

$$= 1068$$

We have already simulated 100 one-week periods; hence 968 additional periods must be simulated.

TABLE 6.2 Selected Values of the Standard Normal Distribution

z	0	1	2	3	4	5	6	7	8	9
0.0	0.0000	0.0040	0.0080	0.0120	0.0160	0.0199	0.0239	0.0279	0.0319	0.0359
0.1	0.0398	0.0438	0.0478	0.0517	0.0557	0.0596	0.0636	0.0675	0.0714	0.0754
0.2	0.0793	0.0832	0.0871	0.0910	0.0948	0.0987	0.1026	0.1064	0.1103	0.1141
0.3	0.1179	0.1217	0.1255	0.1293	0.1331	0.1368	0.1406	0.1443	0.1480	0.1517
0.4	0.1554	0.1591	0.1628	0.1664	0.1700	0.1736	0.1772	0.1808	0.1844	0.1879
0.5	0.1915	0.1950	0.1985	0.2019	0.2054	0.2088	0.2123	0.2157	0.2190	0.2224
0.6	0.2258	0.2291	0.2324	0.2357	0.2389	0.2422	0.2454	0.2486	0.2518	0.2549
0.7	0.2580	0.2612	0.2642	0.2673	0.2704	0.2734	0.2764	0.2794	0.2823	0.2852
0.8	0.2881	0.2910	0.2939	0.2967	0.2996	0.3023	0.3051	0.3078	0.3106	0.3133
0.9	0.3159	0.3186	0.3212	0.3238	0.3264	0.3289	0.3315	0.3340	0.3365	0.3389
1.0	0.3413	0.3438	0.3461	0.3485	0.3508	0.3531	0.3554	0.3577	0.3599	0.3621
1.1	0.3643	0.3665	0.3686	0.3708	0.3729	0.3749	0.3770	0.3790	0.3810	0.3830
1.2	0.3849	0.3869	0.3888	0.3907	0.3925	0.3944	0.3962	0.3980	0.3997	0.4015
1.3	0.4032	0.4049	0.4066	0.4082	0.4099	0.4115	0.4131	0.4147	0.4162	0.4177
1.4	0.4192	0.4207	0.4222	0.4236	0.4251	0.4265	0.4279	0.4292	0.4306	0.4319
1.5	0.4332	0.4345	0.4357	0.4370	0.4382	0.4394	0.4406	0.4418	0.4429	0.4441
1.6	0.4452	0.4463	0.4474	0.4484	0.4495	0.4505	0.4515	0.4525	0.4535	0.4545
1.7	0.4554	0.4564	0.4573	0.4582	0.4591	0.4599	0.4608	0.4616	0.4625	0.4633
1.8	0.4641	0.4649	0.4656	0.4664	0.4671	0.4678	0.4686	0.4693	0.4699	0.4706

1.9	0.4713	0.4719	0.4726	0.4732	0.4738	0.4744	0.4750	0.4756	0.4761	0.4767
2.0	0.4772	0.4778	0.4783	0.4788	0.4793	0.4798	0.4803	0.4808	0.4812	0.4817
2.1	0.4821	0.4826	0.4830	0.4834	0.4838	0.4842	0.4846	0.4850	0.4854	0.4857
2.2	0.4861	0.4864	0.4868	0.4871	0.4875	0.4878	0.4881	0.4884	0.4887	0.4890
2.3	0.4893	0.4896	0.4898	0.4901	0.4904	0.4906	0.4909	0.4911	0.4913	0.4916
2.4	0.4918	0.4920	0.4922	0.4925	0.4927	0.4929	0.4931	0.4932	0.4934	0.4936
2.5	0.4938	0.4940	0.4941	0.4943	0.4945	0.4946	0.4948	0.4949	0.4951	0.4952
2.6	0.4953	0.4955	0.4956	0.4957	0.4959	0.4960	0.4961	0.4962	0.4963	0.4964
2.7	0.4965	0.4966	0.4967	0.4968	0.4969	0.4970	0.4971	0.4972	0.4973	0.4974
2.8	0.4974	0.4975	0.4976	0.4977	0.4977	0.4978	0.4979	0.4979	0.4980	0.4981
2.9	0.4981	0.4982	0.4982	0.4983	0.4984	0.4984	0.4985	0.4985	0.4986	0.4986
3.0	0.4987	0.4987	0.4987	0.4988	0.4988	0.4989	0.4989	0.4989	0.4990	0.4990
3.1	0.4990	0.4991	0.4991	0.4991	0.4992	0.4992	0.4992	0.4992	0.4993	0.4993
3.2	0.4993	0.4993	0.4994	0.4994	0.4994	0.4994	0.4994	0.4995	0.4995	0.4995
3.3	0.4995	0.4995	0.4995	0.4996	0.4996	0.4996	0.4996	0.4996	0.4996	0.4997
3.4	0.4997	0.4997	0.4997	0.4997	0.4997	0.4997	0.4997	0.4997	0.4997	0.4998
3.5	0.4998	0.4998	0.4998	0.4998	0.4998	0.4998	0.4998	0.4998	0.4998	0.4998
3.6	0.4998	0.4998	0.4999	0.4999	0.4999	0.4999	0.4999	0.4999	0.4999	0.4999
3.7	0.4999	0.4999	0.4999	0.4999	0.4999	0.4999	0.4999	0.4999	0.4999	0.4999
3.8	0.4999	0.4999	0.4999	0.4999	0.4999	0.4999	0.4999	0.4999	0.4999	0.4999
3.9	0.5000	0.5000	0.5000	0.5000	0.5000	0.5000	0.5000	0.5000	0.5000	0.5000

Blocking

The validity of the method for determining run length (n) presented in the last section requires that the individual Y_i's at least be symmetrical about the mean $(\overline{Y})$, and preferably be normally distributed. This symmetry requirement can be satisfied in many simulation problems, but it cannot, in general, be guaranteed. We now consider a variation of this method, known as *blocking*, which makes use of random variates that are always approximately normally distributed.

The procedure is based upon the use of several consecutive short runs rather than one long run to establish a confidence interval. Specifically, consider m short runs (that is, m *blocks*), where each block contains n simulated values of the performance criterion. Let

$\overline{Y}_i =$ the calculated mean value of the system performance criterion for each block, $i = 1, 2, \ldots, m$. (Hence, each $\overline{Y}_i$ will be averaged over n_i independently generated values of the performance criterion.)

$\overline{Y} =$ the average of the calculated block means. Thus

$$\overline{Y} = \frac{1}{m} \sum_{i=1}^{m} \overline{Y}_i \qquad (6.10)$$

$d =$ the standard deviation of the calculated block means about the average, that is,

$$d = \left\{ \left[\frac{1}{m} \sum_{i=1}^{m} \overline{Y}_i^2 \right] - (\overline{Y})^2 \right\}^{1/2} \qquad (6.11)$$

$\mu =$ the true mean value (unknown) of the system performance criterion.

We can again establish a confidence interval for μ using the t distribution. Now, however, the appropriate statistic for a given confidence level $(1 - \alpha)$ is

$$-t_{m-1,\ 1-\alpha/2} < \left(\frac{\overline{Y} - \mu}{d} \right) < t_{m-1,\ 1-\alpha/2} \qquad (6.12)$$

or

$$\left(\overline{Y} - t_{m-1,\ 1-\alpha/2}\ d \right) < \mu < \left(\overline{Y} + t_{m-1,\ 1-\alpha/2}\ d \right) \qquad (6.13)$$

This equation states that the true mean (μ) falls within the interval

$$\overline{Y} \pm t_{m-1,\ 1-\alpha/2}\ d \tag{6.14}$$

with $100\,(1-\alpha)$ percent confidence.

Equations (6.12) to (6.14) are based upon the use of block averages, which tend to be normally distributed (because of the central limit theorem). Therefore we are more likely to satisfy the required normality condition when using one of these equations than when using Eq. (6.5) or (6.6). This is the reason for favoring a blocking procedure. It should be understood, however, that the number of blocks (m) will usually range from ten to twenty in a typical simulation run. The normal approximation to the t distribution cannot be used under these conditions.

Example 6.6

A proposed business venture is being studied by means of simulation, using a blocking technique. Each simulation consists of ten successive blocks, with 100 values of the present worth generated for each block. The following block averages were obtained:

Block number	Present worth (millions of dollars)
1	15.6
2	22.7
3	11.8
4	17.0
5	20.9
6	13.5
7	13.1
8	12.7
9	18.8
10	21.0

Determine the limits of μ that correspond to a 95 percent confidence level (where μ represents the true mean value of the present worth).

The overall sample mean and the standard deviation of the block means can be obtained using Eqs. (6.10) and (6.11). Thus

$$\overline{PW} = \frac{1}{10} \sum_{i=1}^{10} \overline{PW_i}$$

$$= \frac{1}{10}(15.6 + 22.7 + 11.8 + 17.0 + 20.9 + 13.5 + 13.1 + 12.7 + 18.8 + 21.0)$$

$$= 16.71$$

and

$$d = \left\{ \left[\frac{1}{10} \sum_{i=1}^{10} (\overline{PW_i})^2 \right] - (\overline{PW})^2 \right\}^{1/2}$$

$$= \left\{ \frac{1}{10} (15.6^2 + 22.7^2 + 11.8^2 + 17.0^2 + 20.9^2 + 13.5^2 + 13.1^2 + 12.7^2 \right.$$

$$\left. + \quad 18.8^2 + 21.0^2) - 16.71^2 \right\}^{1/2} = 3.76$$

We have 9 degrees of freedom in this example. Therefore, the appropriate value of $t_{m-1,\ 1-\alpha/2}$ can be obtained from Table 6.1 as $t_{9,0.975} = 2.26$. We can now obtain the desired limits from Eq. (6.14), as

$$PW \pm t_{9,0.975}\ d = 16.71 \pm (2.26)\ (3.76)$$
$$= 16.71 \pm 8.50$$

Therefore

$$8.21 < \mu < 24.92$$

We conclude that the true mean value for PW falls between \$8.21 million and \$24.92 million with a 95 percent confidence level.

It should be noted that Eqs. (6.12) to (6.14) do not explicitly involve the number of simulated values per block (n_i), or the standard deviation (s_i) of these values about each block average. The variability in the calculated block averages (and consequently d) will, however, decrease as n_i increases. This will affect the size of the confidence interval when the number of blocks (m) and the confidence level ($1 - \alpha$) are fixed.

Now suppose that a simulation consisting of m blocks, with n simulated values per block, has already been carried out. We can easily calculate values for $\overline{Y_i}$, $\overline{Y}$, and d, and a corresponding confidence interval, using the procedure described above. If the confidence interval is too large, then the simulation will have to be repeated (or at least restarted) using a larger number of random variates. Usually the block size (n) will be increased rather than the number of blocks (m). The new value for n can be obtained in the following manner.

Since the block averages will tend to be normally distributed, we know that their variance about the true average is given by σ^2/n, where σ represents the true standard deviation of the random variates. As a rule σ will be unknown. We can estimate σ, however, by utilizing the expression $d^2 = \sigma^2/n$. To do so, we write

$$\sigma^2 = n_1 d_1^2 = n_2 d_2^2 \tag{6.15}$$

where n_1 represents the original (known) block size and d_1 represents the corresponding standard deviation. Thus the quantity $n_2 d_2^2$ can easily be obtained, where n_2 represents the new block size and d_2 represents the new standard deviation. Also d_2 can be estimated by writing

$$t_{m-1,\ 1-\alpha/2}\ d_2 = \theta \tag{6.16}$$

where $\overline{Y} \pm \theta$ is the desired confidence interval. Now Eq. (6.16) can be solved directly for d_2, and n_2 can then be obtained as

$$n_2 = n_1(d_1^2/d_2^2) \tag{6.17}$$

Example 6.7

Again consider the situation described in Ex. 6.6. Determine how many values of the present worth will have to be determined within each block so that the true mean value falls within $\pm5\%$ of the calculated mean at a 95 percent confidence level.

If we base our calculations on the previously determined sample mean ($\overline{PW} = 16.71$), then

$$\theta = 0.05\ \overline{PW} = 0.05\ (16.71) = 0.8355$$

Since m remains equal to 10, we can write

$$t_{m-1,\ 1-\alpha/2} = t_{9,0.975} = 2.26$$

as before. Hence from Eq. (6.16),

$$2.26\ d_2 = 0.8355$$

which yields $d_2 = 0.3697$

We have already established that $n_1 = 100$ and $d_1 = 3.76$. Therefore Eq. (6.17) yields

$$n_2 = 100\ (3.76)^2/(0.3697)^2 = 10{,}344$$

We conclude that the desired confidence interval will be obtained with ten blocks provided at least 10,344 values are simulated within each block.

The reader is referred to the texts by Fishman (1973, 1978) for additional information on the use of confidence intervals in conjunction with run length.

Warm-up Period

There is one additional consideration when running a dynamic model (for example, a facility utilization model or a queuing model), in which the random events occur sequentially. This concerns the effect of the *initial conditions* that are required in order to initiate the computation. These conditions are chosen arbitrarily, though they attempt to describe an initial state that is representative of actual system performance. However the statistical information that is accumulated during the simulation should be based upon *steady-state behavior*, that is, it should be *independent* of the particular initial conditions. We must therefore include a *warm-up period* within the simulation run. This allows the effects of the initial conditions to dissipate before beginning to collect statistical information on system performance.

There is no particular criterion for establishing the length of the warm-up period, so that its length is more or less arbitrary. Typically a warm-up period may be anywhere from 5 percent to 20 percent of the remaining (steady-state) run.

Example 6.8

A particular system involving an airline check-in counter is being simulated. It has been established that 782 simulated passengers are required in order that the mean waiting time falls within ± 10 percent of the calculated mean with 95 percent confidence.

When carrying out the actual computation, however, a total of 900 passengers were simulated. The first 100 passengers were not included when determining a distribution of waiting times. Thus the run included a 100-passenger warm-up period in addition to an 800-passenger "steady-state" simulation.

The decision to include the first 100 passengers in the warm-up period was, of course, arbitrary. Some numerical experimentation with different-length periods may be required in order to establish a more appropriate size for the warm-up period.

Randomization

Randomization refers to the initialization of the random number generator by specifying a new seed. Each seed will result in the generation of a different, unique sequence of random variates. Therefore with a given simulation model, it is possible to carry out many independent simulations by this procedure. The blocking technique discussed above makes use of this concept, since the seed that is used for a particular block (other than the first) is simply the final value of the random number generator from the preceding block.

When carrying out an actual simulation study the question of whether to specify a new seed or utilize the old (previous) seed arises quite often. In some situations a new seed is desirable, and in other situations it is not. Some guidelines are outlined:

1. A *new* seed should be specified under the following circumstances:
    a. When carrying out multiple runs of the same problem (for example, when utilizing a blocking technique), in order to generate multiple sample means and variances, etc.
    b. When restarting a previous problem, in order to obtain a larger sample size (and perhaps a correspondingly smaller confidence interval) before calculating a sample mean and variance, etc.
    In such situations it is convenient to use the last random number generated in the previous run as the seed for the new run, though the new seed can also be specified arbitrarily if desired.

2. A new seed should *not* be provided (that is, the *same* seed should be used repeatedly) under the following circumstances:
   a. When debugging a simulation program
   b. When carrying out multiple runs with different input parameters (parametric analysis)
   c. When faced with any other situation requiring a constant, reproducible sequence of random variates.

The desirability of a reproducible random number sequence is readily apparent for debugging purposes. This is less clear when conducting parametric studies, although this procedure does eliminate random statistical variations between runs. Moreover, this procedure tends to reduce the variance of the difference between successive runs if the random number sequence exhibits positive autocorrelation [see Emshoff and Sisson (1970) for a further discussion of this last point].

Finally, the reader is referred to a classical paper by Conway (1963) for a further discussion of various computational strategies, including the use of blocking techniques and the question of whether or not to randomize.

6.6 VALIDATING THE MODEL

In order to determine the validity of a simulation model we must first recognize the following likely sources of error:

1. The data.
   This refers to both the *accuracy* of the data and the *type* of data (that is, the particular parameters that were measured).
2. The model itself.
   Here we are primarily concerned with the assumptions that were used to build the model, and the validity of the cause-and-effect relationships among the variables.
3. Implementation of the model.
   This is largely a matter of programming accuracy.
4. Interpretation of the results.

Each of these items should be examined carefully if the accuracy of the model is considered to be in question. Each item is fundamentally distinct, however, which precludes the use of simple standardized procedures for error detection. Thus, the analyst must examine each item carefully and critically, applying sound judgment, common sense, and attention to detail. A good deal of time and patience may be required for this phase of the work.

Perhaps the easiest way to assess the validity of a model is to determine (that is, simulate) the expected behavior of a system whose performance characteristics are known. Comparisons can then be made between the simulated and the known results.

There are several ways in which the performance characteristics of the system can be determined. These include the use of theoretical predictions, hand calculations, and historical data.

Theoretical predictions can be used only for simple, idealized models. We have already encountered predictions of this nature in our discussion of inventory systems in Chap. 5 (see Sec. 5.4). Some additional theoretical results for idealized queuing systems are presented in the next chapter. Such results may be more useful than the reader may realize, since a simple, idealized model can often be formulated as a special case of a more general and a more complicated model. Comparisons of this nature should always be made whenever possible. One good test of a complicated model is, of course, its ability to predict the performance of a simplified subsystem.

The use of hand calculations also tends to be limited to simple systems with a small number of random events, since it is generally impractical to carry out an extensive amount of computation manually. On the other hand, some occasional hand calculations can provide considerable insight into the details of the computational procedures as well as establish a basis of comparison for the computerized simulation model. The examples in Chap. 5 contain several brief, illustrative hand calculations (others will be seen in Chap. 7). Such calculations can be very useful, particularly for debugging purposes.

Historical data generally refer to information that has been obtained for an existing system whose behavior is well understood. The assessment of a simulation model based upon such historical data tends to be much more subjective than the other two methods. Nevertheless, historical assessments are frequently of greater value because they can be applied to more realistic and complex types of situations. For many problems the successful simulation of known past performance is a critical test that usually enhances one's confidence in the validity of the simulation model.

There is also some value in using a simulation model to determine the expected behavior of a system whose performance characteristics are *not* known. This can be tricky, but common sense will often indicate areas in which the simulation model is in need of modification. Results of this nature can be a useful supplement to the controlled comparisons previously described.

Finally it should again be mentioned that the development of a valid simulation model is a repeated, cyclical procedure. Several cycles are usually required in order to obtain an accurate model which contains an adequate level of detail. Typically each cycle will involve some modification of the basic underlying assumptions, the specific cause-and-effect relationships, or the supporting data base. It is usually the validation step that determines whether or not the model has undergone a sufficient degree of refinement. Another cycle will be initiated if the desired degree of refinement has not been attained.

6.7 USE OF THE SIMULATION MODEL

Once a simulation model has been fully developed, it can be used in a number of different ways. These include

1. The detailed analysis of a particular policy.
2. Sensitivity analysis.
3. Comparisons between alternative policies.
4. Cost–benefit assessments.

Let us consider each of these items in some detail.

The detailed analysis of a particular policy involves the determination of the pertinent state variables, and the evaluation of the system performance criterion. The latter is customarily expressed as a statistical distribution function, from which an expected value and a standard deviation can be obtained. This is, of course, the most obvious and most important single application of a simulation study. The various simulation models developed elsewhere in this book are all oriented toward this type of application (though these models can also be used for other types of applications, as will be subsequently discussed).

Another important consideration related to the behavior of a system is the change in the performance criterion resulting from a change in one of the input quantities (that is, a decision variable or a system parameter). This is known as *sensitivity analysis.* Such studies are somewhat awkward to carry out with a simulation model, since we must carry out a complete simulation, then change an input quantity by some small, known amount (say, 10 percent), and then repeat the entire simulation. The information that is obtained often justifies the effort, however, because we are able to identify the particular input parameters that are capable of causing wide variations in the system performance. Such parameters must be accurately measured and carefully controlled. Hence we can obtain some indication of where additional data-gathering activities, and perhaps process control efforts, may be warranted.

The comparison of alternative policies is a logical extension of the previous two types of applications, since we can determine the sensitivity of the system performance criterion to various policy parameters as well as obtain a detailed analysis of the system behavior under different operating conditions. Ultimately this type of comparison is of greatest value to decision makers, since it allows them to choose the best of many different alternatives. A decision maker can then assess the cost of operating the system in this manner, and relate that cost to the benefits that the system provides.

PROBLEMS

6.1. Examine a computer program that was written to solve an actual simulation problem, such as those described at the end of Chap. 5. Consider whether or not ample attention has been given to the following programing characteristics:
  (a) Generality
  (b) Modularity
  (c) Internal documentation
  (d) Appearance of the output data
  (e) Printing of input data to identify the problem
  If necessary, rewrite the program to accommodate each characteristic.

6.2. For the situation described in Ex. 6.3, determine the limit of μ corresponding to a 90 percent confidence level. Solve by means of
  (a) the t distribution
  (b) the standard normal distribution
  Compare the results obtained using each method. Also, compare with the limits obtained in Exs. 6.3 and 6.4, for a 95 percent confidence level.

6.3. Again consider the situation described in Ex. 6.3.
  (a) Estimate the number of simulated 1-week periods (that is, the value for n) required so that $0.616 < \mu < 0.636$ with 90 percent confidence.
  (b) Repeat the calculation for a 99 percent confidence level. Compare with the result obtained in part **a**.

6.4. For the situation described in Ex. 6.6, determine
  (a) the limits of μ that correspond to a 98 percent confidence level
  (b) the confidence level that is associated with the interval $15.71 < \mu < 17.71$
  (c) the block size (that is, the value of n) required so that $15.71 < \mu < 17.71$ with 98 percent confidence.

6.5. The operation of a supermarket checkout counter is being analyzed by simulating the checkout time for 500 consecutive customers. An average checkout time of 4.79 minutes and a standard deviation of 1.65 minutes were obtained. Determine
  (a) The limits of μ corresponding to a 95 percent confidence level (where μ represents the true mean checkout time)
  (b) the number of consecutive customers that must be simulated in order that the mean checkout time fall within ±0.1 min with 99 percent confidence.

Assume that the checkout times are normally distributed when carrying out the calculations.

6.6. Suppose that the 500 checkout times obtained in Prob. 6.5 were broken up into ten blocks, with fifty events per block, resulting in the following block averages ($\overline{CT}_i$) and standard deviations (s_i).

i	$\overline{CT}_i$ (min)	s_i (min)
1	4.92	1.63
2	5.13	1.71
3	4.62	1.74
4	4.94	1.71
5	4.64	1.56
6	4.84	1.77
7	4.42	1.77
8	4.52	1.46
9	4.91	1.71
10	4.93	1.48

Use these block data to determine
(a) the limits of μ corresponding to a 95 percent confidence level (where μ represents the true mean checkout time)
(b) the block size that would be required in order that the mean checkout time fall within ± 0.1 min with 99 percent confidence
Compare your answers with the results obtained in Prob. 6.5. What conclusions can be reached concerning the pros and cons of using blocked data?

6.7. From the results obtained in Prob. 3.15, determine the limits of the average yearly profit corresponding to a 95 percent confidence level.

6.8. Rerun Prob. 3.15 using a large enough number of simulated events so that the average yearly profit is known to within ± 2 percent of the calculated mean with 99 percent confidence.

6.9. Using the results obtained in Prob. 5.3, determine a confidence interval for the expected recovery time corresponding to a 98 percent level of confidence. From a practical standpoint, has a large enough number of simulated events been carried out to obtain a reasonably precise estimate of the mean recovery time?

6.10. Solve the inventory control problem described in Ex. 5.3, with an initial inventory level of 60 units, and values of $ROP = 40$ and $Q = 50$ as decision variables. Simulate a large enough number of 1-year time periods so that the expected value of the total yearly inventory cost falls within ± 10 percent of the calculated mean with 90 percent confidence.

6.11. Use the results obtained in Prob. 5.16 to determine a confidence interval for the mean present worth. Simulate a large enough number of 10-year periods so that the expected value of the present worth falls within ± 5 percent of the calculated mean with 95 percent confidence.

7

QUEUING APPLICATIONS

Perhaps the most important type of problem in discrete-event simulation is that which involves a *queue* (that is, a *waiting line*). There are many practical applications of queuing problems. For example, banks, doctors' offices, supermarkets, car washes, airports, and gasoline stations all involve situations in which customers may have to wait in line for service.

Furthermore, there are many problems in which some other kind of entity must wait to be processed. Orders entering a job shop, for example, must wait for their turn before being processed. The same is true of machines waiting to be repaired, jobs entering a computer system, and so on. Such entities need not necessarily be processed on a first-in, first-out basis.

In this chapter we will consider techniques for simulating the more common types of queuing situations. We begin with the simplest and most common situation, consisting of a single waiting line followed by a single server. Two different methods are presented for simulating such problems. We then extend one of these models to describe two other important situations, namely, the case where there are multiple servers (and multiple waiting lines) in series, and the problem of multiple servers in parallel with a common waiting line. The chapter concludes with a discussion of some computational procedures that should be used when implementing these models.

7.1 THE SINGLE-CHANNEL SINGLE-STATION QUEUE

We begin our study of queuing models with the single-channel, single station queue. This model consists of a single server which accommodates customers sequentially, on a one-at-a-time basis. Thus newly arriving customers either enter the server directly, if it is idle, or else they must wait in line until it is their turn to enter. A waiting customer will enter the server as soon as the previous customer leaves unless, of course, no one is waiting to enter, in which case the server becomes idle. Each customer enters and leaves the queue (that is, the waiting line) on a first-in first-out (FIFO) basis.

Figure 7.1 shows a schematic representation of the single-channel single-station queue. Within this figure the box marked "Q" represents the queue, and the box marked "S" represents the server.

Let us proceed by simulating a succession of customers passing through the system. Thus we will refer to this model as a *next-customer* model of the single-channel single-station queue. Our ultimate objective will be to obtain distributions for the waiting-time, the queue length, and the time spent within the system. Certain related statistics, such as the fraction of time the server is idle, the maximum queue length, and the maximum time required to pass through the system, can also be obtained.

We now define the following variables:

$A_i =$ arrival time of the ith customer (that is, the time when the ith customer enters the system)

$B_i =$ the time when the ith customer enters the server

$D_i =$ departure time of the ith customer (that is, the time when the ith customer leaves the server)

$AT_i =$ time interval between the arrival of the $(i - 1)$st and the ith customers (a random variate)

$ST_i =$ time required to serve the ith customer (a random variate)

$WT_i =$ waiting time for the ith customer (that is, the length of time the ith customer spends in the queue, waiting to enter the server)

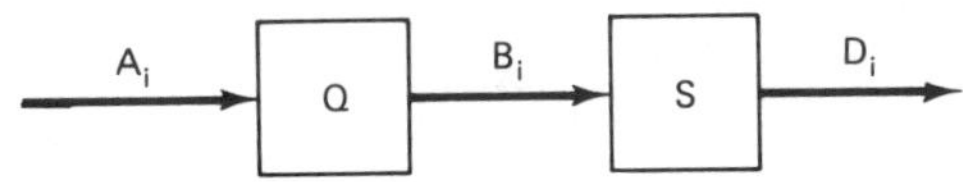

Figure 7.1

IT_i = idle time associated with the ith customer (that is, the length of time the server is idle prior to serving the ith customer)

N = total number of customers to be simulated (thus $i = 1, 2, \ldots, N$)

The single-letter variables (A_i, B_i, and D_i) are state variables that represent *clock times*, whereas the double-letter variables (AT_i, ST_i, WT_i, and IT_i) are either randomly generated or calculated quantities that represent *time intervals*. These time intervals must all be nonnegative, that is,

$$AT_i, ST_i, WT_i, IT_i \geq 0$$

Furthermore a customer cannot be waiting in the queue if the server is idle, and the server cannot be idle if customers are waiting. Therefore for each customer (each i), either WT_i or IT_i must equal zero. (It is possible, of course, that at some particular time there will be no customers in the system; hence WT_i and IT_i will both equal zero.)

Since AT_i and ST_i are random variates whose values will be generated by the computer, we are left with five unknown quantities: A_i, B_i, D_i, IT_i, and WT_i. The mathematical model presented below consists of five independent equations. Thus we will have five equations in five unknowns, which can readily be solved.

The *consecutive arrivals* to the system can be described by the equation

$$A_i = A_{i-1} + AT_i \tag{7.1}$$

and the *consecutive departures* can be expressed as

$$D_i = D_{i-1} + IT_i + ST_i \tag{7.2}$$

The arrival and departure of each customer can be related to one another by considering all of the events experienced by each customer (that is, a *customer history*). Thus we can write

$$D_i = A_i + WT_i + ST_i \tag{7.3}$$

The restriction that IT_i and WT_i cannot both be positive can be expressed mathematically as

$$(IT_i)(WT_i) = 0 \tag{7.4}$$

Finally the *consecutive arrival of customers to the server* can be expressed as

$$B_i = A_i + WT_i \tag{7.5}$$

Notice that a newly arriving customer will not have to wait in line if the server is idle, that is, if $A_i \geq D_{i-1}$. Hence we set $WT_i = 0$ and $B_i = A_i$. A nonzero value will then have to be calculated for IT_i. If the new customer arrives before the previous customer has departed, however (that is, if $A_i < D_{i-1}$), then we set $IT_i = 0$ and calculate a nonzero value for WT_i.

In order to simulate the behavior of the system with these equations, we must specify an appropriate set of initial conditions (that is, values A_1, B_1, D_1, IT_1, and WT_1). The most convenient way to do this is to allow the first customer to arrive at time level zero. Then $A_1 = 0$ and $D_1 = ST_1$ (where ST_1 is randomly generated). Also note that B_1, IT_1, and WT_1 will also equal zero. Thus we have a complete set of initial conditions.

For each successive customer ($i = 2, 3, \ldots, N$), we generate a value for AT_i and for ST_i, and then calculate values for A_i, B_i, D_i, IT_i, and WT_i using Eqs. (7.1) through (7.5). It is important, however, that the calculations be carried out in the correct order. Figure 7.2 presents a flowchart of the computational procedure. Notice that Eq. (7.4) is never used as such, but is replaced by an equivalent conditional branch.

Example 7.1

The manager of a supermarket wishes to study the operation of a checkout counter. The arrival times of customers to the checkout counter are known to be exponentially distributed, with a mean interarrival time of 2.3 min. The checkout times (service times) are normally distributed, with a mean of 1.8 minutes and a standard deviation of 0.5 minutes. If the checkout counter is busy the customers must wait in line until the counter becomes free.

The operation of the checkout counter can be simulated using the procedure presented in Fig. 7.2. In order to demonstrate the manner in which the calculations are carried out, however, let us calculate a detailed history for the first few simulated customers, as subsequently outlined.

Observe that the second and fourth columns (labeled AT_i and ST_i) contain random variates representing the customer interarrival times and the service times, respectively. The remaining columns (except, of course, columns 1 and 5) contain calculated quantities.

The table should be examined on a row-by-row basis, since each row contains information that is associated with a different customer. Notice that $WT_i = 0$ and $IT_i > 0$ for the fourth, fifth, and eighth customers. For all other customers (except the first), $IT_i = 0$ and $WT_i > 0$. Thus eight of the twelve simulated customers were required to wait before entering the server.

Finally it is interesting to compare this problem with the facility utilization problem presented in Ex. 5.1. The two problems are identical except that waiting is not considered in the earlier problem.

Once a simulated history has been determined for all of the customers, it is quite straightforward to obtain distributions for the customer waiting times and the total times spent within the system (the latter quantity is simply $T_i = D_i - A_i$). The individual values can be

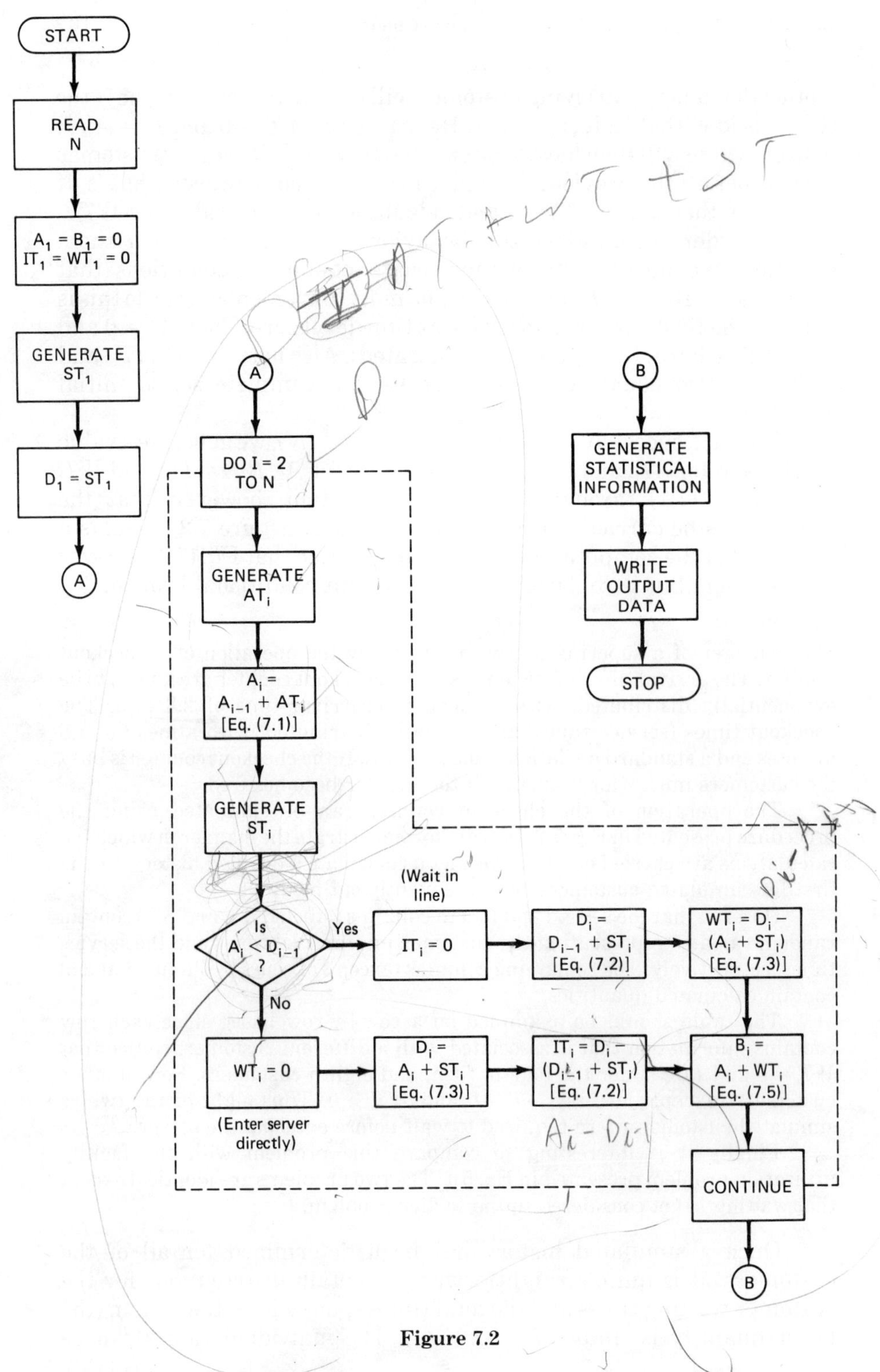

Figure 7.2

i	AT_i^*	A_i	$\bar{ST}_i^*$	Is $A_i < D_{i-1}$?	IT_i	D_i	WT_i	B_i
1		0	1.840		0	1.840	0	
2	1.470	1.470	3.465	Yes	0	5.305	0.370	1.840
3	2.415	3.885	1.766	Yes	0	7.071	1.420	5.305
4	4.363	8.248	1.414	No	1.177	9.662	0	8.248
5	3.483	11.731	0.796	No	2.069	12.527	0	11.731
6	0.071	11.802	1.184	Yes	0	13.711	0.725	12.527
7	1.175	12.977	1.934	Yes	0	15.645	0.734	13.711
8	2.942	15.919	2.303	No	0.274	18.222	0	15.919
9	0.542	16.461	2.857	Yes	0	21.079	1.761	18.222
10	1.503	17.964	2.192	Yes	0	23.271	3.115	21.079
11	3.701	21.665	2.326	Yes	0	25.597	1.606	23.271
12	1.991	23.656	1.971	Yes	0	27.568	1.941	25.597

*Random variates.

grouped using the procedure described in Sec. 4.11, and the corresponding means and standard deviations can be determined as

$$\overline{WT} = \frac{1}{N} \sum_{i=1}^{N} WT_i \tag{7.6}$$

$$\overline{T} = \frac{1}{N} \sum_{i=1}^{N} (D_i - A_i) \tag{7.7}$$

$$s_{WT} = \left[\left\{ \frac{1}{N} \sum_{i=1}^{N} WT_i^2 \right\} - (\overline{WT})^2 \right]^{1/2} \tag{7.8}$$

$$s_T = \left[\left\{ \frac{1}{N} \sum_{i=1}^{N} (D_i - A_i)^2 \right\} - (\overline{T})^2 \right]^{1/2} \tag{7.9}$$

where $\overline{T}$ represents the mean value of the total time spent in the system. Also, the fraction of time the server is idle can easily be obtained by summing the idle times and dividing by the total simulated time, that is,

$$F = \left(\sum_{i=1}^{N} IT_i \right) / (D_N - A_1) \tag{7.10}$$

The desired statistical information for the queue length is more difficult to obtain, since queue length must be expressed as a function of time. This can be accomplished by merging the arrivals and departures associated with the queue, as described in the next section.

7.2 DETERMINING QUEUE LENGTH VERSUS TIME

Let us now examine the behavior of the queue in greater detail. We will assume that the arrivals and departures associated with the queue (A_i and B_i, $i = 1, 2, \ldots, N$) are known. Our present objective will be to merge the arrivals and departures, in chronological order, into a single list of events. We will also determine the *status* of the queue (that is, the number of customers in the queue) after each event. This information will allow us to obtain the desired distribution of queue length versus time.

In order to develop the computational strategy for carrying out the merge, we must use different subscripts for the arrivals and the departures. This will allow us to compare the arrival of one customer (for example, the jth customer) with the departure (from the queue) of another customer (for example, the kth customer) in order to determine which event occurs first. We therefore define the following symbols:

A_j = arrival time of the jth customer (that is, the time when the jth customer enters the queue), $j = 1, 2, \ldots, N$

B_k = departure time of the kth customer (that is, the time when the kth customer leaves the queue and enters the server), $k = 1, 2, \ldots, N$

T_i = time when the ith event occurs (the ith event will be either an arrival or a departure), $i = 1, 2, \ldots, 2N$

Q_i = the status of the queue (that is, the number of customers in the queue) after the ith event.

We wish to form a single, chronological list of all events (including all arrivals and all departures), and a corresponding list of the queue length following each event. (Note that each list will contain $2N$ elements.) To do so, we proceed as follows:

1. For the first event ($i = 1$), set $T_1 = A_1$ and $Q_1 = 1$.

2. Set $j = 2$ and $k = 1$ (We are now going to compare the arrival of the second customer with the departure of the first customer, to see which event occurs first.)

3. Compare A_j with B_k

   a. If $A_j < B_k$, then the jth customer arrives before the kth customer leaves. Hence set

   $$T_{i+1} = A_j$$
   $$Q_{i+1} = Q_i + 1$$
   $$i = i + 1$$
   $$j = j + 1 \text{ (prepare for next arrival)}$$

   b. If $A_j > B_k$, then the kth customer leaves before the jth customer arrives. Hence set

   $$T_{i+1} = B_k$$
   $$Q_{i+1} = Q_i - 1$$
   $$i = i + 1$$
   $$k = k + 1 \text{ (prepare for next departure)}$$

   c. If $A_j = B_k$, then the jth customer arrives at the same time that the kth customer departs. Hence set

   $$T_{i+1} = A_j$$

$$Q_{i+1} = Q_i + 1$$

$$T_{i+2} = B_k = A_j = T_{i+1}$$

$$Q_{i+2} = Q_{i+1} - 1 = Q_i$$

$$i = i + 2$$

$$j = j + 1 \text{ (prepare for next arrival)}$$

$$k = k + 1 \text{ (prepare for next departure)}$$

4. Step 3 is repeated successively until j exceeds N.
   a. If $j > N$ and $k \leq N$, the last customer has arrived but there are still customers in the queue. Hence the next event must be a departure. Therefore set

$$T_{i+1} = B_k$$

$$Q_{i+1} = Q_i - 1$$

$$i = i + 1$$

$$i = k + 1 \text{ (prepare for next departure)}$$

   b. If $j > N$ and $k > N$, then the last customer has departed. The merge will therefore be completed.

A flowchart of the entire merge procedure is shown in Fig. 7.3.

Once the merge has been completed we will have determined the status of the queue as a function of time. The mean queue length can then be obtained as a time-weighted average. Thus

$$\overline{Q} = \sum_{i=1}^{2N-1} \left[\frac{T_{i+1} - T_i}{T_{2N} - T_1} \right] Q_i$$

$$= \frac{1}{T_{2N} - T_1} \sum_{i=1}^{2N-1} (T_{i+1} - T_i)\, Q_i \tag{7.11}$$

Similarly the standard deviation can be determined as

$$s_Q = \left\{ \sum_{i=1}^{2N-1} \left[\frac{T_{i+1} - T_i}{T_{2N} - T_1} \right] Q_i^2 - (\overline{Q})^2 \right\}^{1/2}$$

$$= \left[\left\{ \frac{1}{T_{2N} - T_1} \sum_{i=1}^{2N-1} (T_{i+1} - T_i)\, Q_i^2 \right\} - (\overline{Q})^2 \right]^{1/2} \tag{7.12}$$

In addition, the individual Q_i's can be grouped using the procedure presented in Sec. 4.11.

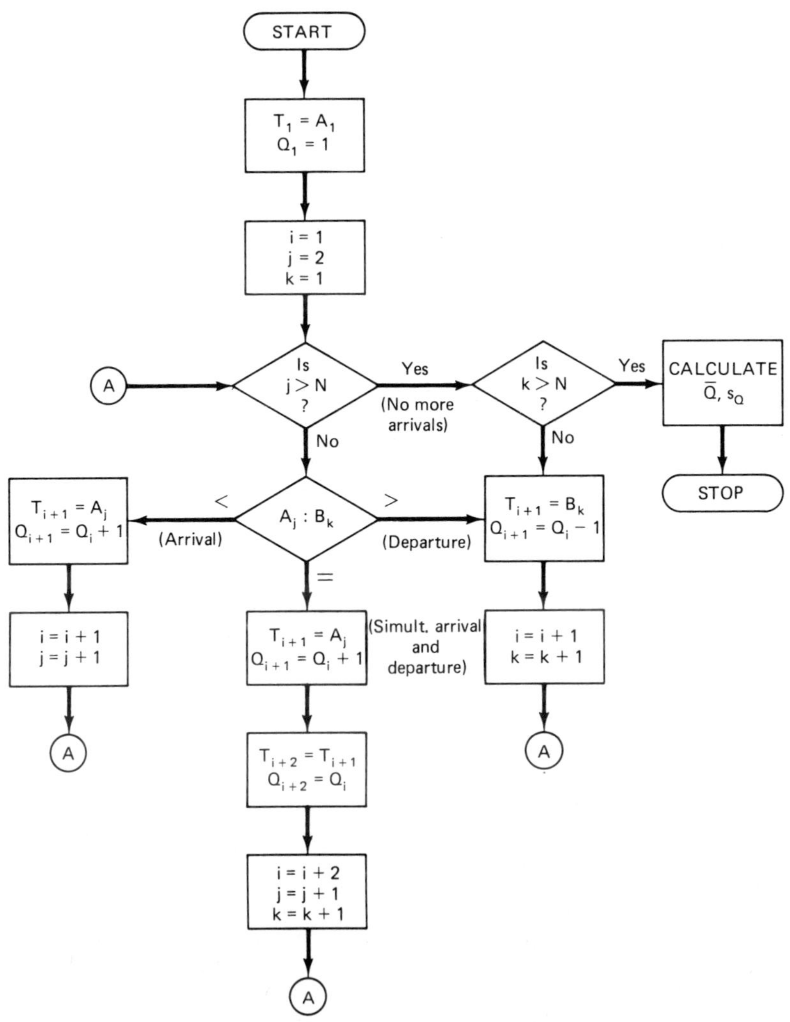

Figure 7.3

Example 7.2

Determine the mean, the standard deviation, the relative frequencies, and the cumulative distribution of the waiting time and the queue length for the twelve simulated customers in Ex. 7.1.

The individual waiting times have already been determined in Ex. 7.1. Hence the mean waiting time can be determined from Eq. (7.6) as

$$\overline{WT} = \frac{1}{12}(0 + 0.370 + 1.420 + 0 + 0 + 0.725 + 0.734 + 0 + 1.761 + 3.115$$
$$+ 1.606 + 1.941)$$
$$= 0.973 \text{ min}$$

Similarly, the standard deviation can be obtained from Eq.(7.8). Thus

$$
s_{WT} = \left[\frac{1}{12} (0^2 + 0.370^2 + 1.420^2 + 0^2 + 0^2 + 0.725^2 + 0.734^2 + 0^2 + 1.761^2 \right.
$$
$$
\left. + 3.115^2 + 1.606^2 + 1.941^2) - 0.973^2 \right]^{1/2}
$$
$$
= 0.958 \text{ min}
$$

The relative frequencies can be determined simply by counting the number of waiting times that fall into each time interval, and the cumulative distribution is obtained by calculating the partial sums of the relative frequencies. Thus if we choose 1-minute time intervals, we obtain

Time interval (min)	Number of events	Relative frequency	Cumulative distribution
0–1	7	7/12 = 0.583	0.583
1–2	4	4/12 = 0.333	0.916
2–3	0	0/12 = 0.000	0.916
3–4	1	1/12 = 0.083	0.999
	12		

In order to determine the queue length as a function of time, we must merge the arrivals (A_i) and the departures from the queue (B_i). The results, obtained using the procedure previously described, are as follows:

i	T_i	$(T_{i+1} - T_i)$ (min)	Q_i	i	T_i	$(T_{i+1} - T_i)$ (min)	Q_i
1	0 (A1)	0	1	13	12.977 (A7)	0.734	1
2	0 (B1)	1.470	0	14	13.711 (B7)	2.208	0
3	1.470 (A2)	0.370	1	15	15.919 (A8)	0	1
4	1.840 (B2)	2.045	0	16	15.919 (B8)	0.542	0
5	3.885 (A3)	1.420	1	17	16.461 (A9)	1.503	1
6	5.305 (B3)	2.943	0	18	17.964 (A10)	0.258	2
7	8.248 (A4)	0	1	19	18.222 (B9)	2.857	1
8	8.248 (B4)	3.483	0	20	21.079 (B10)	0.586	0
9	11.731 (A5)	0	1	21	21.665 (A11)	1.606	1
10	11.731 (B5)	0.071	0	22	23.271 (B11)	0.385	0
11	11.802 (A6)	0.725	1	23	23.656 (A12)	1.941	1
12	12.527 (B6)	0.450	0	24	25.597 (B12)		0

The mean queue length can now be obtained from Eq. (7.11). Thus

$$
\overline{Q} = \left(\frac{1}{25.596 - 0} \right) \left[(0)(1) + (1.470)(0) + (0.370)(1) + \ldots + (0.385)(0) \right.
$$
$$
\left. + (1.941)(1) \right] = 0.456 \text{ customers}
$$

Similarly the standard deviation can be determined using Eq. (7.12), resulting in

$$s_Q = \left\{ \left(\frac{1}{25.596 - 0} \right) \left[(0)\,(1)^2 + (1.470)\,(0)^2 + (0.370)\,(1)^2 + \ldots + (0.385)\,(0)^2 \right.\right.$$
$$\left.\left. + (1.941)\,(1)^2 \right] - (0.456)^2 \right\}^{1/2} = 0.528 \text{ customers}$$

We can group the merged data by determining the fraction of time that the queue holds one customer, two customers, etc. These calculations are summarized as follows:

Queue length (customers)	Number of events	Total time (min)	Relative frequency	Cumulative distribution
0	11	14.183	14.183/25.597 = 0.554	0.554
1	12	11.156	11.156/25.597 = 0.436	0.990
2	1	0.258	0.258/25.597 = 0.010	1.000
	24	25.597		1.000

A realistic simulation should, of course, be based upon a larger number of simulated events. The present supermarket checkout problem was therefore repeated for 500 simulated customers, resulting in a mean waiting time of 3.39 minutes, a mean queue length of 1.54 customers, and a value of 0.191 for the fraction of idle time.

7.3 THE SINGLE-CHANNEL SINGLE-STATION QUEUE— ANOTHER APPROACH

Although the previous next-customer model of the single-channel single-station queue is easy to understand and easy to implement on a computer, there are certain drawbacks associated with its use. First there is the need to merge the arrivals with the departures from the queue, in order to obtain the queue length as a function of time. Second and more troublesome, is the requirement that several large lists (containing the state variables A_i, B_i, and D_i) must be stored within the computer's memory. This storage requirement can be costly, and it becomes increasingly severe as the number of simulated events (that is, the number of customers) increases, or if the model is extended to a multiple-channel or a multiple-station situation. Finally the model does not permit any variation in the first-in, first-out priority rule that is applied to the queue.

Let us now consider a different approach to the single-channel single-station queue, based upon the occurrence of consecutive random events (that is, arrivals and departures) rather than a flowthrough of consecutive customers. Hence we will refer to this as a next-event model.

This approach is more complicated than the next-customer model, but it is more flexible and computationally more efficient. The greater flexibility involves the arrivals to and departures from the queue, which need not be on a first-in, first-out basis. Furthermore all of the desired statistical information can be obtained without storing large arrays in the computer's memory.

The basic idea is to generate the time of the next arrival and the time of the next departure, and then compare them to determine which event occurs next. In order to develop a quantitative description of the procedure, however, we must make use of the following symbols, in addition to those defined in the last two sections.

$$T0 = \text{time that the last event occurred}$$

$DT =$ time interval between the occurrence of the last event and the occurrence of the present event

$S =$ status of the server prior to the present event
$S = 0$ if the server is idle
$S = 1$ if the server is busy

$Q =$ queue length prior to the present event (note that this variable is defined differently from that in Sec. 7.2)

$QIN =$ time when a customer enters the queue

$IT =$ cumulative idle time (note that this variable is defined differently from that in Sec. 7.1)

$POINTER =$ an integer that indicates the next customer to leave the queue

The following three subscripts are also required:

i refers to arrivals to the system ($i = 1, 2, \ldots, N$)
j refers to departures from the system ($j = 1, 2, \ldots, N$)
k refers to the order of the customers waiting in the queue ($k = 1, 2, \ldots, Q$)

Let us assume that values for A_i (the time of the next arrival) and D_j (the time of the next departure) have been generated. The computation then proceeds as follows:

1. Test for idle server:
   If the server is idle (that is, if $S = 0$) then the next event must be an arrival. (A current value for D_j will not even be available in this special case). Hence proceed to step 3.

2. Compare time of next arrival with time of next departure:
   If the server is busy (that is, if $S = 1$) then A_i and D_j must be compared to determine which event occurs next.
   a. If $A_i \leq D_j$, then the next event is an arrival. Hence proceed to step 3.
   b. If $A_i > D_j$, then the next event will be a departure. Therefore proceed to step 4.
3. Arrival is next:
   a. The time increment between the last event and the current event is determined as $DT = A_i - T0$, and the time of the last event is updated $(T0 = A_i)$. The value of DT is then grouped in accordance with the corresponding queue length (Q).
   b. A test is made to determine the status of the server.
      i. If the server is idle $(S = 0)$, then the current arrival enters directly into the server. Therefore update the status of the server $(S = 1)$, update the preceding idle time $(IT = IT + DT)$, prepare for the departure of the current customer by updating the departure counter $(j = j + 1)$ and generating a new service time (ST_j), and, finally, calculate the time of the next departure as $D_j = B_j + ST_j$, where $B_j = A_i$.
      ii. If the server is busy $(S = 1)$, then the current arrival must enter the queue. Therefore update the queue length $(Q = Q + 1)$, record the time that the new arrival enters the queue, as $(QIN_{k=Q} = A_i)$, and determine who will leave the queue by assigning a value to POINTER (see below).
   c. Prepare for the next arrival by updating the arrival counter $(i = i + 1)$, generating a new interarrival time (AT_i), and calculating the time of the next arrival as $A_i = A_{i-1} + AT_i$.
   d. A test is then made to determine if the maximum number of arrivals has already been simulated (that is, if the new value of i exceeds N). If so, the time of the next arrival is set to some very large number (say, 10^{38}). This forces the next event to be a departure.
   e. A test is made to determine if a departure occurred at the same time as the last arrival (that is, if $T0 = D_j$). If so we proceed directly to step 4. Otherwise return to step 1, in preparation for the next comparison.
4. Departure is next:
   a. The time increment between the last event and the current event is now determined as $DT = D_j - T0$, and the time of the last event is updated $(T0 = D_j)$. The value of DT is then grouped in accordance with the corresponding queue length (Q).
   b. A test is made to determine if this is the final departure $(j = N)$. If so, the desired statistical information is generated and

written out, and the computation ceases. Otherwise, the computation continues as subsequently outlined.

c. A test is made to determine the status of the queue.

   i. If the queue is empty ($Q = 0$), then the server must become idle. Therefore set $S = 0$ and proceed to step 3 (because the next event must be an arrival).

   ii. If the queue is not empty ($Q > 0$), then the next customer enters the server as the current customer departs. The departure counter is therefore incremented ($j = j + 1$), the time that this next customer enters the server is recorded as $B_j = D_{j-1}$, and the queue length is decreased accordingly ($Q = Q - 1$). The waiting time for the next customer is then determined as $WT_j = B_j - QIN_{k = \text{POINTER}}$, and the value of WT_j is then added to the grouped waiting-time data, as subsequently described. Finally the waiting time in the queue is compressed (see later), a new service time is generated (ST_j), and the next departure time is calculated as $D_j = B_j + ST_j$. The computation then returns to step 2, for the next comparison.

The handling of the queue requires some further explanation. Every new customer that enters the queue will start out in the last position (that is, $k = Q$) and gradually work forward ($k = Q - 1$, $Q - 2, \ldots, 1$) as others leave. The next customer to leave will be indicated by the POINTER, which is reset whenever a new customer enters.

If the queue priority is first-in, first-out (which is usually the case when people are waiting in line), then the POINTER is always set equal to 1 when a new customer enters. This means that the customer at the front of the line will be the next to leave. As soon as a customer does leave, everyone in the queue moves up one position (that is, the queue is compressed). This is accomplished by changing the index associated with each of the entry times. Thus

$$QIN\ (1) = QIN\ (2)$$
$$QIN\ (2) = QIN\ (3)$$
$$\cdot$$
$$\cdot$$
$$\cdot$$
$$QIN\ (Q) = QIN\ (Q + 1)$$

where Q now represents the number of people waiting after the most recent departure from the queue. When a customer leaves the queue, that customer's waiting time can then be determined as $WT = B - QIN\,(1)$.

In the more general case, the POINTER can be set to something other than 1. (If the queue priority is last-in, first-out, for example, then the POINTER will be set equal to Q.) The waiting time for the customer leaving is always determined as $WT = B - QIN\,(POINTER)$.

The desired statistical information will consist of queue length and waiting time distributions, and a determination of the fraction of time the server is idle. The queue length distribution is obtained by grouping the DT versus Q data. In other words, an interval is defined for each expected value of Q (that is, $Q = 0, 1, 2, \ldots,$ etc.) and the total amount of simulated time corresponding to each value of Q is obtained. From this grouped data we can determine the relative frequencies and the cumulative distribution of queue lengths, as well as the mean queue length and the standard deviation (see Sec. 4.11).

Similarly the waiting time distribution is obtained by defining several adjacent waiting time intervals (that is, WT between 0 and 0.5 minutes, 0.5 and 1.0 minutes, etc.) and the total number of customers falling into each interval is obtained. These grouped data allow us to determine the relative frequencies and the cumulative distribution of waiting times, as well as the mean waiting time and the standard deviation. (Actually, *two* mean waiting times are usually defined: one based upon all N customers, and the other based only upon those customers who must wait before entering the server.)

A detailed flowchart of the entire procedure is shown in Fig. 7.4.

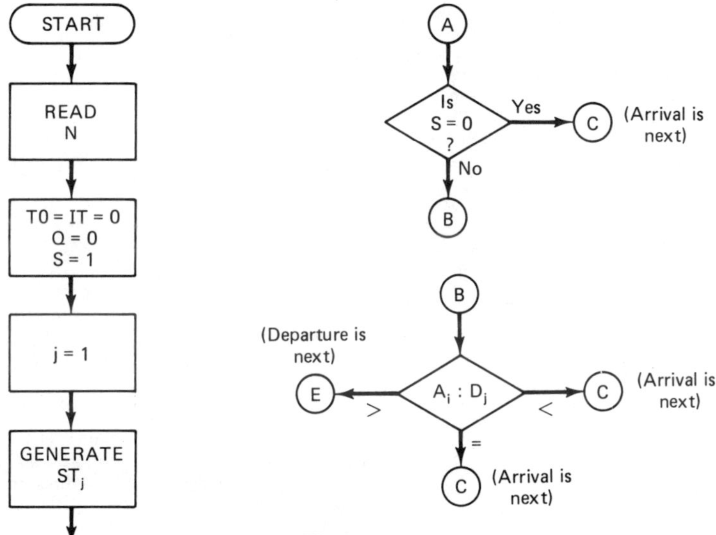

Figure 7.4

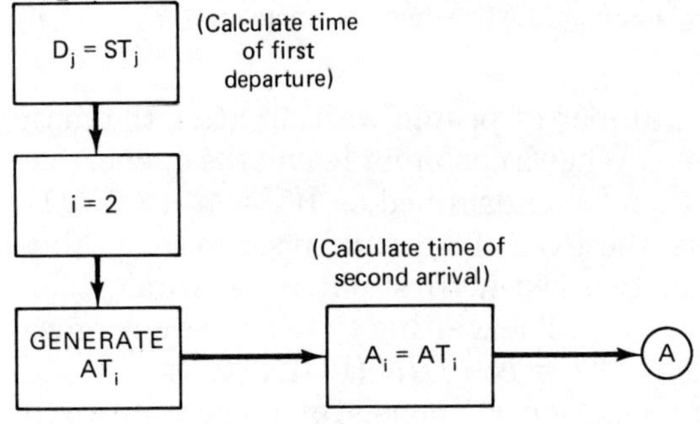

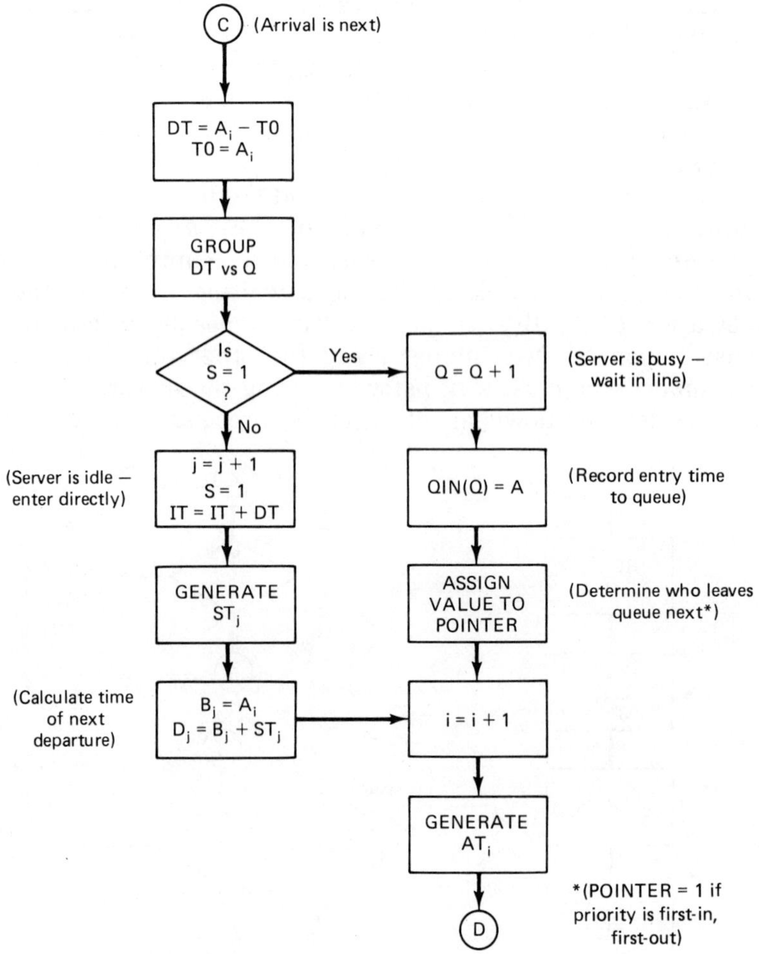

Figure 7.4 *(continued)*

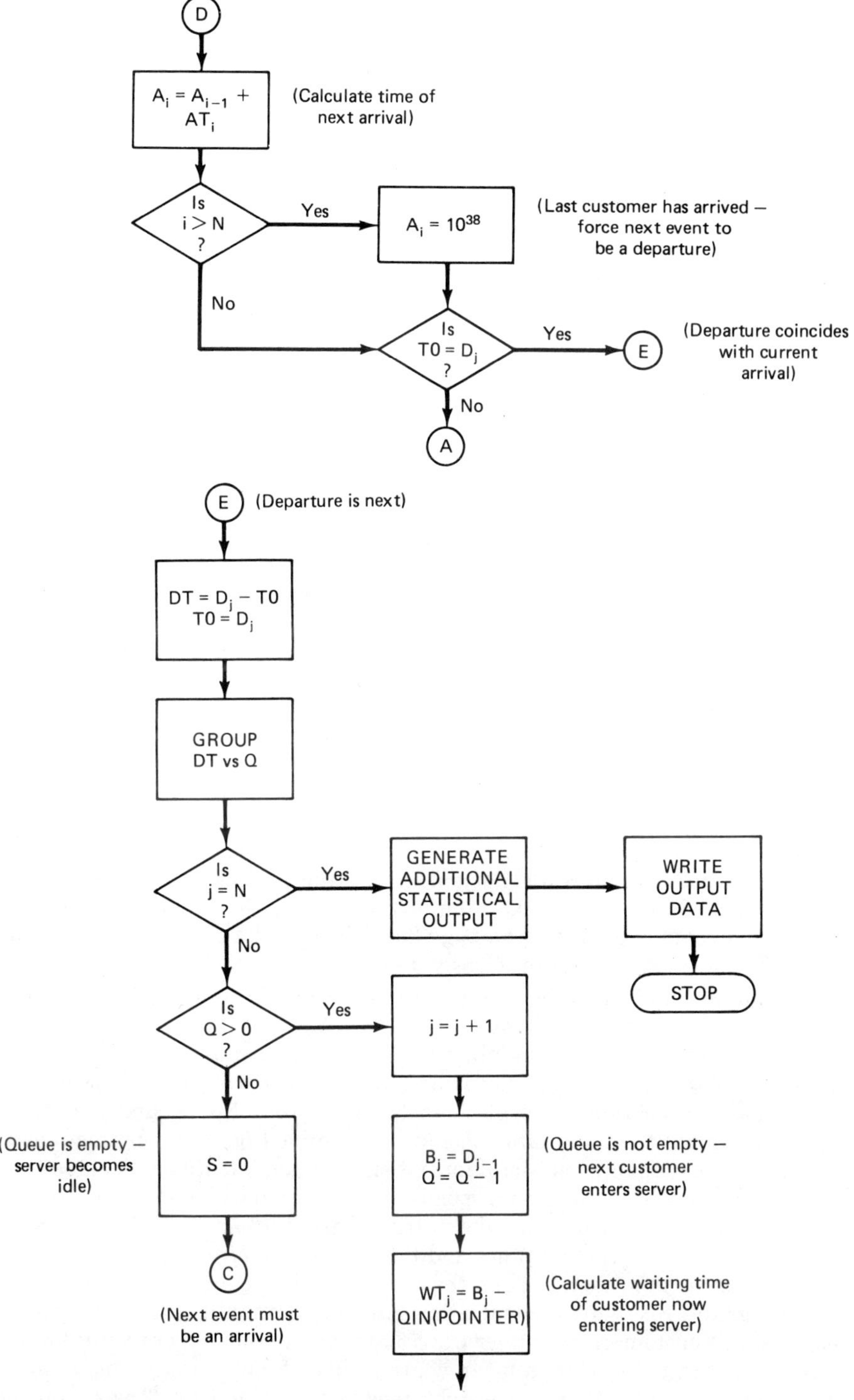

Figure 7.4 *(continued)*

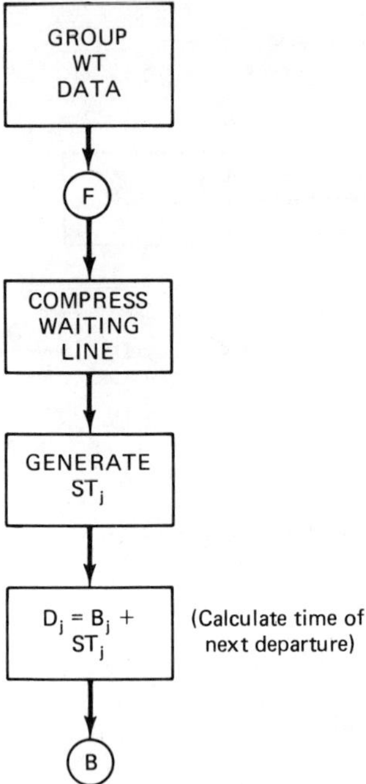

Figure 7.4 *(continued)*

Example 7.3

In order to illustrate the manner in which the calculations are carried out, let us again consider the problem described in Ex. 7.1 (that is, exponential arrivals with a mean of 2.3 minutes, and normal service times with a mean of 1.8 minutes and a standard deviation of 0.5 minutes). A detailed history of the first few simulated events is presented on page 203. The calculations were carried out using the next-event model previously described.

Column S indicates the status of the server and column Q indicates the queue length prior to each event.

Notice that the columns labeled AT_i and ST_j contain random variates representing the customer interarrival times and the service times respectively. The remaining columns (except those labeled Event No., i, and j) contain calculated quantities. The table should be examined on a row-by-row basis, however, since each row contains information associated with a particular event, and the ordering of the rows corresponds to the order in which the events occur.

It must be emphasized that the statistical distributions (and the associated mean values) of the waiting times and the queue lengths are generated *as the events occur* in this model. Thus the thirty simulated events shown in the preceding table result in a mean waiting time of 2.24 minutes and a mean queue length of 1.23 customers. Moreover when 500 simulated customers were allowed to pass through the system, a mean waiting time of 3.50 minutes and a mean queue length of 1.52 customers were obtained. Also a value of 0.223 was obtained for the fraction of idle time.

Event			Arrivals			Departures					
No.	S	Q	i	AT_i^*	A_i	j	B_j	ST_j^*	D_j	IT_j	WT_j
1	0	0	1		0						
2	1	0				1	0	1.71	1.71	0	0
3	0	0	2	2.06	2.06						
4	1	0	3	1.32	3.38						
5	1	1				2	2.06	2.41	4.47	0.35	0
6	1	0	4	2.37	6.11						
7	1	1	5	0.78	6.89						
8	1	2				3	4.47	2.70	7.17	0.35	1.09
9	1	1	6	2.46	9.35						
10	1	2				4	7.17	2.25	9.42	0.35	1.06
11	1	1	7	1.30	10.65						
12	1	2				5	9.42	1.53	10.95	0.35	2.53
13	1	1	8	0.62	11.27						
14	1	2	9	0.81	12.08						
15	1	3				6	10.95	1.55	12.50	0.35	1.60
16	1	2				7	12.50	1.98	14.48	0.35	1.85
17	1	1	10	2.74	14.82						
18	1	2	11	0.30	15.12						
19	1	3	12	0.14	15.26						
20	1	4				8	14.48	1.39	15.88	0.35	3.21
21	1	3	13	0.73	15.99						
22	1	4				9	15.88	1.21	17.08	0.35	3.80
23	1	3				10	17.08	1.73	18.81	0.35	2.24
24	1	2				11	18.81	0.88	19.69	0.35	3.69
25	1	1	14	5.68	21.67						
26	1	2				12	19.69	2.44	12.13	0.35	4.43
27	1	1				13	22.13	1.47	23.60	0.35	6.14
28	1	0				14	23.60	1.61	25.21	0.35	1.93
29	0	0	15	5.74	27.41						
30	1	0				15	27.41	1.82	29.23	2.55	0

*Random variates.

7.4 SOME THEORETICAL RESULTS FOR THE SINGLE-CHANNEL SINGLE-STATION QUEUE

The single-channel single-station queue has been studied theoretically for certain conditions that are of practical interest. Some of the results that are obtained are subsequently summarized. These predictions can be of great value when debugging a simulation program whose underlying assumptions correspond to those used in the theoretical analysis. Moreover, many simulation models that employ different assumptions can easily be modified (temporarily) for debugging purposes. Thus the theoretical results presented below can be used to check a wide variety of queuing programs.

Consider a single-channel, single-station queue where the interarrival times are exponentially distributed with mean λ (the minimum, x_0, is also assumed to equal zero), and the service times are exponentially distributed with mean μ. Let

$$\beta = \frac{\text{expected service time}}{\text{expected interarrival time}} = \frac{\mu}{\lambda} \qquad (7.13)$$

The following results can then be obtained provided $\beta < 1$ (that is, $\lambda > \mu$) (Cooper 1972; Hillier and Lieberman 1980):

(a) Expected waiting time $= \mu\, \beta/(1 - \beta)$ $\qquad (7.14)$

(b) Expected time spent in the system
(waiting time + service time) $= \mu/(1 - \beta)$ $\qquad (7.15)$

(c) Expected queue length $= \beta^2/(1 - \beta)$ $\qquad (7.16)$

(d) Expected nonempty queue length
$= 1/(1 - \beta)$ $\qquad (7.17)$

(e) Expected fraction of time the server is idle
$= (1 - \beta)$ $\qquad (7.18)$

If $\beta > 1$, then the system will be unstable (customers arrive faster than they depart). The queue will continue to grow with time under these conditions. Moreover, the queue length will oscillate with time if $\beta = 1$. These undesirable situations should be avoided if at all possible.

Now consider the case where the interarrival times are exponentially distributed with mean λ as before, but the service times are governed by a two-parameter distribution having mean μ and standard deviation σ (for example, a normal distribution). We again define β by Eq. (7.13). The following results can be obtained provided $\beta < 1$:

(a) Expected waiting time $= \dfrac{\sigma^2/\lambda + \beta^2\lambda}{2(1-\beta)}$ (7.19)

(b) Expected time spent in the system

(waiting time + service time) $= \dfrac{\sigma^2/\lambda + \beta^2\lambda}{2\,(1-\beta)} + \mu$ (7.20)

(c) Expected queue length $= \dfrac{\sigma^2/\lambda^2 + \beta^2}{2\,(1-\beta)}$ (7.21)

(d) Expected fraction of time the server is idle
$= (1-\beta)$ (7.22)

Example 7.4

Consider once again the supermarket check-out counter described in Exs. 7.1 to 7.3 (that is, exponential arrivals with a mean of 2.3 minutes, and normal service times with a mean of 1.8 minutes and a standard deviation of 0.5 min). Calculate the following theoretically expected values and compare with the simulated results obtained in the previous examples:

a. Expected waiting time
b. Expected queue length
c. Expected fraction of time the server is idle.

We first obtain a value for β, using Eq. (7.13). Thus

$$\beta = 1.8/2.3 = 0.783$$

The expected waiting time can now be obtained from Eq. (7.19) as

$$\overline{WT} = \frac{(0.5)^2/(2.3) + (0.783)^2\,(2.3)}{(2)\,(1-0.783)} = 3.50 \text{ min}$$

Similarly the expected queue length can be determined from Eq. (7.21) as

$$Q = \frac{(0.5)^2/(2.3)^2 + (0.783)^2}{(2)\,(1-0.783)} = 1.52 \text{ customers}$$

Finally the use of Eq. (7.22) results in the following value for the fraction of time the server is idle.

$$F = (1 - 0.783) = 0.217$$

The following table summarizes the theoretical results obtained in this example and the simulated results described in the last two examples.

	Theoretical prediction	Next-customer model	Next-event model
Number of simulated customers	—	500	500
Mean waiting time (minutes)	3.50	3.39	3.50
Mean queue length (number of customers)	1.52	1.54	1.52
Fraction of time the server is idle	0.217	0.191	0.213

The agreement appears reasonable, thus enhancing our confidence in the simulated results.

7.5 THE SINGLE-CHANNEL MULTISTATION QUEUE

Let us again consider the next-customer model discussed in Sec. 7.1. Now, however, we will extend the analysis to the multistation case, where each customer enters several different servers, one after the other. A waiting line will be allowed to form in front of each server. Thus we are considering several stations in series, where each station consists of a queue followed by a server, as indicated in Fig. 7.5.

We will again simulate a succession of customers passing through the system. Our objective will be to obtain distributions for pertinent system performance criteria, such as the overall time spent in the system, the waiting time for each station (in front of each server), the queue length for each station, and the fraction of time each server is idle.

The analysis is a logical extension of the single-station case considered in Sec. 7.1. Before developing the necessary equations however we must first redefine the basic system variables in terms of a double-subscript notation. Therefore let

$$A_{i,j} = \text{arrival time of the } i\text{th customer to the } j\text{th station}$$
$$B_{i,j} = \text{the time when the } i\text{th customer enters the } j\text{th server}$$

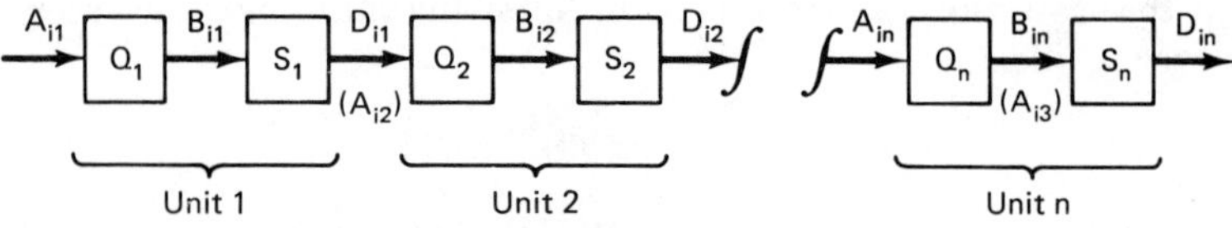

Figure 7.5

$$D_{i,j} = \text{departure time of the } i\text{th customer from the } j\text{th server}$$

$AT_i =$ time interval between the arrival of the $(i - 1)$st and the ith customers to the system (a random variate)

$ST_{i,j} =$ time required to serve the ith customer in the jth server (a random variate)

$WT_{i,j} =$ waiting time for the ith customer at the jth station

$IT_{i,j} =$ idle time associated with the ith customer at the jth server (that is, the length of time the jth server is idle prior to serving the ith customer)

$N =$ total number of customers $(i = 1, 2, \ldots, N)$

$n =$ total number of stations $(j = 1, 2, \ldots, n)$

As before the single-letter variables $(A_{i,j}, B_{i,j}, \text{and } D_{i,j})$ are state variables that represent *clock times*, whereas the double-letter variables $(AT_i, ST_{i,j}, WT_{i,j}, \text{and } IT_{i,j})$ are either randomly generated or calculated quantities that represent *time intervals*. These time intervals must all be nonnegative. Thus

$$AT_i, \ ST_{i,j}, \ WT_{i,j}, \ IT_{i,j} \geq 0$$

Moreover for any given i and j, either WT_{ij} or IT_{ij} (or both) must equal zero.

We can now analyze the detailed behavior of the system. The consecutive arrivals to the first station (that is, the *system arrivals*) can be expressed as

$$A_{i,1} = A_{i-1,1} + AT_i \tag{7.23}$$

and the *consecutive arrivals to the remaining stations* $(j = 2, 3, \ldots, n)$ can be written

$$A_{i,j} = D_{i,j-1} \tag{7.24}$$

Similarly the *consecutive departures from each station* $(j = 1, 2, \ldots, n)$ can be expressed as

$$D_{i,j} = D_{i-1,j} + IT_{i,j} + ST_{i,j} \tag{7.25}$$

For each customer $(i = 1, 2, \ldots, N)$, we can write a detailed *customer history* as

$$D_{i,j} = A_{i,1} + \sum_{k=1}^{j} (WT_{i,k} + ST_{i,k}) \tag{7.26}$$

where

$$j = 1, 2, \ldots, n$$

The restriction that $WT_{i,j}$ and $IT_{i,j}$ cannot both be greater than zero, for each value of i and j, can be expressed mathematically as

$$(IT_{i,j})\,(WT_{i,j}) = 0 \qquad (7.27)$$

Finally the *consecutive arrivals to the jth server* $(j = 1, 2, \ldots, n)$ can be expressed as

$$B_{i,j} = A_{i,j} + WT_{i,j} \qquad (7.28)$$

This last equation can also be written in terms of a system arrival (similar to the customer history) as

$$B_{i,j} = A_{i,1} + \sum_{k=1}^{(j-1)} (WT_{i,k} + ST_{i,k}) + WT_{i,j} \qquad (7.29)$$

where

$$j = 1, 2, \ldots, n$$

The initial conditions can be handled most conveniently by allowing the first customer $(i = 1)$ to arrive at time level zero. Thus we have

Station 1 $(j = 1)$		Station 2 $(j = 2)$		...	Station n $(j = n)$	
$A_{1,1}$	$= 0$	$A_{1,2}$	$= D_{1,1}$	...	$A_{1,n}$	$= D_{1,n-1}$
$B_{1,1}$	$= 0$	$B_{1,2}$	$= A_{1,2}$	...	$B_{1,n}$	$= A_{1,n}$
$D_{1,1}$	$= ST_{1,1}$	$D_{1,2}$	$= B_{1,2} + ST_{1,2}$	...	$D_{1,n}$	$= B_{1,n} + ST_{1,n}$
$WT_{1,1}$	$= 0$	$WT_{1,2}$	$= 0$	...	$WT_{1,n}$	$= 0$
$IT_{1,1}$	$= 0$	$IT_{1,2}$	$= B_{1,2}$	...	$IT_{1,n}$	$= B_{1,n}$

For each successive customer $(i = 2, 3, \ldots, N)$, we generate random values for the interarrival time and the n different service times, and we then solve for the remaining variables in essentially the same manner as outlined in Sec. 7.1. It is necessary, however, that the calculations for the different stations be carried out in their *natural order*, that is, first station 1, then station 2, etc.

The required distributions can be obtained quite easily once all N customers have been allowed to pass through the system. In particular, the distributions for the total time spent in the system and the waiting time for each server can be obtained by the grouping procedure discussed in Sec. 4.12. The computation of the fraction of time that each server is idle is also straightforward, using Eq. (7.10). In order to obtain the distributions of the queue lengths, however, the arrival and departure times for each queue must be merged, as described in Sec. 7.2.

Figure 7.6 presents a flowchart of the overall procedure.

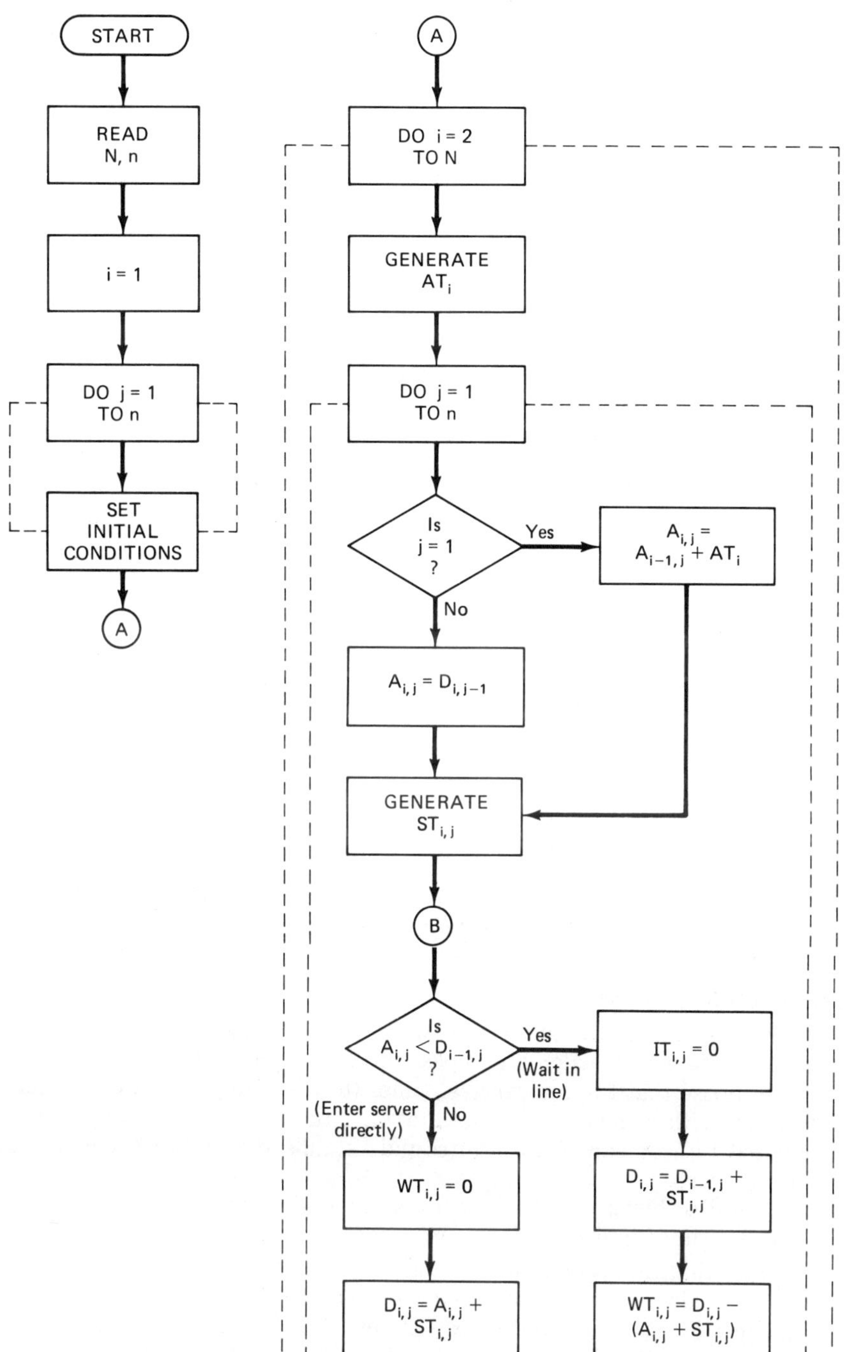

START
READ
N, n
i = 1
DO j = 1
TO n
SET
INITIAL
CONDITIONS
A
A
DO i = 2
TO N
GENERATE
AT_i
DO j = 1
TO n
Is
j = 1
?
Yes
No
$A_{i,j} = A_{i-1,j} + AT_i$
$A_{i,j} = D_{i,j-1}$
GENERATE
ST_{i,j}
B
Is
$A_{i,j} < D_{i-1,j}$
?
Yes
(Wait in
line)
(Enter server
directly)
No
$IT_{i,j} = 0$
$WT_{i,j} = 0$
$D_{i,j} = D_{i-1,j} + ST_{i,j}$
$D_{i,j} = A_{i,j} + ST_{i,j}$
$WT_{i,j} = D_{i,j} - (A_{i,j} + ST_{i,j})$

Figure 7.6

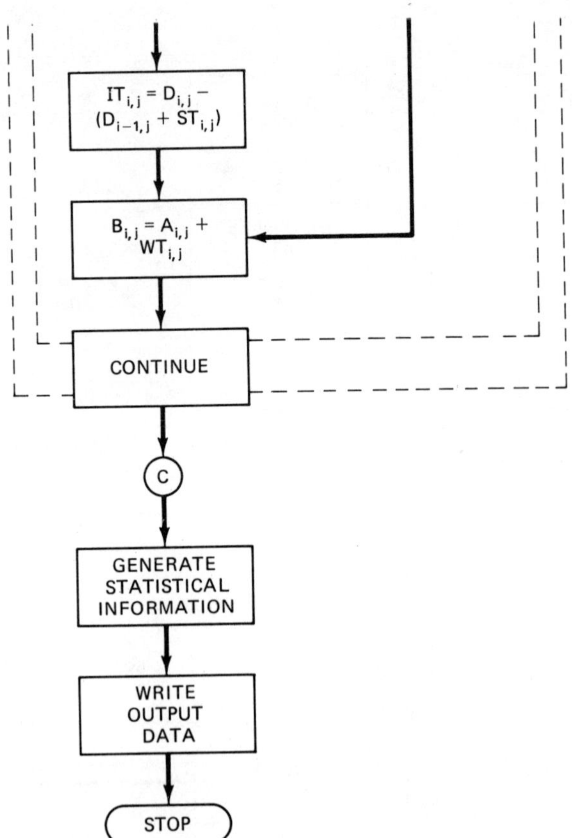

Figure 7.6 *(continued)*

Example 7.5

In order to illustrate the manner in which the calculations are carried out, consider the first few customers to pass through a single-channel, two-station queue. Suppose that the interarrival times (to the system) are exponentially distributed with a mean of 2.3 minutes. Let the service times be normally distributed with a mean of 1.8 minutes and a standard deviation of 0.5 minutes for each server.

A detailed history of the first few simulated customers, based upon the procedure outlined in Fig. 7.6, is given on following page.

Notice that the second, fourth, and eleventh columns (labeled AT_i, $ST_{i,1}$, and $ST_{i,2}$) contain random variates that represent the interarrival times, and the service times for stations 1 and 2, respectively. All of the remaining numerical columns (except the first) contain calculated quantities, obtained by the procedure shown in Fig. 7.6.

The table should be examined on a row-by-row basis, since each row contains information that is associated with a different customer. Notice that

	Station 1 ($j = 1$)								Station 2 ($j = 2$)						
i	AT_i^*	$A_{i,1}$	$ST_{i,1}^*$	Is $A_{i,1} < D_{i-1,1}$?	$IT_{i,1}$	$D_{i,1}$	$WT_{i,1}$	$B_{i,1}$	$A_{i,2}$	$ST_{i,2}^*$	Is $A_{i,2} < D_{i-1,2}$?	$IT_{i,2}$	$D_{i,2}$	$WT_{i,2}$	$B_{i,2}$
1		0	1.711		0	1.711	0	1.711	1.711	2.437		0	4.148	0	1.711
2	2.466	2.466	2.462	No	0.755	4.927	0	2.466	4.927	2.770	No	0.780	7.697	0	4.927
3	3.053	5.519	1.537	No	0.591	7.056	0	5.519	7.056	1.891	Yes	0	9.588	0.641	7.697
4	6.726	12.245	1.568	No	5.189	13.813	0	12.245	13.813	1.992	No	4.225	15.805	0	13.813
5	2.691	14.936	2.370	No	1.123	17.306	0	14.936	17.306	0.910	No	1.502	18.217	0	17.306
6	2.402	17.337	1.372	No	0.031	18.710	0	17.337	18.710	1.084	No	0.493	19.794	0	18.710
7	0.767	18.140	2.093	Yes	0	20.803	0.606	18.710	20.803	1.222	No	1.009	22.025	0	20.803
8	0.112	18.216	1.234	Yes	0	22.036	2.587	20.803	22.036	1.936	No	0.011	23.972	0	22.036
9	0.430	18.646	1.526	Yes	0	23.562	3.390	22.036	23.562	0.942	Yes	0	24.914	0.410	23.972
10	2.170	20.816	1.587	Yes	0	25.150	2.747	23.562	25.150	2.152	No	0.236	27.302	0	25.150

*Random variates.

four customers ($k = 2$, 4, 5, and 6) did not experience waiting at either station, whereas one customer ($k = 9$) was required to wait at both stations. The remaining customers experienced some waiting at one of the stations.

When simulating an actual system the individual values are normally not printed in this manner (except when debugging), since we are interested primarily in mean queue lengths and waiting times, and their associated distributions. Such values are, of course, obtained from the individual quantities shown above. For example, when 500 simulated customers were allowed to pass through the present system, it was found that the average value for total time in the system is 7.33 minutes per customer. In addition, the following averages were obtained for each of the stations.

	Station 1	Station 2
Waiting time	2.78 min	0.97 min
Queue length	1.19 customers	0.41 customers
Fraction of time the server is idle	0.243	0.221

Recall that the queue lengths were obtained by merging the values of $A_{i,j}$ and $B_{i,j}$ for each station and then obtaining a time-weighted average, as described in Sec. 7.2.

7.6 THE MULTICHANNEL SINGLE-STATION QUEUE

Another queuing model that is of considerable practical importance is the multichannel single-station queue. Here we have several servers in parallel with a common waiting line. Each incoming customer enters the waiting line, works forward, and then enters the first available server after reaching the front of the line. If a server is idle when the customer enters the system, then the customer will proceed directly to the server without waiting. If two or more servers are idle, then the customer will choose a server randomly, or in accordance with some other selection rule. Figure 7.7 illustrates this particular system.

In order to quantify the model, we must introduce a set of symbols based upon the following three subscripts:

i refers to a particular customer within a given channel

j refers to a particular channel (server)

k refers to a particular customer entering the *system*

The actual variables are defined as follows:

A_k = arrival time of the kth customer to the system

B_k = time when the kth customer leaves the waiting line

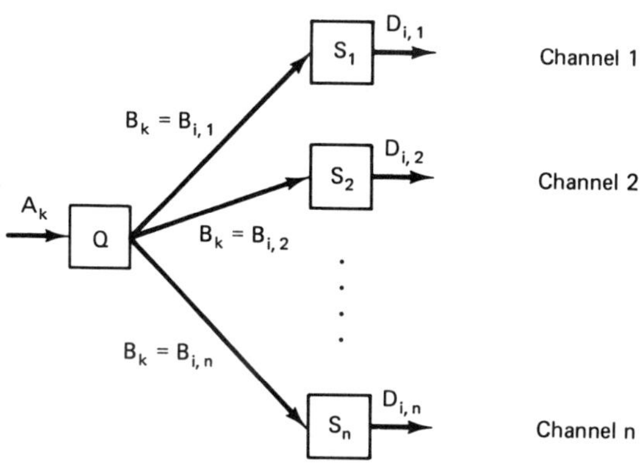

Figure 7.7

$B_{i,j}$ = time when the ith customer enters the jth channel (server)

$D_{i,j}$ = departure time of the ith customer from the jth channel

LD_j = time of the last (most recent) departure from the jth channel

AT_k = time interval between the arrival of the $(k-1)$st and the kth customers to the system (a random variate)

$ST_{i,j}$ = time required to serve the ith customer in the jth channel (a random variate)

WT_k = waiting time for the kth customer

$IT_{i,j}$ = idle time associated with the ith customer in the jth channel (that is, the length of time the jth server is idle prior to serving the ith customer)

N = total number of customers passing through the system $(k = 1, 2, \ldots, N)$

n = total number of channels (servers) $(j = 1, 2, \ldots, n)$

Once again the single-letter variables (A_k, B_k, $B_{i,j}$, and $D_{i,j}$) are state variables that represent *clock times*, and the double-letter variables (AT_k, $ST_{i,j}$, WT_k, and $IT_{i,j}$) are calculated or randomly generated quantities that represent *time intervals*. (Note that LD_j is an exception to this rule.) These time intervals must all be nonnegative, that is,

$$AT_k, ST_{i,j}, WT_k, IT_{i,j} \geq 0$$

Furthermore for a given i and j, either $IT_{i,j}$ or the corresponding WT_k (or both) must equal zero.

   The following system of equations can now be written to describe the multichannel single-station queue. We begin with the consecutive *arrivals to the system.*

$$A_k = A_{k-1} + AT_k \tag{7.30}$$

The consecutive *departures from the queue* can be expressed as

$$B_k = A_k + WT_k \tag{7.31}$$

Each customer will *enter a channel* (that is, a server) as soon as he or she leaves the queue. Thus

$$B_{i,j} = B_k \tag{7.32}$$

At this point it is not clear, however, which channel (that is, which value of j) is selected by the kth customer. This point will be further discussed later.

The consecutive *departures from each channel* can be expressed as

$$D_{i,j} = D_{i-1,j} + IT_{i,j} + ST_{i,j} \tag{7.33}$$

Also, the time of the *last departure* from each channel is simply the largest value of $D_{i,j}$ for that channel, that is,

$$LD_j = \operatorname*{Max}_i \left\{ D_{i,j} \right\} \tag{7.34}$$

We can also write a *customer history* for each channel as

$$D_{i,j} = B_{i,j} + ST_{i,j} \tag{7.35}$$

Finally the relationship between $IT_{i,j}$ and the associated WT_k can be expressed as

$$(IT_{i,j})(WT_k) = 0 \tag{7.36}$$

Once again we see the need for a correspondence between the channel index (j) and the customer index (k).

In order to initiate the computation, we must specify an appropriate set of initial conditions for each of the channels. A convenient way to do this is to arbitrarily assign the first customer to the first channel, the second customer to the second channel, etc., until each channel has been assigned one customer. Thus we have

Channel 1 $(j = 1)$	Channel 2 $(j = 2)$		Channel n $(j = n)$
$A_1 = 0$	$A_2 = AT_2$	....	$A_n = A_{n-1} + AT_n$
$B_{1,1} = B_1 = 0$	$B_{1,2} = B_2 = A_2$	....	$B_{1,n} = B_n = A_n$
$D_{1,1} = ST_{1,1}$	$D_{1,2} = B_{1,2} + ST_{1,2}$	....	$D_{1,n} = B_{1,n} + ST_{1,n}$
$LD_1 = D_{1,1}$	$LD_2 = D_{1,2}$	....	$LD_n = D_{1,n}$
$WT_1 = 0$	$WT_2 = 0$	....	$WT_n = 0$
$IT_{1,1} = 0$	$IT_{1,2} = B_{1,2}$	....	$IT_{1,n} = B_{1,n}$

For each successive customer ($k = n + 1,\ n + 2, \ldots,\ N$), we determine the arrival time, A_k, using Eq. (7.30), and then compare this arrival time with the earliest of the most recent departures. This latter quantity can be determined as

$$LD_{j^*} = \operatorname*{Min}_{j} \left\{ LD_j \right\} \qquad (7.37)$$

(Note that j^* indicates the first channel to become vacant.)

If $A_k < LD_{j^*}$ then all of the servers will be busy, requiring the customer to wait until channel j^* becomes vacant. Thus we can write

$$
\begin{aligned}
B_k \quad &= B_{i,j^*} = LD_{j^*} \\
WT_k &= B_k - A_k \qquad \text{(from Eq. 7.31)} \\
D_{i,j^*} &= B_{i,j^*} + ST_{i,j^*} \quad \text{(Eq. 7.35)} \\
IT_{i,j^*} &= 0 \qquad\qquad\quad \text{(Eq. 7.36)}
\end{aligned}
$$

We then update the value of LD_{j^*} that is,

$$LD_{j^*} = D_{i,j^*}$$

Now suppose that $A_k \geq LD_{j^*}$. At least one channel will then be vacant when the kth customer arrives. If *only* one channel is vacant (indicated by the condition that $A_k \geq LD_{j^*}$ but A_k less than all other LD_j), then the customer will immediately enter channel j^*. Hence

$$
\begin{aligned}
B_k \quad &= B_{i,j^*} = A_k \\
WT_k &= 0 \qquad\qquad\qquad\quad \text{(Eq. 7.36)} \\
D_{i,j^*} &= B_{i,j^*} + ST_{i,j^*} \quad \text{(Eq. 7.35)} \\
IT_{i,j^*} &= B_{i,j^*} - D_{i-1,j^*} \quad \text{(from Eq. 7.33)}
\end{aligned}
$$

and we again update LD_{j^*} as

$$LD_{j^*} = D_{i,j^*}$$

Finally consider the case where two or more channels are vacant when the kth customer arrives. Let us refer to these as $j_1{}^*, j_2{}^*, \ldots,$ etc. These values can be placed in a "vacancy list," and one of them selected at random (or by some other selection process). Now suppose that j^{**} refers to the vacant channel that is selected by the kth customer. We then write

$$
\begin{aligned}
B_k \quad &= B_{i,j^{**}} = A_k \\
WT_k &= 0 \qquad\qquad\qquad\quad \text{(Eq. 7.36)} \\
D_{i,j^{**}} &= B_{i,j^{**}} + ST_{i,j^{**}} \quad \text{(Eq. 7.35)}
\end{aligned}
$$

$$IT_{i,j^{**}} = B_{i,j^{**}} - D_{i-1,j^{**}} \text{ (from Eq. 7.33)}$$

$$LD_{j^{**}} = D_{i,j^{**}}$$

as before.

After all N customers have passed through the system, it is quite simple to obtain distributions for the waiting time and the total time in the system using the grouping procedure presented in Sec. 4.11. Furthermore a distribution of queue lengths can be obtained by merging the successive A_k's and B_k's, as described in Sec. 7.2.

Figure 7.8 presents a flowchart showing the procedure for simulating the N successive customers.

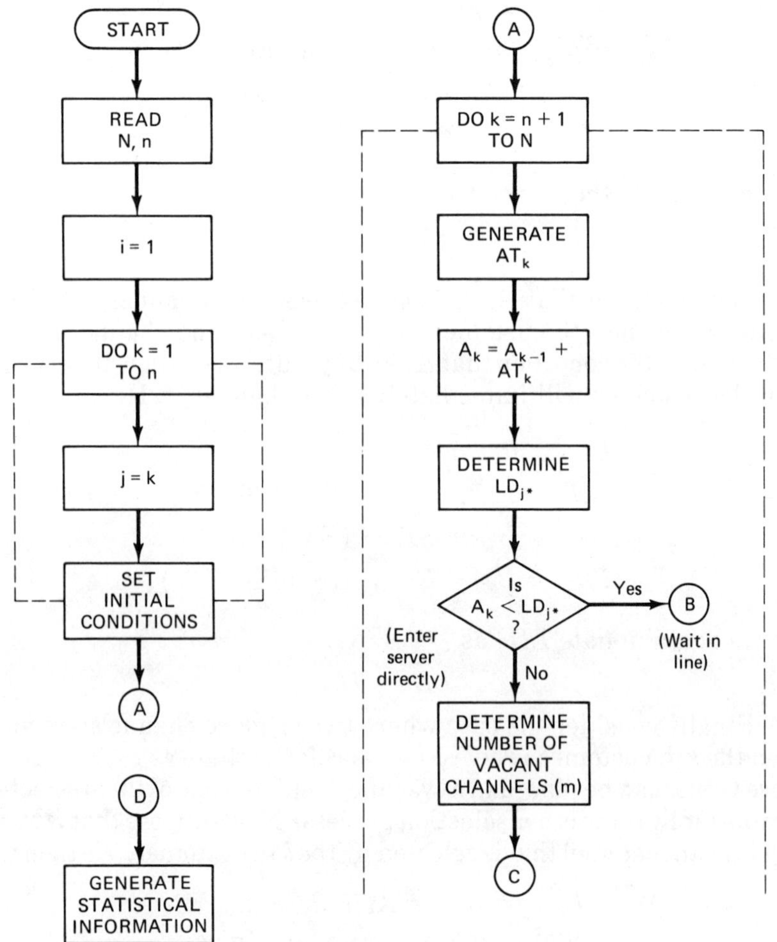

Figure 7.8

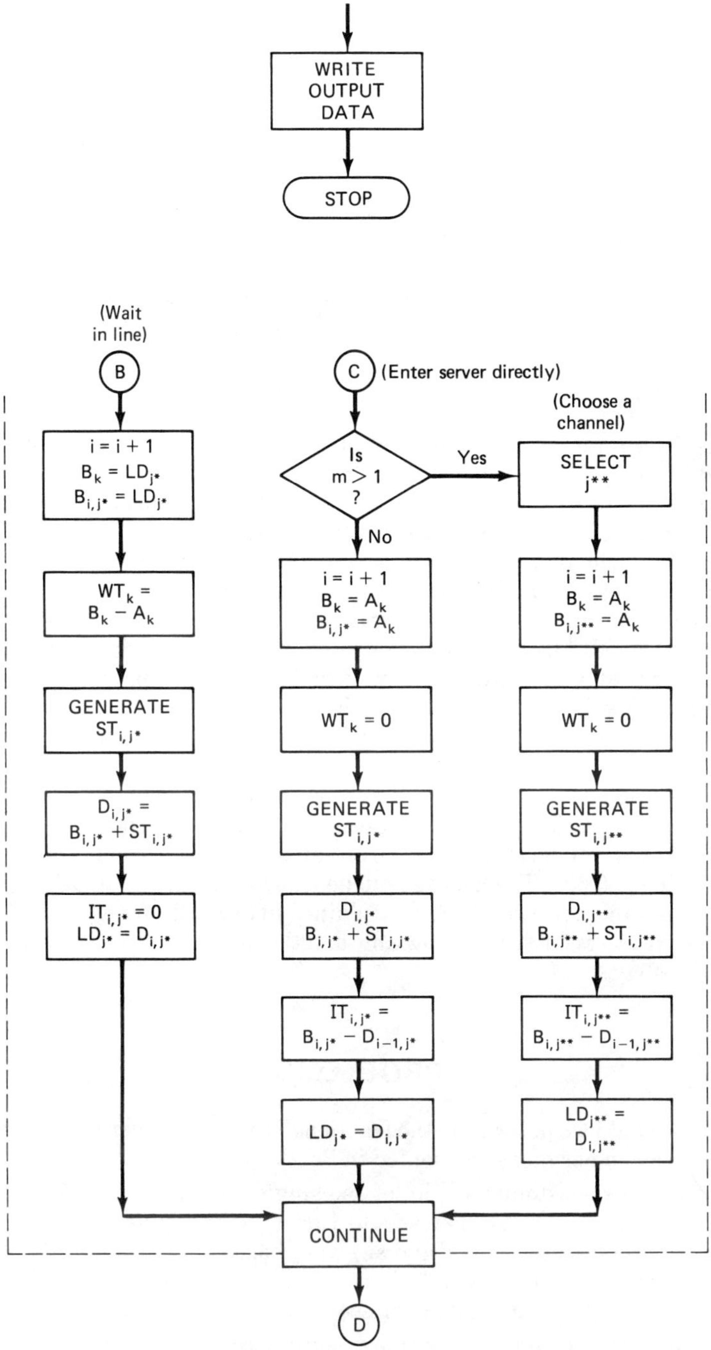

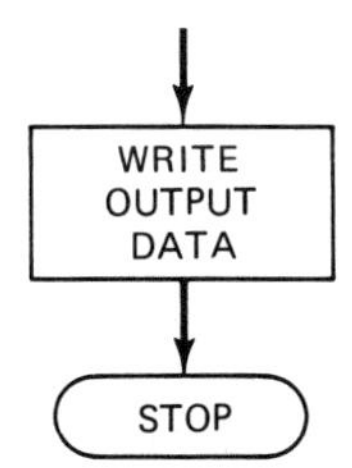

Figure 7.8 *(continued)*

Example 7.6

Let us illustrate the manner in which the calculations are carried out by considering a two-channel queue with a common waiting line. Suppose that the customer interarrival times are exponentially distributed with a mean of 3 minutes. Let the service times be normally distributed with a mean of 4 minutes and a standard deviation of 2 minutes for each server. Assume that a channel is selected at random if both channels are empty when a customer enters the system.

A detailed history of the first few simulated customers is given on the following page. The individual calculations were carried out using the procedure shown in Fig. 7.8.

Observe that the columns labeled AT_k, $ST_{i,1}$, and $ST_{i,2}$ contain random variates that represent the interarrival times, and the service times for the two channels. The remaining subscripted quantities represent calculated values, obtained by the computational procedure given in Fig. 7.8.

Again the table should be studied on a row-by-row basis, since each row corresponds to a different customer. Notice that customers 1, 3, 5, 6, 7, 9, and 11 enter the first channel, whereas the remaining customers enter the second channel. The choice of channels was arbitrary (and therefore randomly selected) for customers 7 and 11, since both channels were vacant when each of these customers arrived. All of the other customers chose whichever channel was free or whichever first became available. Note that only three of the customers ($k = 5$, 6, and 10) were required to wait for a channel to become available.

If we were to calculate some overall statistics for these twelve customers, we would find that the average waiting time per customer is $0.715/12 = 0.060$ minutes, and the average waiting time for those customers who experienced some waiting is $0.715/3 = 0.238$ minutes. Also the fraction of idle time for the first channel is $10.967/34.895 = 0.314$, and for the second channel it is $18.156/35.863 = 0.506$. The average queue length would have to be obtained by merging the values of A_k and B_k, as explained in Sec. 7.2. (We should, of course, obtain *distributions* for the waiting time and the queue length, in addition to the mean values.)

PROBLEMS

7.1. How would we determine the fraction of time the server is idle, using the next-customer model presented in Sec. 7.1?

7.2. In the next-customer model of the single-channel single-station queue presented in Sec. 7.1, how should Eqs. (7.6) and (7.8) be modified if the calculated values for $\overline{WT}$ and s_{WT} were based only upon those customers who were required to wait before entering the server? Are these statistics as meaningful as those determined by Eqs. (7.6) and (7.8)?

7.3. Write a self-contained computer subprogram to carry out the merge procedure described in Sec. 7.2. Allow the two arrays A and B to be input items, along with N, the size of each array. Let the arrays Q and T be output items.

		System arrivals						Channel 1 ($j=1$)					Channel 2 ($j=2$)				
k	AT_k	A_k	LD_j	j	Is $A_k < LD_j$	B_k	WT_k	i	$B_{i,1}$	$ST_{i,1}$	$D_{i,1}$	$IT_{i,1}$	i	$B_{i,2}$	$ST_{i,2}$	$D_{i,2}$	$IT_{i,2}$
1		0				0	0	1	0	3.367	3.367	0					
2	1.917	1.917				1.917	0						1	1.917	2.237	4.154	1.917
3	2.150	4.067	3.367	1	No	4.067	0	2	4.067	4.898	8.965	0.700					
4	3.691	7.758	4.154	2	No	7.758	0						2	7.758	3.402	11.160	3.604
5	0.704	8.462	8.965	1	Yes	8.965	0.503	3	8.965	1.043	10.008	0					
6	1.533	9.995	10.008	1	Yes	10.008	0.013	4	10.008	0.225	10.233	0					
7	5.093	15.088	10.233	1†	No	15.088	0	5	15.088	5.953	21.041	4.855					
8	3.837	18.925	11.160	2	No	18.925	0						3	18.925	6.504	25.429	7.765
9	4.544	23.469	21.041	1	No	23.469	0	6	23.469	3.604	27.073	2.428					
10	1.761	25.230	25.429	2	Yes	25.429	0.199						4	25.429	2.355	27.784	0
11	4.827	30.057	27.073	1†	No	30.057	0	7	30.057	4.838	34.895	2.984					
12	2.597	32.654	27.784	2	No	32.654	0						5	32.654	3.209	35.863	4.870
							0.715					10.967					18.156

AT_k, $ST_{i,1}$, and $ST_{i,2}$ are random variates.

†Channel selected randomly.

7.4. Solve the supermarket check-out problem given in Ex. 7.1. Obtain distributions and mean values for the waiting time, the total time in the system, and the queue length, and determine the fraction of time the server is idle. Consider a large enough number of customers so that the true mean waiting time falls within ± 5 percent of the calculated mean waiting time with 90 percent confidence. Include a warm-up period.

7.5. Repeat Prob. 7.4 using the following input parameters, rather than those given in Ex. 7.1:

> Interarrival times: exponentially distributed, with a mean of 2.2 minutes.
>
> Service times: normally distributed, with a mean of 2.0 minutes and a standard deviation of 0.7 minutes.

Compare the results with those obtained in Prob. 7.4. Also, compare the calculated means with the corresponding theoretical predictions given in Sec. 7.4.

7.6. Repeat Prob. 7.4 using the following input parameters rather than those given in Ex. 7.1:

> Interarrival times: exponentially distributed, with a mean of 2.8 minutes.
>
> Service times: normally distributed, with a mean of 2.8 minutes and a standard deviation of 1.2 minutes.

Compare the results with those obtained in Prob. 7.4.

7.7. Consider the operation of a bank that has a separate waiting line in front of each teller's window. In this situation each window can be considered a single-channel, single-station queue with exponentially distributed interarrival times and exponentially distributed service times. Suppose that λ is the mean interarrival time, and μ is the mean service time. Use the next-customer model to simulate the operation of a teller's window under the following conditions:

(a) $\lambda = 5$ min, $\mu = 1.5$ min
(b) $\lambda = 3$ min, $\mu = 2$ min
(c) $\lambda = 2.5$ min, $\mu = 2.3$ min
(d) $\lambda = 2.5$ min, $\mu = 2.5$ min
(e) $\lambda = 2.3$ min, $\mu = 2.5$ min

Simulate a large enough number of events so that the true mean waiting time falls within ± 10 percent of the calculated mean with 90 percent confidence. Include a warm-up period. When possible, compare the calculated means with the corresponding theoretical predictions given in Sec. 7.4.

7.8. Repeat Prob. 7.7 using the next-event model of the single-channel, single-station queue.

7.9. Consider once again the bank teller's window described in Prob. 7.7. Simulate the operation of the teller's window for an 8-hour period using the following time-dependent mean values.

Time of day	λ	μ
9–10 A.M.	4.5	3
10–11 A.M.	4.2	3
11–12 Noon	3.4	3
12–1 P.M.	2.8	3
1–2 P.M.	3.2	3
2–3 P.M.	3.7	3.5
3–4 P.M.	4.4	3.5
4–5 P.M.	3.9	3.5

Determine a confidence interval for the mean waiting time corresponding to a 90 percent confidence level.

7.10. Consider once again the emergency medical facility described in Ex. 5.1. Simulate this problem twice, once *with* and once *without* the restriction that any newly arriving patient will leave (that is, will not wait) if the facility is busy. Obtain distributions and mean values for the total time in the system, the waiting time, and the queue length, and determine the fraction of time that the facility is in use. Compare the results obtained for both simulations and discuss their significance. In each case, consider a large enough number of customers so that the true mean total time in the system falls within ±10 percent of the calculated mean with 90 percent confidence. Include a warm-up period. Compare the calculated means with the corresponding theoretical predictions given in Sec. 7.4.

7.11. A one-chair barbershop can be modeled as a single-channel single-station queue. Suppose that the customer interarrival times are exponentially distributed with a mean of 20 minutes, and the time required to cut a customer's hair is normally distributed with a mean of 18 minutes and a standard deviation of 6 minutes.

    (a) Simulate the operation of the barbershop for a 10-hr. day.

    (b) Repeat the simulation for the same set of conditions described above except that there will be a 30 percent probability that a customer will leave if he or she has waited for 10 minutes and the barber's chair is still occupied.

    (c) Repeat the simulation once again. Now, however, let there be a 50 percent probability that a newly arriving customer will leave immediately (that is, not enter the barbershop) if two or more customers are already waiting. (Note that this problem can best be handled by the next-event model.)

7.12. An airline check-in counter in a major airport is open from 6:00 A.M. to 11:00 P.M. every weekday. At least one check-in window will always be open during these hours. Additional check-in windows will be opened as the need arises, during busy periods of the day. As many as six windows can be open at any time. Each window will have its own waiting line.

    Suppose that the passenger interarrival times are exponentially distributed, with mean values that fluctuate during the day as indicated in

Fig. 7.9. Also suppose the check-in times (service times) are distributed in accordance with the following empirical distribution:

Time interval (min.)	Relative frequency
0–0.5	0.159
0.5–1.0	0.238
1.0–1.5	0.212
1.5–2.0	0.132
2.0–2.5	0.095
2.5–3.0	0.063
3.0–4.0	0.042
4.0–5.0	0.032
5.0–7.0	0.016
7.0–10.0	0.011
	1.000

Simulate the behavior of the system for various operating policies and recommend a satisfactory operating policy based upon the simulated results. (An operating policy in this situation will be a schedule indicating how many windows will be open at different times during the day.) Estimate the validity of your results through the use of confidence intervals.

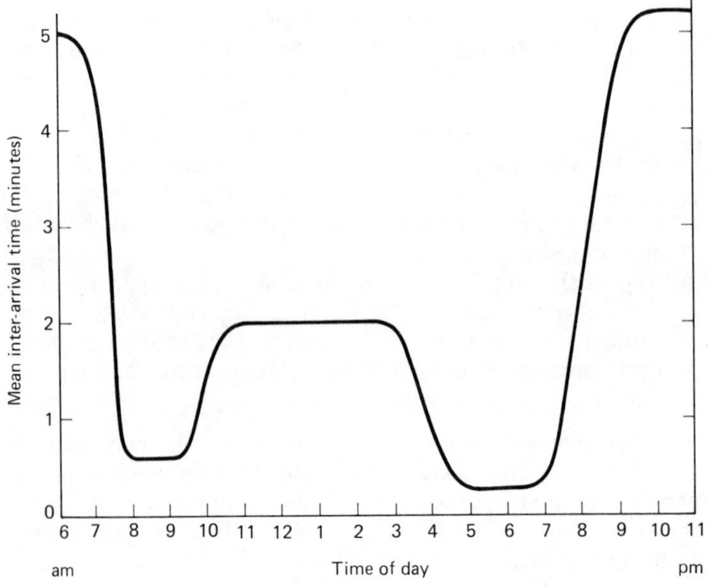

Figure 7.9

7.13. Suppose that the next-event model of the single-channel, single-station queue will be used to simulate the processing of orders within a job shop. Assume that an *estimated* service time is determined when each order arrives. Discuss how the next-event model should be modified to accommodate the following queue priorities:

   (a) For those orders that must wait for service, the order with the smallest estimated service time is the first to leave the queue.

   (b) The orders which must wait for service are divided into two groups: short jobs and long jobs, based upon the estimated service time. A long job will not be started until all short jobs waiting for service have been processed. Within each group, however, the orders will be processed on a first-in, first-out basis.

7.14. Use the model developed for Prob. 7.13a to simulate the operation of a job shop under the following conditions:

   New orders: interarrival times are exponentially distributed, with a mean of 3.7 days.

   Estimated service time per order: determined by the following empirical distribution:

Estimated service time (days)	Relative frequency
0–0.5	0.0113
0.5–1.0	0.0282
1.0–1.5	0.0452
1.5–2.0	0.0734
2.0–2.5	0.1073
2.5–3.0	0.1357
3.0–3.5	0.1695
3.5–4.0	0.1525
4.0–4.5	0.1243
4.5–5.0	0.0791
5.0–5.5	0.0509
5.5–6.0	0.0226
	1.0000

   Deviation between actual service time and estimated service time, per order: normally distributed, with a mean of 0 and a standard deviation of ($0.5\times$ estimated time).

(Note that the deviations can be either positive or negative, but the *actual* service time (which is the sum of the estimated service time and the corresponding deviation) must be nonnegative.)

   Simulate a large enough number of orders so that the true mean time spent in the shop falls within ±5 percent of the calculated mean with 95 percent confidence. Include a warm-up time.

7.15. Simulate the job shop described in Prob. 7.14, using the model developed for Prob. 7.13b rather than 7.13a. Let the cutoff between "short" and "long" orders be 2 days. Develop a separate set of statistics for each group.

7.16. Repeat Prob. 7.15 using several different values for the cutoff that separates short and long orders. Based upon the simulated results, recommend an appropriate cutoff value.

7.17. The arrival of passengers at an airline counter is known to be Poisson-distributed, with a mean of 4.3 passengers per minute. It is also known that the number of passengers serviced per minute is normally distributed, with a mean of 4.8 and a standard deviation of 1.6.

(a) If the check-in counter is considered to be a single-channel, single-station queue, determine how the passenger arrivals and departures can be simulated on the basis of equal, successive time increments. (Note that the nature of the arrival and service data suggests the use of successive time increments rather than successive passengers.)

(b) Simulate a large enough period of time so that the mean queue length falls within ± 5 percent of the calculated queue length with 90 percent confidence. Include a warm-up period in your calculations.

(c) Why are the results requested in part b based upon queue length rather than passenger waiting time?

7.18. Simulate the behavior of the two-station queuing system described in Ex. 7.5. Obtain distributions and mean values for the following system performance criteria:

(a) Total time spent in the system

(b) Waiting time for each station

(c) Queue length for each station

Also, determine the fraction of time each server is idle. Consider a large enough number of customers so that the true mean time spent in the system falls within ± 10 percent of the calculated mean with 95 percent confidence. Include a warm-up period.

7.19. Write a complete computer program to simulate a single-channel n-station queuing system, using the computational procedure shown in Fig. 7.6. Allow the particular value for n to be specified as an input parameter, up to a maximum value of $n = 5$. Include a provision for determining distributions and mean values for the following system performance criteria:

(a) Total time spent in the system

(b) Waiting time for each station

(c) Queue length for each station

Also, determine the fraction of time each server is idle. Include a provision for a warm-up period.

7.20. Use the program written for Prob. 7.19 to simulate the following situation:

A job shop manufactures a highly specialized product line. Each item must be custom built to its own individual specifications. Four distinct steps are required in the manufacturing process: rough machining, finishing, assembling, and testing. These steps are carried out sequentially. The processing of an order may be delayed prior to each step, depending upon the availability of the processing facility. When such delays do occur, the orders enter and leave the various waiting lines on a first-in, first-out basis.

Suppose that the time between arrivals of successive orders is exponentially distributed with a mean interarrival time of 3.2 days. Also, suppose that each of the service facilities can be characterized in the following manner:

Rough machining: exponentially distributed service times, with a mean of 2.7 days and a minimum of 1 day.

Finishing: normally distributed service times. For each order, the expected value will be 80 percent of the time required for rough machining, and the standard deviation will be 45 percent of the rough machining time.

Assembling: normally distributed service times, with a mean of 3 days and a standard deviation of 0.7 days.

Testing: times distributed in accordance with the following empirical distribution:

Testing time (hr)	Relative frequency
4–6	0.40
6–10	0.30
10–16	0.20
16–24	0.10

(Assume 1 working day = 8 hours.)

Consider a large enough number of orders so that the true mean manufacturing time falls within ±10 percent of the calculated mean with 95 percent confidence.

Use the simulated results to identify potential bottlenecks within the system and suggest appropriate remedial action.

7.21. A computer center consists of three remote job-entry stations, each containing a card reader and a line printer, supported by a single central processor. Each job arrives at one of the three card readers where it must wait to be read into the system. It then enters a common queue, where it waits to be processed by the central processor. Once the job has been processed, it is routed to a line printer (at the same remote job-entry station where it was read in), where it enters a queue until the line printer is available. After the job has been printed, it is placed on a shelf where it can be picked up by its owner. Figure 7.10 shows a diagram of the system.

Simulate the operation of the computer center, using the following information:

1. Job arrivals: exponentially distributed, with the following mean interarrival time for each remote job-entry station

Queue CR1
Card reader #1
Queue CR2
Card reader #2
Queue CR3
Card reader #3
Queue CP
Central processor
Queue LP1
Line printer #1
Queue LP2
Line printer #2
Queue LP3
Line printer #3

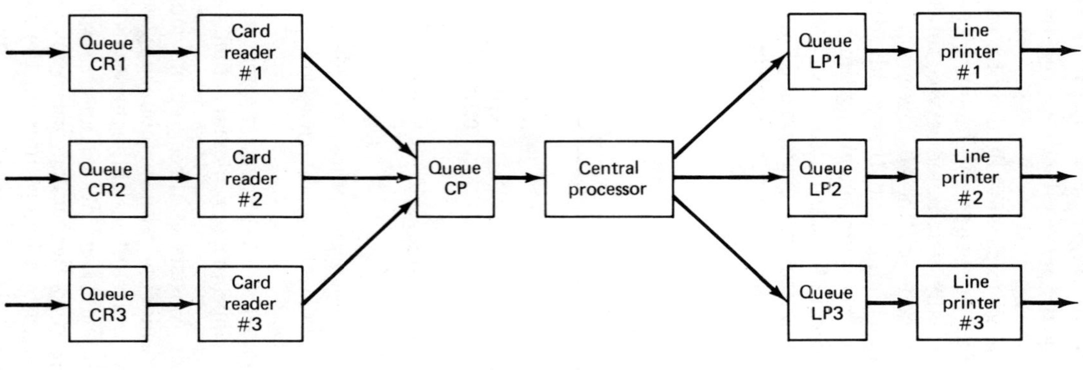

Figure 7.10

station 1: 70 sec
station 2: 15 sec
station 3: 35 sec

2. Time required to read a job: normally distributed, with a mean of 3 seconds and a standard deviation of 1 second.
3. Time required to process a job: exponentially distributed, with a mean of 2 seconds.
4. Time required to print a job: normally distributed, with a mean of 7 seconds and a standard deviation of 2.1 seconds.

Assume that each of the queues operates on a first-in, first-out basis, and that a job leaves the system once it leaves the line printer. Simulate a large enough number of jobs so that the true mean time in the system falls within ±10 percent of the calculated mean with 95 percent confidence.

Based upon the simulated results, would you recommend adding a fourth remote job-entry station?

7.22. Simulate the behavior of the two-channel queuing system described in Ex. 7.6. Obtain distributions and mean values for the total time spent in the system, the waiting time, and the queue length. Also, determine the fraction of time each server is idle. Consider a large enough number of customers so that the true mean waiting time falls within ±10 percent of the calculated mean waiting time with 90 percent confidence. Include a warm-up period.

7.23. Write a complete computer program to simulate an n-channel single-station queuing system with a common waiting line, using the computational procedure shown in Fig. 7.8. Allow the particular value for n to be specified as an input parameter, up to a maximum value of $n = 6$. Include a provision for determining distributions and mean values for the total time spent in the system, the waiting time, and the queue length. Also, determine the fraction of time each server is idle. Include a provision for a warm-up period.

7.24. Use the program written for Prob. 7.23 to simulate the following situation:

A barbershop has three chairs (three barbers). The time between arrivals of successive customers is exponentially distributed, with a mean of 6 minutes. The time required to cut a customer's hair is normally distributed, with means and standard deviations that vary from one barber to another, as indicated in the following table:

Barber	Mean (μ)	Standard deviation (σ)
1	18 min	3 min
2	22 min	4 min
3	15 min	2 min

(a) Simulate the operation of the barbershop for a 12-hour day.
(b) Suppose that the mean could be reduced from $\mu = 22$ minutes to $\mu = 18$ minutes for barber 2. Would this have a significant impact on the operation of the barbershop?

   (c) Suppose that barber 3 calls in sick. A substitute barber is available on a temporary basis, though this barber is less efficient than the original barber. If $\mu = 25$ minutes and $\sigma = 4$ minutes for the substitute, what effect will this have on the operation of the barbershop?

7.25. Use the program written for Prob. 7.23 to simulate the following situation:

On a typical business day, the operation of a suburban bank can be characterized as follows:

   Number of tellers: five

   Customer interarrival times: exponentially distributed, with a mean of 0.5 minutes

   Service times: normally distributed, with a mean of 2 minutes and a standard deviation of 0.7 minutes

The bank makes use of a common waiting line, where the customer at the head of the line goes to the first available teller.

Simulate the operation of this bank. Consider a large enough number of customers so that the true mean time spent in the bank falls within ±10 percent of the calculated mean with 95 percent confidence. Include a warm-up period.

7.26. Suppose that the bank described in Prob. 7.25 has a new manager who wants to abandon the common waiting time, replacing it with a separate waiting line in front of each teller's window. Simulate the operation of the bank under these conditions. Compare the results with those obtained in Prob. 7.25 (common waiting line).

Which system would you recommend, based upon the two sets of simulated results?

8

SPECIAL-PURPOSE SIMULATION LANGUAGES

Over the past several years a number of special-purpose languages have been developed to facilitate the programming of discrete-event simulation problems. These languages vary considerably in their complexity and their generality, although they share some common characteristics. In particular, each of these languages includes various "macro" instructions that are more or less equivalent to lengthy sequences of instructions, or even entire subroutines, in a general-purpose language such as FORTRAN. These macros provide mechanisms for the automatic implementation of certain essential activities, such as the sequencing of random events and the generation of pertinent statistics. Therefore these special-purpose languages tend to simplify the programming requirements of discrete-event models. On the other hand, the manner in which certain key variables interact may not be as well understood or controlled when using these special-purpose languages. Thus there is a trade-off between the convenience and programming simplicity that is offered by a special-purpose simulation language, and the insight and flexibility that result from the use of a general-purpose language.

In this chapter we will examine the overall characteristics of a few of the more commonly used special-purpose languages. In particular, we will consider GPSS, SIMULA, GASP, and SLAM. It should be understood, however, that we are only scratching the surface with each of these languages. More extensive study would be required in order to develop any practical degree of proficiency with any of them.

8.1 GPSS

GPSS (General-Purpose Simulation System) is actually a process-oriented, special-purpose language developed by International Business Machines for solving discrete-event simulation problems. The language has evolved through several different versions over a period of years. The more recent versions (for example, GPSS/360 and GPSS V) are widely used in the United States.

Programming with GPSS is relatively simple for certain kinds of problems, though the language lacks the flexibility that is found in some of the more recent simulation languages.

The underlying idea of GPSS is that a sequence of *transactions* (for example, customers) is generated randomly, in accordance with a prescribed distribution function. Each transaction proceeds through the system where it encounters various *process entities*, such as *queues*, *facilities* (for example, service areas), or *storage areas* (for example, multicustomer service areas). Random delays may be associated with these entities. In some cases (for example, at a facility or a storage area) the delays will be generated directly, in accordance with a given distribution function. Such delays usually represent various kinds of service times. In other cases, however (for example, at queues), the delays may represent times spent waiting for a busy entity to become vacant. In either case, the delays are expressed as multiples of an arbitrary time unit, as defined by the programmer.

The language includes eight different random number generators, which may be sampled in whatever order the programmer may wish.

The creation of a GPSS program usually begins with a *block diagram* (flowchart), which indicates how each transaction will be processed. The individual blocks are then translated into equivalent GPSS statements. Each of these statements may correspond to a *group* of statements in a true general-purpose language, such as FORTRAN. Thus it is possible to represent complicated discrete-event models with a relatively small number of GPSS statements. Moreover many of the logical intricacies that are inherent in most discrete-event problems are automatically taken care of in GPSS. This results in a considerable simplification of the overall programming task.

Certain statistics that describe the overall behavior of the system are automatically gathered during the course of the simulation. These statistics are then printed out after the simulation has been completed. Thus GPSS programs generate a certain amount of output data automatically, without the need to include explicit output statements within the program. Additional output can be generated, however, through the use of such statements.

Example 8.1

Let us once again consider the supermarket checkout problem described in Ex. 7.1. (We are considering a single-channel, single-station queue with exponentially distributed interarrival times and normally distributed service times. Recall that the mean interarrival time is 2.3 minutes. Also the mean service time is 1.8 minutes with a standard deviation of 0.5 minutes.)

Figure 8.1 shows a GPSS block diagram for this problem, based upon an arbitrarily chosen time unit of 0.01 minutes. Thus the GENERATE block generates a new transaction (that is, a new customer) with an exponentially distributed interarrival time whose mean is 230 time units. The QUEUE block causes the transaction to enter the queue (that is, the waiting area) called LINE, and the SEIZE block causes the transaction to enter the server called CNTR at the proper time, based upon a first-in first-out queuing priority. DEPART allows the transaction to leave the queue once the server becomes available. The ADVANCE block is used to produce the actual time delay within the server, as defined by the variable $V1$. (Though not apparent in the block diagram, $V1$ will define a normally distributed random variate with a mean of 180 time units and a standard deviation of 50 time units.) RELEASE causes the transaction to leave the server once the service has been completed, and TABULATE is used to create a statistical distribution called TIME of the total time spent in the system. Finally the TERMINATE block allows the transaction to leave the system and at the same time decreases the transaction counter by 1 unit. (The transaction counter proceeds backwards, from some specified maximum value to zero.)

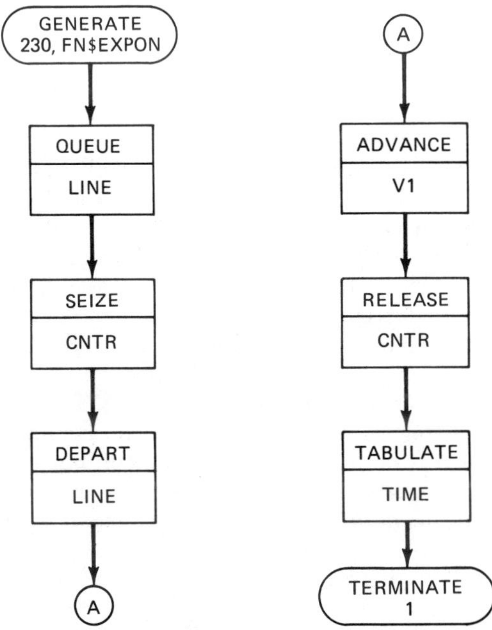

Figure 8.1

```
* SINGLE-CHANNEL SINGLE-STATION QUEUE (SUPERMARKET CHECKOUT COUNTER)
*
* EXPONENTIAL INTER-ARRIVAL TIMES, NORMAL SERVICE TIMES
*
          SIMULATE
*
 EXPON FUNCTION   RN1,BE              DEFINE EXPONENTIAL DISTRIBUTION
  NORM FUNCTION   RN2,BN              DEFINE NORMAL DISTRIBUTION
    1  VARIABLE   180+50*FN$NORM      DEFINE SERVICE TIME
*
          GENERATE   230,FN$EXPON     CUSTOMER ARRIVES
          QUEUE      LINE             CUSTOMER ENTERS QUEUE
          SEIZE      CNTR             CUSTOMER SEIZES SERVER
          DEPART     LINE             CUSTOMER LEAVES QUEUE
          ADVANCE    V1               CUSTOMER RECEIVES SERVICE
          RELEASE    CNTR             CUSTOMER LEAVES SERVER
          TABULATE   TIME             TABULATE TOTAL TIME IN SYSTEM
          TERMINATE  1                CUSTOMER DEPARTS FROM SYSTEM
*
  TIME TABLE      M1,100,100,40       DEFINE DISTRIBUTION OF TOTAL TIME IN SYSTEM
  WAIT QTABLE     LINE,100,100,40     DEFINE DISTRIBUTION OF WAITING TIMES
*
          START      100              BEGIN WARMUP PERIOD
*
          RESET
*
          START      1000             BEGIN ACTUAL RUN
*
          END
```

Figure 8.2

Figure 8.2 presents a complete GPSS program for this problem. The two numerical columns on the left are not a part of the actual GPSS *source program*, but are generated by the computer when the source program is processed. The left-most column simply numbers all of the lines consecutively, for easy reference; the next column (numbered 1 through 8) identifies and numbers the GPSS *block statements*, that is, those statements that correspond directly to the individual blocks shown in Fig. 8.1.

The GPSS source program is printed to the right of these two columns of numbers. This program consists of several different types of statements, including the numbered *block statements* (GENERATE, QUEUE, SEIZE, . . . , TERMINATE); two *table definition statements* (TABLE and QTABLE), which define two statistical distributions; and several *control statements* (SIMULATE, START, RESET, and END). Most of these statements include annotations, which occupy the right-hand portion of the line.

Let us now consider the purpose of each line within the program. The first line is a *comment* that serves as a program heading. The remaining lines that begin with asterisks (for example, lines 2, 4, 6, 10, etc.) are blank comments (blank lines) that merely delineate various sectors of the program.

The SIMULATE statement (line 5) causes the program to execute and to generate certain output data after the program has been *assembled* (that is, translated into machine language). This statement may appear anywhere within the GPSS program, though it is customarily placed ahead of the block statement, as in Fig. 8.2.

Line 7 accesses library function BE, which generates exponentially distributed random variates having a mean of 1. This function is referred to as EXPON. The first of the eight available random number generators ($RN1$) is used with this function. Similarly, line 8 accesses library function *BN*, which generates standard normal random variates. This function is called NORM. The second sequence of random numbers (*RN*2) will be used with this function. In line 9 we define a *variable* (referred to as *V*1) which is used in conjunction with NORM to generate the required normally distributed service times.

Lines 11 to 18 contain the block statements, which are the heart of the program. (Again note that the second column contains the consecutive block numbers 1 to 8.) We have already discussed the purpose of each block. It should be pointed out, however, that certain of these statements are always used in pairs. In particular the GENERATE and TERMINATE statements are used together, as are QUEUE and DEPART, and SEIZE and RELEASE. Thus the block portion of the program should be thought of in terms of the following sets of statement pairs.

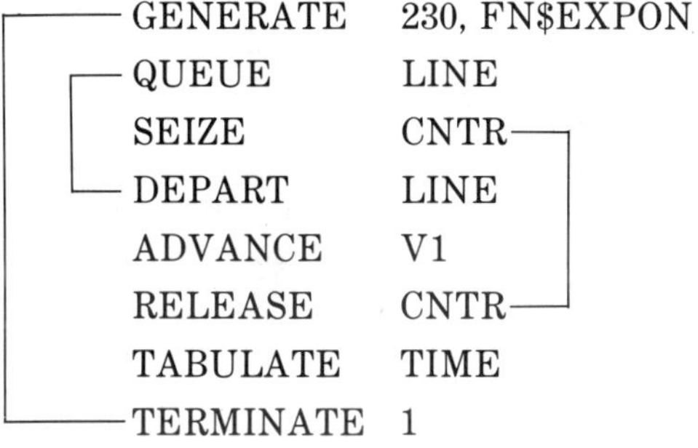

Notice the way the QUEUE–DEPART pairing overlays the SEIZE–RELEASE pairing. This is required in order to sequence the queue and the server properly.

The TABLE and QTABLE statements (lines 20 and 21) are used to define statistical distributions of total time in the system (called *M*1) and waiting time in the queue. Each distribution will be permitted as many as forty intervals, each interval having a width of 100 time units.

Line 23 contains a control statement (START 100) which is used to initiate a warm-up period consisting of 100 transactions. All statistics are RESET to zero (line 25) following the warm-up period. The actual run, consisting of 1000 transactions, is then initiated in line 27 (START 1000). Finally the END statement (line 29) identifies the physical end of the program.

Notice that the program does not contain any input/output statements. The required input data are included within the structure of the program, and the output is generated automatically when the program is executed.

Figures 8.3a through c illustrate some of the output that is generated by this program. The upper portion of Fig. 8.3a, for example, summarizes the information obtained for the queue. Thus we see that the maximum queue length was thirteen customers, the average was 1.74 customers, and the average waiting time was 388.80 time units (3.89 minutes). Also, the average waiting time for those customers *who experienced some waiting* was 485.27 time units (4.85 minutes).

The lower portion of Fig. 8.3a is concerned with the server. We see, for example, that the average utilization is 0.8088 (which implies that the fraction of

QUEUE	MAXIMUM CONTENTS	AVERAGE CONTENTS	TOTAL ENTRIES	ZERO ENTRIES	PERCENT ZEROS	AVERAGE TIME/TRAN	$AVERAGE TIME/TRAN	TABLE NUMBER	CURRENT CONTENTS
LINE	13	1.74	1001	199	19.880	388.80	485.27	2	1

FACILITY	AVERAGE UTILIZATION	NUMBER ENTRIES	AVERAGE TIME/TRANS	SEIZING TRANS. NO.	PREEMPTING TRANS. NO.
CNTR	0.8088	1000	180.55	0	0

(a)

Figure 8.3

TABLE TIME

ENTRIES IN TABLE 1000		MEAN ARGUMENT 570.252		STANDARD DEVIATION 451.434		SUM OF ARGUMENTS 570252.000	NON-WEIGHTED
UPPER LIMIT	OBSERVED FREQUENCY	PER CENT OF TOTAL	CUMULATIVE PERCENTAGE	CUMULATIVE REMAINDER	MULTIPLE OF MEAN	DEVIATION FROM MEAN
100	5	0.50	0.5	99.5	0.175	-1.042
200	199	19.90	20.4	79.6	0.351	-0.820
300	135	13.50	33.9	66.1	0.526	-0.599
400	144	14.40	48.3	51.7	0.701	-0.377
500	92	9.20	57.5	42.5	0.877	-0.156
600	78	7.80	65.3	34.7	1.052	0.066
700	67	6.70	72.0	28.0	1.228	0.287
800	51	5.10	77.1	22.9	1.403	0.509
900	49	4.90	82.0	18.0	1.578	0.730
1000	33	3.30	85.3	14.7	1.754	0.952
1100	25	2.50	87.8	12.2	1.929	1.173
1200	32	3.20	91.0	9.0	2.104	1.395
1300	14	1.40	92.4	7.6	2.280	1.617
1400	14	1.40	93.8	6.2	2.455	1.838
1500	7	0.70	94.5	5.5	2.630	2.060
1600	6	0.60	95.1	4.9	2.806	2.281
1700	9	0.90	96.0	4.0	2.981	2.503
1800	6	0.60	96.6	3.4	3.156	2.724
1900	9	0.90	97.5	2.5	3.332	2.946
2000	6	0.60	98.1	1.9	3.507	3.167
2100	9	0.90	99.0	1.0	3.683	3.389
2200	4	0.40	99.4	0.6	3.858	3.610
2300	2	0.20	99.6	0.4	4.033	3.832
2400	1	0.10	99.7	0.3	4.209	4.053
2500	1	0.10	99.8	0.2	4.384	4.275
2600	1	0.10	99.9	0.1	4.559	4.496
2700	1	0.10	100.0	0.0	4.735	4.718

REMAINING FREQUENCIES ARE ALL ZERO

(b)

Figure 8.3 *(continued)*

```
TABLE WAIT

ENTRIES IN TABLE        MEAN ARGUMENT           STANDARD DEVIATION            S
       1000               389.702                    449.029

          UPPER     OBSERVED     PER CENT     CUMULATIVE        CUMULATIV
          LIMIT     FREQUENCY    OF TOTAL     PERCENTAGE        REMAINDE
            100        316        31.60          31.6              68.4
            200        140        14.00          45.6              54.4
            300        110        11.00          56.6              43.4
            400         80         8.00          64.6              35.4
            500         67         6.70          71.3              28.7
            600         50         5.00          76.3              23.7
            700         52         5.20          81.5              18.5
            800         35         3.50          85.0              15.0
            900         22         2.20          87.2              12.8
           1000         31         3.10          90.3               9.7
           1100         18         1.80          92.1               7.9
           1200         15         1.50          93.6               6.4
           1300          9         0.90          94.5               5.5
           1400          5         0.50          95.0               5.0
           1500          8         0.80          95.8               4.2
           1600          9         0.90          96.7               3.3
           1700          5         0.50          97.2               2.8
           1800          8         0.80          98.0               2.0
           1900          8         0.80          98.8               1.2
           2000          5         0.50          99.3               0.7
           2100          3         0.30          99.6               0.4
           2200          1         0.10          99.7               0.3
           2300          1         0.10          99.8               0.2
           2400          1         0.10          99.9               0.1
           2500          1         0.10         100.0               0.0
REMAINING FREQUENCIES ARE ALL ZERO
```

(c)

Figure 8.3 *(continued)*

time the server is idle is 0.1912). We also see that the average service time is
180.55 time units (1.81 minutes). This value is, of course, essentially the same as
the specified value of 1.80 minutes.

Figure 8.3b shows the distribution of total time in the system. Notice that
the mean value is 5.70 minutes, and the standard deviation is 4.51 minutes
(shown at the top of the figure). Both the relative frequencies and the cumulative
distribution are given in the columnar data (columns 3 and 4, respectively).

Similarly, Fig. 8.3c presents the distribution of waiting times. At the top of
the figure we see that the mean waiting time is 3.897 minutes, with a standard
deviation of 4.49 minutes. The tabulated values of the relative frequency and the
cumulative distribution are again shown beneath these values, in the third and
fourth columns.

A few of the more pertinent statistics for this simulation are summarized
below. It is interesting to compare these values with the results presented for this
same problem in Ex. 7.4.

Number of simulated customers	1000
Mean waiting time (minutes)	3.89
Mean queue length (number of customers)	1.74
Fraction of time the server is idle	0.191

For additional information about GPSS, the reader is referred to
the textbooks by Bobillier, Kahan, and Probst (1976); Gordon (1975);
Maisel and Gnugnoli (1972); and Schriber (1974).

8.2 SIMULA

SIMULA is a general-purpose language that includes some special provisions for handling discrete-event simulation problems. The language is an outgrowth of ALGOL-60 (a general-purpose language that is used extensively in Europe) which was developed at the Norwegian Computing Center in Oslo, Norway (Birtwistle, Dahl, Myhrhaug, and Nygaard 1979).

Conceptually, SIMULA is somewhat like GPSS in the sense that a sequence of active system entities (for example, customers) is generated randomly and allowed to interact with passive system components in accordance with prescribed distribution functions. In practice, however, SIMULA is altogether different from GPSS. For example, SIMULA introduces *objects* to represent customers, servers, queues. *Like* objects (that is, objects that refer to the same type of entity) are represented collectively within a *class*. Of particular importance is the *process class*, in which the scheduling of events (for example, the assignment of a starting time and a stopping time for each event) may occur. A system clock is used to synchronize actions that are scheduled between active and passive objects.

Closely associated with the class construct is the *procedure*, which resembles a subroutine in FORTRAN. Both a class and a procedure are considered to consist of a data structure and a corresponding group of action statements. The two differ in the sense that a class retains its data structure between successive calls (that is, *activations*), whereas a procedure does not.

Example 8.2

Figure 8.4 contains a SIMULA program which simulates our familiar single-channel single-station queue described in Ex. 7.1. Notice that the program is considerably longer than the corresponding GPSS program shown in Fig. 8.2. The program consists, essentially, of three process classes (CUSTOMERS, SERVER, and CUSTARRIVAL), two procedures (PNORMA and REPORT), several initialization statements, and a short executive routine. Let us consider the overall purpose of each of these major program components.

The first few lines of the program provide miscellaneous initial information and define several of the program variables to be either integer or real. This information is followed by PROCESS CLASS CUSTOMERS—an indented sequence of statements which continues through the line containing END $$$ OF CLASS CUSTOMERS $$$. This process class has several functions; first, it assigns initial values to certain "local" variables. It also controls the passage of customers through the queue (even if the queue is empty), and generates most of the statistics for the queue. In addition, this process class generates service times and controls the entry into the server. It is called by PROCESS CLASS CUSTARRIVAL (see the following).

```
OPTIONS(/LIST);
BEGIN
SIMULATION CLASS EXECUTINGJOBS(N);INTEGER N;
        BEGIN
        REF(HEAD)RESTINGPLACE,WAITINGLINE;
        REAL THROUGHTIME,SIMPERIOD,MAXTIMEIN,GOTOIDLE,IDLETIME,P;
        REAL MEAN,STDDEV,THROUGHTIMESQ,SIGMA,ENDTIME,AVGLINELENGTH;
        REAL TIMEINQUEUE,ACTIVETIME,GOTOACTIVE;
        INTEGER NROFCUSTOMERS,MAXLENGTH,U,I,NUMCUSTIN,MESS,GLOBQLGTH;
        INTEGER QMARK,QTIMEMARK,QTIMELOOK;
        PROCESS CLASS CUSTOMERS;
                BEGIN
                REAL ENTRYTIME,ELAPSEDTIME,RUNTIME,INITIALRT,LENGTH,LOCTIMEINQ;
                INTEGER QLENGTH;
                LOCTIMEINQ:=0.0;
                ENTRYTIME:=TIME;
                RUNTIME:= PNORMA;
                INITIALRT:=RUNTIME;
                QLENGTH:=WAITINGLINE.CARDINAL;
                GLOBQLGTH:=QLENGTH;
                NUMCUSTIN:=NUMCUSTIN+1;
                IF MAXLENGTH < QLENGTH
                    THEN  MAXLENGTH:=QLENGTH;
                IF NOT RESTINGPLACE.EMPTY
                    THEN
                      BEGIN
                          ACTIVATE RESTINGPLACE.FIRST;
                      END;
                PASSIVATE;
                ELAPSEDTIME:=TIME-ENTRYTIME;
                IF MAXTIMEIN < ELAPSEDTIME
                    THEN MAXTIMEIN:=ELAPSEDTIME;
                NROFCUSTOMERS:= NROFCUSTOMERS + 1;
                THROUGHTIME := THROUGHTIME + ELAPSEDTIME;
                THROUGHTIMESQ:=THROUGHTIMESQ + ELAPSEDTIME**2;
                TIMEINQUEUE:=TIMEINQUEUE+LOCTIMEINQ;
                END  $$$ OF CLASS CUSTOMERS $$$;
            PROCESS CLASS SERVER;
                BEGIN
                REF(CUSTOMERS) SERVED;
                WHILE TRUE DO
                        BEGIN
                        OUT;
                        WHILE NOT WAITINGLINE.EMPTY DO
                          BEGIN  REAL  TESTTIME;
                          GOTOACTIVE:=TIME;
                          QMARK:=WAITINGLINE.CARDINAL;
                          QTIMEMARK:=TIME-QTIMELOOK;
                          AVGLINELENGTH:=QMARK*QTIMEMARK + AVGLINELENGTH;
                          QTIMELOOK:=TIME;
                          SERVED:- WAITINGLINE.FIRST;
                          SERVED.OUT;
                          SERVED.QLENGTH:=SERVED.QLENGTH-1;
                          SERVED.LOCTIMEINQ:=TIME-SERVED.ENTRYTIME;
                          GLOBQLGTH:=GLOBQLGTH-1;
                          TESTTIME:= SERVED.RUNTIME;
                          HOLD(TESTTIME);
                          ACTIVETIME:=TIME-GOTOACTIVE+ACTIVETIME;
                          SERVED.RUNTIME :=0;
                          ACTIVATE SERVED;
                            END;
                          BEGIN
                          GOTOIDLE:=TIME;
                          WAIT(RESTINGPLACE);
                          IDLETIME:=TIME-GOTOIDLE+IDLETIME;
                          END;
                          END
                    END  $$$ OF CLASS SERVER $$$;
            REAL PROCEDURE PNORMA;
                BEGIN
                INTEGER TEMP1,ITEM;
```

Figure 8.4

```
          REAL TEMP2,SUM;
          ITEM:=0;
          SUM:=0;
          WHILE ITEM < 12 DO
                    BEGIN
                    ITEM:=ITEM+1;
                    SUM:=SUM+UNIFORM(0.0,1.0,MESS);
                    END;
          SUM:=SUM-6;
          PNORMA:=MEAN+STDDEV*SUM;
END  $$ OF PROCEDURE PNORMA  $$;
PROCESS CLASS CUSTARRIVAL;
          BEGIN
          REF(CUSTOMERS)X;
          WHILE NUMCUSTIN < 1000 DO
          BEGIN
                    QMARK:=WAITINGLINE.CARDINAL;
                    QTIMEMARK:=TIME-QTIMELOOK;
                    AVGLINELENGTH:=QMARK*QTIMEMARK + AVGLINELENGTH;
                    QTIMELOOK:=TIME;
                    X:-NEW CUSTOMERS;
                    X.INTO(WAITINGLINE);
                    ACTIVATE X;
                    HOLD(NEGEXP(P,U));
                    END;
          ENDTIME:=TIME;
          END  $$$ OF CLASS CUSTARRIVAL $$$;
PROCEDURE REPORT;
          BEGIN
          OUTTEXT("     SINGLE SERVER/SINGLE QUEUE");
          OUTIMAGE ; OUTIMAGE;
          OUTTEXT("NO. OF CUSTOMERS =");
          OUTINT(NROFCUSTOMERS,26);
          OUTIMAGE;
          OUTTEXT("AVE. ELAPSED TIME IN WHOLE SYSTEM=");
          OUTFIX(THROUGHTIME/NROFCUSTOMERS,2,10);
          OUTIMAGE;
          OUTTEXT("STANDARD DEVIATION OF THROUGHTIME=");
          SIGMA:=(THROUGHTIMESQ/NROFCUSTOMERS-(THROUGHTIME/NROFCUSTOMERS)**2);
          SIGMA:=SQRT(SIGMA);
          OUTFIX(SIGMA,3,10);
          OUTIMAGE;
          OUTTEXT("MEAN TIME IN QUEUE = ");
          OUTFIX(TIMEINQUEUE/NROFCUSTOMERS,2,23);
          OUTIMAGE;
          OUTTEXT("MAX.QUEUE LENGTH=");
          OUTINT(MAXLENGTH,27);
          OUTIMAGE;
          OUTTEXT("LONGEST TIME IN SYSTEM=");
          OUTFIX(MAXTIMEIN,2,21);
          OUTIMAGE;
          OUTTEXT("AVE. IDLE TIME PER CUSTOMER=");
          OUTFIX(IDLETIME/NROFCUSTOMERS,2,16);
          OUTIMAGE;
          OUTTEXT("FRACTION OF TIME IDLE= ");
          OUTFIX(IDLETIME/(IDLETIME+ACTIVETIME),2,21);
          OUTIMAGE;
          OUTTEXT("ESTIMATE OF MEAN QUEUE LENGTH= ");
          OUTFIX(AVGLINELENGTH/(ACTIVETIME+IDLETIME),2,13);
          OUTIMAGE;
          END  $$$ OF REPORT $$$;
QTIMELOOK:=0;
P:=1/2.3;
U:=1;
MEAN:=1.8;
STDDEV:=.5;
MESS:=1;
NROFCUSTOMERS:=0;
IDLETIME:=0.0;
NUMCUSTIN:=0;
THROUGHTIME:=0.0;
```

Figure 8.4 *(continued)*

```
        THROUGHTIMESQ:=0.0;
        GOTOIDLE:=0;
        TIMEINQUEUE:=0.0;
        AVGLINELENGTH:=0.0;
        RESTINGPLACE:=NEW HEAD;
        WAITINGLINE:=NEW HEAD;
        NEW SERVER.INTO(RESTINGPLACE);
        END $$$ OF SIMULATION CLASS EXECUTINGJOBS $$$;
EXECUTINGJOBS(1) BEGIN
        ACTIVATE NEW CUSTARRIVAL;
                HOLD(1000000);
                REPORT
                END;
OUTIMAGE; OUTIMAGE;
OUTTEXT(" THE END OF THE SIMULATION ");
END $$$$ OF THE PROGRAM $$$$;
```

Figure 8.4 *(continued)*

Following PROCESS CLASS CUSTOMERS is another indented sequence of statements, beginning with PROCESS CLASS SERVER and continuing through END $$$ OF CLASS SERVER $$$. This portion of the program simulates the passage of customers through the server. It also generates statistics for the server and a value for the mean queue length. This routine is called by PROCESS CLASS CUSTOMERS.

The next group of statements, beginning with REAL PROCEDURE PNORMA and continuing through END $$ OF PROCEDURE PNORMA $$, generates normally distributed random variates using the method based upon the central limit theorem (see Sec. 4.7). This routine is also called by PROCESS CLASS CUSTOMERS. (It should be noted that most versions of SIMULA contain a library function that generates normally distributed random variates, so that this routine would not be needed.)

PROCESS CLASS CUSTARRIVAL assigns values to certain "global" variables and generates new arrivals into the system. This routine references PROCESS CLASS CUSTOMERS (see above), which then handles the entry into the queue. CUSTARRIVAL is called by the system executive routine, as described below.

Following PROCESS CLASS CUSTARRIVAL is a relatively long sequence of output statements, referred to as PROCEDURE REPORT. Finally, there is a sequence of initialization statements (ranging from QTIMELOOK: = 0 through END $$$ OF SIMULATION CLASS EXECUTING JOBS $$$), followed by a brief executive routine (starting with EXECUTINGJOBS(1) BEGIN and continuing through the remainder of the program).

Figure 8.5 shows the output that is generated by this program. This particular simulation is based upon 1000 successive customers, using the same numerical values presented earlier (cf. Ex. 7.1). Notice that the values obtained for the mean waiting time (3.30), the mean queue length (1.50), and the fraction of time the server is idle (0.19) are consistent with the simulated results presented earlier as well as with the theoretical predictions in Sec. 7.4 (see Ex. 7.4).

It should be mentioned that the required statistical data are not generated automatically in SIMULA. The programmer must provide both the statistical calculations and the output routines.

```
              SINGLE SERVER/SINGLE QUEUE

      NO. OF CUSTOMERS =                          1000
      AVE. ELAPSED TIME IN WHOLE SYSTEM=          5.09
      STANDARD DEVIATION OF THROUGHTIME=          3.451
      MEAN TIME IN QUEUE =                        3.30
      MAX.QUEUE LENGTH=                             11
      LONGEST TIME IN SYSTEM=                     18.73
      AVE. IDLE TIME PER CUSTOMER=                0.41
      FRACTION OF TIME IDLE=                      0.19
      ESTIMATE OF MEAN QUEUE LENGTH=              1.50
```

Figure 8.5

8.3 GASP

GASP is a FORTRAN-based, event-oriented, special-purpose language for solving discrete-event simulation problems. The language is actually a series of FORTRAN subroutines that are linked together by a special executive program. Newer versions of the language (that is, GASP IV) can accommodate continuous as well as discrete simulation models.

Within the GASP framework, every real entity is referred to as an *element*. Such elements can be either *permanent* (for example, a server) or *temporary* (for example, a customer passing through the server). One or more *attributes* can be associated with each element. *Events* cause changes to occur to one or more elements, thus allowing the elements to interact with each other.

GASP incorporates a number of macro instructions that allow several fundamental operations to be carried out automatically. These operations include

- (a) generation of random variates
- (b) time movement and control
- (c) file maintenance
- (d) generation of statistical output
- (e) debugging aids (that is, tracing, monitoring, and dumping)

We will see how some of these features can be utilized in the following example.

Example 8.3

Let us return to the single-channel, single-station queuing problem described in Ex. 7.1. Figure 8.6a contains a brief user-supplied main program that initializes a few key variables and then references subroutine GASP. The latter is the true executive routine for the GASP system. Subroutine GASP is linked with two user-supplied programs called EVNTS and OTPUT. These programs are shown in Figs. 8.6b and c, respectively. EVNTS then references the user-supplied subroutines ARRVL, ENDSV, and ENDSM. These three subroutines are shown in Figs. 8.6d to f, respectively. OTPUT provides supplementary output information upon completion of the simulation.

```
C
      COMMON XISYS,BUS,NARRIV
C
C*****SET NCRDR TO THE CARD READER NUMBER AND
C*****SET NPRNT TO THE PRINTER NUMBER.
C
      NCRDR=5
      NPRNT=6
      XISYS=0.
      BUS=0.
      NARRIV=0
C  XL HAS BEEN REPLACED BY PARAM(1,1)
C  XMU HAS BEEN REPLACED BY PARAM(2,1)
      CALL GASP(NSET)
      CALL EXIT
C
```

(a)

```
C
      COMMON XISYS,BUS,NARRIV
      GO TO (1,2,3),I
1     CALL ARRVL(NSET)
      RETURN
2     CALL ENDSV(NSET)
      RETURN
3     CALL ENDSM(NSET)
C
```

(b)

```
      COMMON XISYS,BUS,NARRIV
C
C*****COMPUTE THEORETICAL AND SIMULATED VALUES OF PERFORMANCE MEASURES
C*****FOR THE SYSTEM.
C
      ETISS=SUMA(1,1)/SUMA(1,3)
      EIDTS=(SSUMA(2,1)-SSUMA(2,2))/(SSUMA(2,1))
      EWTS=SUMA(2,1)/SUMA(2,3)
      BETA=PARAM(2,1)/PARAM(1,1)
      EIDTC=1.-BETA
      EWTC=(PARAM(2,4)**2/PARAM(1,1)+PARAM(1,1)*BETA**2)/(2.*(1.-BETA))
      ETISC=PARAM(2,1)+EWTC
      YA=ETISS/(SSUMA(1,2)/SSUMA(1,1))
      YS=ETISS-EWTS
      WRITE(NPRNT,85)
85    FORMAT(/46X,15HSIMULATED VALUE,4X,17HTHEORETICAL VALUE/)
      WRITE(NPRNT,90)EIDTS,EIDTC
90    FORMAT(20X,18HEXPECTED IDLE TIME,11X,F8.3,12X,F8.3)
      WRITE(NPRNT,95)EWTS,EWTC
95    FORMAT(20X,21HEXPECTED WAITING TIME,8X,F8.3,12X,F8.3)
      WRITE(NPRNT,96)ETISS,ETISC
96    FORMAT(20X,23HEXPECTED TIME IN SYSTEM,6X,F8.3,12X,F8.3)
      WRITE(NPRNT,97)YA,PARAM(1,1)
97    FORMAT(20X,21HEXPECTED ARRIVAL TIME,8X,F8.3,12X,F8.3)
      WRITE(NPRNT,98)YS,PARAM(2,1)
98    FORMAT(20X,21HEXPECTED SERVICE TIME,8X,F8.3,12X,F8.3)
C
```

(c)

Figure 8.6

```fortran
      SUBROUTINE ARRVL(NSET)
C*******************************************************************
C
C THIS LINE IS USED TO INCLUDE THE COMMON BLOCK
C  FOR GASP AS DESCRIBED IN THE GASP II MANUAL.
C
C*******************************************************************
      INCLUDE 'PRG:GASP.COM'
      COMMON XISYS,BUS,NARRIV
C
C *** TOTAL NUMBER OF ARRIVALS EXCEEDES MAXIMUN ALLOWED (500).
C *** SCHEDULE END OF SIMULATION EVENT
C
      NARRIV=NARRIV+1
      IF (NARRIV-500)17,17,16
  16  ATRIB(1)=TNOW
      ATRIB(2)=3.
      CALL FILEM(1,NSET)
      RETURN
  17  CONTINUE
C
C*****SINCE ARRVL IS AN ENDOGENOUS EVENT SCHEDULE THE NEXT ARRIVAL
C*****AT TNOW PLUS NUMBER DRAWN FROM AN EXPONENTIAL DISTRIBUTION.  THE
C*****ARRIVAL TIME IS STORED IN ATRIB(1).  THE EVENT CODE FOR AN ARRVL
C*****IS 1. SET ATRIB(2) EQUAL TO 1.
C
      CALL DRAND(ISEED,RNUM)
      ATRIB(1)=TNOW-PARAM(1,1)*ALOG(RNUM)
      ATRIB(2)=1.
      CALL FILEM(1,NSET)
C*****COLLECT THE STATISTICS ON THE NUMBER IN THE SYSTEM SINCE AN
C*****ARRIVAL CAUSES NUMBER IN THE SYSTEM TO CHANGE
C
      CALL TMST(XISYS,TNOW,1,NEST)
      IF(XISYS)7,8,9
  7   CALL ERROR(31,NSET)
      RETURN
C
C*****INCREMENT TNE NUMBER IN THE SYSTEM. SINCE THE NUMBER IN   THE
C*****SYSTEM WAS ZERO THE SERVER WAS NOT BUSY. THE SERVER STATUS WILL
C*****CHANGE DUE TO THE NEW ARRIVAL THEREFORE STATISTICS ON THE TIME
C*****THE SERVER WAS BUSY MUST BE COLLECTED.
C
  8   XISYS=XISYS+1.
      CALL TMST(BUS,TNOW,2,NSET)
C
C*****CHANGE THE STATUS OF THE SERVER TO BUSY. COLLECT STATISTICS ON
C*****THE WAITING TIME OF THE CURRENT ARRIVAL WHICH IS ZERO SINCE THE SERVER
C*****WAS NOT BUSY AT HIS TIME OF ARRIVAL.
C
      BUS=1.
      CALL COLCT(0.,2,NSET)
C
C*****SINCE THE NEW ARRIVAL GOES DIRECTLY INTO SERVICE CAUSE AN END OF
C*****SERVICE EVENT. SET ATRIB(2) EQUAL TO 2 TO INDICATE AN END OF
C*****SERVICE EVENT. SET ATRIB(3) EQUAL TO TNOW THE ARRIVAL TIME OF THE
C*****CUSTOMER.
C
      ATRIB(1)=TNOW+RNORM(2)
      ATRIB(2)=2.
```

(d)

```fortran
      SUBROUTINE ENDSV(NSET)
      INCLUDE 'PRG:GASP.COM'
      COMMON XISYS,BUS,NARRIV
C
C*****COMPUTE TIME IN SYSTEM EQUAL TO CURRENT TIME MINUS ARRIVAL TIME
C*****OF CUSTOMER FINISHING SERVICE. COMPUTE STATISTICS ON THE TIME IN
```

(e)

Figure 8.6 *(continued)*

```
C*****SYSTEM.
C
      TISYS=TNOW-ATRIB(3)
      CALL COLCT(TISYS,1,NSET)
      CALL HISTO(TISYS,2.0,1.0,1)
C
C*****SINCE A CUSTOMER WILL DEPART FROM THE SYSTEM DUE TO THE END OF
C*****SERVICE COLLECT STATISTICS ON NUMBER IN SYSTEM AND DECREMENT THE
C*****NUMBER IN THE SYSTEM BY ONE.
C
      CALL TMST(XISYS,TNOW,1,NSET)
      XISYS=XISYS-1.
C
C*****TEST TO SEE IF CUSTOMERS ARE WAITING FOR SERVICE. IF NONE COLLECT
C*****STATISTICS ON THE BUSY TIME OF THE SERVER AND SET HIS STATUS TO
C*****IDLE BY MAKING BUS EQUAL ZERO. IF CUSTOMERS ARE WAITING FOR SERVICE
C*****REMOVE FIRST CUSTOMER FROM THE QUEUE OF THE SERVER WHICH IS FILE
C*****TWO.
C
      IF(NQ(2))7,8,9
7     CALL ERROR(41,NSET)
      RETURN
8     CALL TMST(BUS,TNOW,2,NSET)
      BUS=0.
      RETURN
9     CALL RMOVE(MFE(2),2,NSET)
C
C*****COMPUTE WAITING TIME OF CUSTOMER AND COLLECT STATISTICS ON
C*****WAITING TIME. PUT CUSTOMER IN SERVICE BY SCHEDULING AN END OF
C*****SERVICE EVENT FOR THE CUSTOMER.
      WT=TNOW-ATRIB(3)
      CALL COLCT(WT,2,NSET)
      ATRIB(1)=TNOW+RNORM(2)
      ATRIB(2)=2.
      CALL FILEM(1,NSET)
      RETURN
      END
      ATRIB(3)=TNOW
      CALL FILEM(1,NSET)
      RETURN
C
C*****INCREMENT THE NUMBER IN THE SYSTEM
9     XISYS=XISYS+1.
C
C*****PUT NEW ARRIVAL IN THE QUEUE WAITING FOR THE SERVER TO BECOME FREE
C*****SET ATRIB(3) EQUAL TO THE ARRIVAL TIME OF THE CUSTOMER.
C
      ATRIB(3)=TNOW
      CALL FILEM(2,NSET)
      RETURN
      END
```

(e)

```
      SUBROUTINE ENDSM(NSET)
      INCLUDE 'PRG:GASP.COM'
      COMMON XISYS,BUS,NARRIV
20    IF(NQ(1))7,8,9
7     CALL ERROR(3,NSET)
C
C*****UPDATE STATISTICS ON NUMBER IN SYSTEM AND STATUS OF SERVER TO END
C*****OF SIMULATION TIME. SET CONTROL VARIABLE TO STOP SIMULATION
8     CALL TMST(XISYS,TNOW,1,NSET)
      CALL TMST(BUS,TNOW,2,NSET)
      MSTOP=-1
      NORPT=0
      RETURN
```

(f)

Figure 8.6 *(continued)*

```
C
C*****REMOVE ALL EVENTS FROM EVENT FILE SO THAT ALL CUSTOMERS ARRIVING
C*****BEFORE END OF SIMULATION TIME ARE INCLUDED IN THE SIMULATION
C*****STATISTICS. ONLY END OF SERVICE EVENT NEED BE PROCESSED. IF
C*****ITEMS ARE IN THE QUEUE OF THE SERVER THEY WILL BE REMOVED IN THE
C*****END OF SERVICE EVENT WHERE ANOTHER END OF SERVICE EVENT WILL BE
C*****CREATED.
C
9       CALL RMOVE(MFE(1),1,NSET)
        TNOW=ATRIB(1)
        IF(ATRIB(2)-2.)20,21,20
21      CALL ENDSV(NSET)
        GOTO 20
        END
```

(f)

Figure 8.6 *(continued)*

Subroutine ARRVL performs a number of required functions such as scheduling arrivals, maintaining the current status of the queue and the server, and gathering various statistics that characterize the performance of the system. For the most part, these functions are carried out by referencing other subroutines (for example, FILEM, DRAND, TMST, ERROR, and COLCT). These subroutines need not be provided by the user—they are included as part of the GASP system.

Similarly subroutine ENDSM updates the statistics on the number of customers in the system and the status of the server. It also references subroutine ENDSV at the end of service. Subroutine ENDSV then computes statistics on the time spent in the system, time spent in the queue, number of customers in the system, and number of customers in the queue. Note that these two subroutines call various other routines (for example, ERROR, TMST, RMOVE, COLCT, HISTO, and FILEM) that are a part of the GASP system. Thus we see that most of the detailed bookkeeping is provided by the GASP system routines; the user-supplied routines essentially control the manner in which the system routines are utilized.

Figure 8.7 shows the output resulting from 500 simulated customers. Parameter 1 refers to the mean interarrival time (2.3), and parameter 2 describes the service time (mean of 1.8, standard deviation of 0.5). Also code 1 refers to the entire system and code 2 refers to the queue.

Three of the computed values are underlined for purposes of discussion and comparison. For example, the value 1.5592, which is labeled "AVERAGE NUMBER IN FILE WAS" (file 2), is actually the mean queue length. This value compares favorably with the theoretical prediction of 1.50. Similarly, near the bottom of the figure, the expected (fraction of) idle time of 0.216 compares favorably with the theoretical prediction of 0.217, and the expected waiting time of 3.46 compares properly with its corresponding theoretical value.

There are two versions of GASP in common use—GASP II and GASP IV. The latter is the more recent. The reader is referred to Pritsker and Kiviat (1969) and Pritsker (1974) for more detailed information.

```
                        **GASP SUMMARY REPORT**

            SIMULATION PROJECT NO.    1  BY   SEPULVEDA

            DATE  4/  7/ 1980                RUN NUMBER      1

    PARAMETER NO.     1      2.3000       0.0000       0.0000       0.0000
    PARAMETER NO.     2      1.8000       0.8000       2.8000       0.5000


                          **GENERATED DATA**
        CODE     MEAN      STD.DEV.      MIN.         MAX.        OBS.

          1     5.1984     4.1355       0.8000      22.2000       500
          2     3.4596     4.1133       0.0000      21.0000       500

                        **TIME GENERATED DATA**
        CODE     MEAN      STD.DEV.      MIN.         MAX.     TOTAL TIME

          1     2.3429     2.4211       0.0000      13.0000    1109.4000
          2     0.7837     0.4117       0.0000       1.0000    1109.4000

                  **GENERATED FREQUENCY DISTRIBUTIONS**
        CODE                    HISTOGRAMS

          1        74   95   86   66   37   33   22   19   13    9    7
                    6    5    5    6    3    1    1    4    4    2    2


                    FILE PRINTOUT, FILE NO.,   1

            AVERAGE NUMBER IN FILE WAS,      0.8738
            STD. DEV.                        0.6025
            MAXIMUM                             5

                    THE FILE IS EMPTY


                    FILE PRINTOUT, FILE NO.,   2

            AVERAGE NUMBER IN FILE WAS,      1.5592
            STD. DEV.                        2.2400
            MAXIMUM                            12

                    THE FILE IS EMPTY

                         SIMULATED VALUE       THEORETICAL VALUE

    EXPECTED IDLE TIME          0.216                 0.217
    EXPECTED WAITING TIME       3.460                 3.490
    EXPECTED TIME IN SYSTEM     5.198                 5.290
    EXPECTED ARRIVAL TIME       2.219                 2.300
    EXPECTED SERVICE TIME       1.739                 1.800
```

Figure 8.7

8.4 SLAM

SLAM (Simulation Language for Alternative Modeling) is a relatively
new, comprehensive simulation language that can be used to simulate
discrete systems, continuous systems, or combinations thereof. Moreover, a discrete system can be described in terms of a model that is either
process oriented or *event oriented*.

In process-oriented models a description is provided of the process
through which the various system entities flow. (Note the similarity to

Figure 8.8

GPSS.) Models of this type are represented by a process-oriented flowchart called a *network structure*. Each SLAM network is comprised of specialized symbols, called *nodes* and *branches*, which represent various process-oriented model elements such as queues, servers, and decision points. The manner in which these symbols are used is illustrated in the following example.

Example 8.4

Consider once again the single-channel, single-station queuing problem described in Ex. 7.1. Figure 8.8 shows a SLAM network diagram for this problem. Notice that the network diagram consists of four circular-type figures plus a small square near the center. The first circle indicates the creation of arrivals to the system; the second represents the queue, and the small square indicates an activity (that is, the server). The oval represents a statistical collection point, and the last circle indicates system departures.

Figure 8.9 shows the corresponding SLAM program. The first three lines and the last line contain control statements. The actual network statements appear on lines 4 through 9. (Note the indentation.) Each of the network statements (except END) corresponds to one of the symbols shown in Fig. 8.8.

In Figure 8.10a we see the output that is generated for this problem, based upon 5000 customers passing through the system. Some of the numerical results have been underlined for clarity and for emphasis. Notice the average queue length of 1.635 customers, the average waiting time of 3.80 min, and the average utilization of 0.773 (corresponding to an average fractional idle time of 0.227). These values compare favorably with the theoretical predictions given in Ex. 7.4. Also, notice the value given for the average time in the system, 5.59 minutes (3.80 minutes waiting time + 1.80 minutes service time).

Figure 8.10b shows a histogram for the total time in the system. This histogram was generated in response to the COLCT statement in line 7 of Fig. 8.9.

```
 1    GEN,B.S.GOTTFRIED,SAMPLE SCSS QUEUE,4,28,82,1;
 2    LIMITS,1,1,100;
 3    NETWORK;
 4         CREATE,EXPON(2.3),,,1;
 5         QUEUE(1);
 6         ACTIVITY/1,RNORM(1.8,0.5);
 7         COLCT,INT(1),TIME IN SYSTEM,48/0.5/0.5;
 8         TERM,5000;
 9         END;
10    FIN;
```

Figure 8.9

```
                    S L A M   S U M M A R Y   R E P O R T

        SIMULATION PROJECT SAMPLE SCSS QUEUE           BY B.S.GOTTFRIED

        DATE  4/28/1982                                RUN NUMBER    1 OF    1


        CURRENT TIME   0.1162E+05
        STATISTICAL ARRAYS CLEARED AT TIME   0.0000E+00
```

STATISTICS FOR VARIABLES BASED ON OBSERVATION

	MEAN VALUE	STANDARD DEVIATION	COEFF. OF VARIATION	MINIMUM VALUE	MAXIMUM VALUE	NUMBER OF OBSERVATIONS
TIME IN SYSTEM	0.5594E+01	0.4847E+01	0.8665E+00	0.1775E+00	0.3180E+02	5000

FILE STATISTICS

FILE NUMBER	ASSOCIATED NODE TYPE	AVERAGE LENGTH	STANDARD DEVIATION	MAXIMUM LENGTH	CURRENT LENGTH	AVERAGE WAITING TIME
1	QUEUE	1.6350	2.5191	17	0	3.7977
2		1.7732	0.4188	3	2	1.4838

SERVICE ACTIVITY STATISTICS

ACTIVITY INDEX	START NODE LABEL/TYPE	SERVER CAPACITY	AVERAGE UTILIZATION	STANDARD DEVIATION	CURRENT UTILIZATION	AVERAGE BLOCKAGE	MAXIMUM IDLE TIME/SERVERS	MAXIMUM BUSY TIME/SERVERS	ENTITY COUNT
1	QUEUE	1	0.7732	0.4188	1	0.0000	22.0811	187.7285	5000

(a)

Figure 8.10

```
                              **HISTOGRAM NUMBER  1**

                                  TIME IN SYSTEM

  OBSV      RELA      CUML         UPPER
  FREQ      FREQ      FREQ     CELL LIMIT   0         20        40        60        80       100
                                            +    +    +    +    +    +    +    +    +    +    +
      4     .001      .001     0.5000E+00   +                                                  +
     74     .015      .016     0.1000E+01   +*                                                 +
    265     .053      .069     0.1500E+01   +***                                               +
    558     .112      .180     0.2000E+01   +******   C                                        +
    541     .108      .288     0.2500E+01   +*****         C                                   +
    396     .079      .368     0.3000E+01   +****              C                               +
    371     .074      .442     0.3500E+01   +****                  C                           +
    302     .060      .502     0.4000E+01   +***                      C                        +
    293     .059      .561     0.4500E+01   +***                         C                     +
    256     .051      .612     0.5000E+01   +***                           C                   +
    209     .042      .654     0.5500E+01   +**                              C                 +
    173     .035      .688     0.6000E+01   +**                               C                +
    175     .035      .723     0.6500E+01   +**                                 C              +
    140     .028      .751     0.7000E+01   +*                                   C             +
    125     .025      .776     0.7500E+01   +*                                     C           +
    111     .022      .799     0.8000E+01   +*                                      C          +
     94     .019      .817     0.8500E+01   +*                                       C         +
     97     .019      .837     0.9000E+01   +*                                        C        +
     59     .012      .849     0.9500E+01   +*                                         C       +
     54     .011      .859     0.1000E+02   +*                                          C      +
     56     .011      .871     0.1050E+02   +*                                          C      +
     48     .010      .880     0.1100E+02   +                                           C      +
     36     .007      .887     0.1150E+02   +                                            C     +
     42     .008      .896     0.1200E+02   +                                            C     +
     42     .008      .904     0.1250E+02   +                                             C    +
     40     .008      .912     0.1300E+02   +                                             C    +
     32     .006      .919     0.1350E+02   +                                             C    +
     52     .010      .929     0.1400E+02   +*                                            C    +
     28     .006      .935     0.1450E+02   +                                              C   +
     34     .007      .941     0.1500E+02   +                                              C   +
     30     .006      .947     0.1550E+02   +                                              C   +
     25     .005      .952     0.1600E+02   +                                               C  +
     20     .004      .956     0.1650E+02   +                                               C  +
     18     .004      .960     0.1700E+02   +                                               C  +
     16     .003      .963     0.1750E+02   +                                               C  +
      9     .002      .965     0.1800E+02   +                                               C  +
      8     .002      .967     0.1850E+02   +                                               C  +
     14     .003      .969     0.1900E+02   +                                               C+
     11     .002      .972     0.1950E+02   +                                               C+
     17     .003      .975     0.2000E+02   +                                               C+
      6     .001      .976     0.2050E+02   +                                               C+
     11     .002      .978     0.2100E+02   +                                               C+
     15     .003      .981     0.2150E+02   +                                               C+
      9     .002      .983     0.2200E+02   +                                               C+
     11     .002      .985     0.2250E+02   +                                               C+
      5     .001      .986     0.2300E+02   +                                               C+
      9     .002      .988     0.2350E+02   +                                               C+
      6     .001      .989     0.2400E+02   +                                               C+
      3     .001      .990     0.2450E+02   +                                                C
     50     .010     1.000        INF       +*                                               C
    ---                                      +    +    +    +    +    +    +    +    +    +    +
   5000                                      0         20        40        60        80       100
```

(b)

Figure 8.10 *(continued)*

Event-oriented models describe changes in state that occur at the time of each event. The changes associated with each type of event are described by user-supplied FORTRAN subroutines. SLAM includes its own utility routines for scheduling events, generating random numbers, manipulating files, collecting pertinent statistics, etc. An executive control program, which advances time and initiates calls to other subroutines, is also included. (Note the similarity to GASP.)

Example 8.5

Figures 8.11a through f contain the user-supplied portion of an event-based SLAM program for our familiar single-channel, single-station queue (see Ex. 7.1). Figure 8.11a defines a number of required variables and arrays and then calls the SLAM executive routine. Subroutine INTLC, shown in Fig. 8.11b, initializes the model, whereas subroutine EVENT, presented in Fig. 8.11c, transforms each event into a call to the appropriate subroutine. Subroutine ARVL, in Fig. 8.11d, contains the logic for each new customer arrival, and subroutine ENDSV, in Fig. 8.11e, presents the end-of-service logic for each new customer. Finally, in Fig. 8.11f we see the required SLAM control statements. The remaining subroutines referenced (that is, SLAM, SCHDL, FILEM, COLCT, and RMOVE), are provided as a part of the SLAM language.

Figure 8.12a shows the output that is generated by this program. The underlined values indicate an average queue length of 1.48 customers, an average waiting time of 4.37 minutes, an average time in the system of 5.22 minutes and an average utilization of 0.782 (corresponding to an average fractional idle time of 0.218). These values compare favorably with the theoretical predictions given in Ex. 7.4.

```
      DIMENSION NSET(5000)
      COMMON/SCOM1/ATRIB(100),DD(100),DDL(100),DTNOW,II,MFA,MSTOP,NCLNR,
     1 NCRDR,NPRNT,NNRUN,NNSET,NTAPE,SS(100),SSL(100),TNEXT,TNOW,XX(100)
      COMMON QSET(5000)
      EQUIVALENCE(NSET(1),QSET(1))
      NCRDR=5
      NPRNT=6
      NTAPE=7
      NNSET=5000
      CALL SLAM
      STOP
      END
```

(a)

```
      SUBROUTINE INTLC
      COMMON/SCOM1/ATRIB(100),DD(100),DDL(100),DTNOW,II,MFA,MSTOP,NCLNR,
     1 NCRDR,NPRNT,NNRUN,NNSET,NTAPE,SS(100),SSL(100),TNEXT,TNOW,XX(100)
      EQUIVALENCE(XX(1),BUSY)
      BUSY=0.
      CALL SCHDL(1,0.,ATRIB)
      RETURN
      END
```

(b)

```
      SUBROUTINE EVENT(I)
      GO TO (1,2),I
    1 CALL ARVL
      RETURN
    2 CALL ENDSV
      RETURN
      END
```

(c)

Figure 8.11

```
      SUBROUTINE ARVL
      COMMON/SCOM1/ATRIB(100),DD(100),DDL(100),DTNOW,II,MFA,MSTOP,NCLNR,
     1 NCRDR,NPRNT,NNRUN,NNSET,NTAPE,SS(100),SSL(100),TNEXT,TNOW,XX(100)
      EQUIVALENCE(XX(1),BUSY)
      CALL SCHDL(1,EXPON(2.3,1),ATRIB)
      ATRIB(1)=TNOW
      IF (BUSY.EQ.0.) GO TO 10
      CALL FILEM(1,ATRIB)
      RETURN
   10 BUSY=1.
      CALL SCHDL(2,RNORM(1.8,0.5,2),ATRIB)
      RETURN
      END
```

(d)

```
      SUBROUTINE ENDSV
      COMMON/SCOM1/ATRIB(100),DD(100),DDL(100),DTNOW,II,MFA,MSTOP,NCLNR,
     1 NCRDR,NPRNT,NNRUN,NNSET,NTAPE,SS(100),SSL(100),TNEXT,TNOW,XX(100)
      EQUIVALENCE(XX(1),BUSY)
      TSYS=TNOW-ATRIB(1)
      CALL COLCT(TSYS,1)
      IF (NNQ(1).GT.0) GO TO 10
      BUSY=0.
      RETURN
   10 CALL RMOVE(1,1,ATRIB)
      CALL SCHDL(2,RNORM(1.8,0.5,2),ATRIB)
      RETURN
      END
```

(e)

```
    1   GEN,B,S,GOTTFRIED,SCSS QUEUE,8/10/81,1;
    2   LIMITS,5,2,200;
    3   STAT,1,TIME IN SYSTEM,50/,5/,5;
    4   TIMST,XX(1),UTILIZATION;
    5   INIT,0,600000;
    6   FIN;
```

(f)

Figure 8.11 *(continued)*

A user-defined histogram of the total time in the system is shown in Fig. 8.12b. This histogram was generated in response to the STAT statement in line 3 of Fig. 8.11f.

SLAM can also simulate other types of discrete systems, including systems that are based upon combined process-oriented and event-oriented viewpoints. In addition, SLAM can be used to simulate continuous systems, or combined continuous-discrete systems. A discussion of these features is beyond the scope of the present text. The reader is referred to Pritsker and Pegden (1979) for more information about SLAM.

```
                            S L A M   S U M M A R Y   R E P O R T

        SIMULATION PROJECT SCSS QUEUE                    BY B.S.GOTTFRIED

        DATE  8/10/1981                                  RUN NUMBER   1 OF    1


        CURRENT TIME   0.6000E+05
        STATISTICAL ARRAYS CLEARED AT TIME   0.0000E+00


        **STATISTICS FOR VARIABLES BASED ON OBSERVATION**

                        MEAN      STANDARD    COEFF. OF   MINIMUM     MAXIMUM     NUMBER OF
                        VALUE     DEVIATION   VARIATION   VALUE       VALUE       OBSERVATIONS

TIME IN SYSTEM      0.5222E+01    0.4008E+01  0.7676E+00  0.0000E+00  0.3447E+02     25973


        **STATISTICS FOR TIME-PERSISTENT VARIABLES**

                        MEAN      STANDARD    MINIMUM     MAXIMUM     TIME        CURRENT
                        VALUE     DEVIATION   VALUE       VALUE       INTERVAL    VALUE

UTILIZATION         0.7818E+00    0.4130E+00  0.0000E+00  0.1000E+01  0.6000E+05  0.1000E+01


        **FILE STATISTICS**

FILE    ASSOCIATED   AVERAGE    STANDARD    MAXIMUM   CURRENT   AVERAGE
NUMBER  NODE TYPE    LENGTH     DEVIATION   LENGTH    LENGTH    WAITING TIME

   1                 1.4785     2.0905        17        0       4.3682
   2                 0.0000     0.0000         0        0       0.0000
   3                 0.0000     0.0000         0        0       0.0000
   4                 0.0000     0.0000         0        0       0.0000
   5                 0.0000     0.0000         0        0       0.0000
   6                 1.7818     0.4130         2        2       2.0580
```

(a)

Figure 8.12

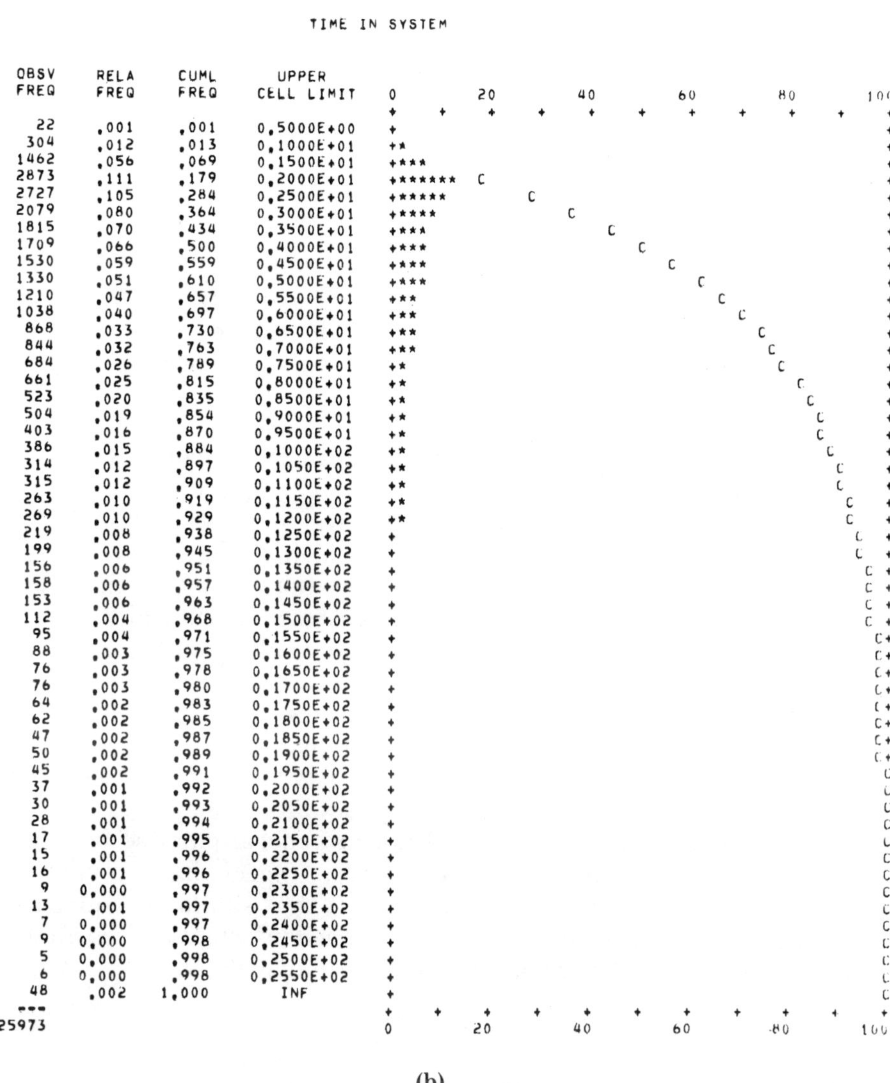

OBSV FREQ	RELA FREQ	CUML FREQ	UPPER CELL LIMIT
22	.001	.001	0.5000E+00
304	.012	.013	0.1000E+01
1462	.056	.069	0.1500E+01
2873	.111	.179	0.2000E+01
2727	.105	.284	0.2500E+01
2079	.080	.364	0.3000E+01
1815	.070	.434	0.3500E+01
1709	.066	.500	0.4000E+01
1530	.059	.559	0.4500E+01
1330	.051	.610	0.5000E+01
1210	.047	.657	0.5500E+01
1038	.040	.697	0.6000E+01
868	.033	.730	0.6500E+01
844	.032	.763	0.7000E+01
684	.026	.789	0.7500E+01
661	.025	.815	0.8000E+01
523	.020	.835	0.8500E+01
504	.019	.854	0.9000E+01
403	.016	.870	0.9500E+01
386	.015	.884	0.1000E+02
314	.012	.897	0.1050E+02
315	.012	.909	0.1100E+02
263	.010	.919	0.1150E+02
269	.010	.929	0.1200E+02
219	.008	.938	0.1250E+02
199	.008	.945	0.1300E+02
156	.006	.951	0.1350E+02
158	.006	.957	0.1400E+02
153	.006	.963	0.1450E+02
112	.004	.968	0.1500E+02
95	.004	.971	0.1550E+02
88	.003	.975	0.1600E+02
76	.003	.978	0.1650E+02
76	.003	.980	0.1700E+02
64	.002	.983	0.1750E+02
62	.002	.985	0.1800E+02
47	.002	.987	0.1850E+02
50	.002	.989	0.1900E+02
45	.002	.991	0.1950E+02
37	.001	.992	0.2000E+02
30	.001	.993	0.2050E+02
28	.001	.994	0.2100E+02
17	.001	.995	0.2150E+02
15	.001	.996	0.2200E+02
16	.001	.996	0.2250E+02
9	0.000	.997	0.2300E+02
13	.001	.997	0.2350E+02
7	0.000	.997	0.2400E+02
9	0.000	.998	0.2450E+02
5	0.000	.998	0.2500E+02
6	0.000	.998	0.2550E+02
48	.002	1.000	INF
25973			

(b)

Figure 8.12 *(continued)*

8.5 OTHER SIMULATION LANGUAGES

A number of other special-purpose simulation languages have been developed over the past several years, some of which have had a reasonably widespread user acceptance. These include SIMSCRIPT, Q-GERT, and SIMPL/1. SIMSCRIPT is perhaps the best known of this group. This is a comprehensive programming language that encompasses the following five distinct levels:

1. A simple, easy-to-use instruction set intended for teaching purposes.
2. A general-purpose instruction set comparable to FORTRAN IV.
3. A more comprehensive general-purpose instruction set comparable to ALGOL or PL/1.
4. A set of instructions that facilitate the representation of discrete-event simulation models in terms of sets, entities, and attributes.
5. A specialized set of instructions that will provide for time advance, allow the scheduling of events, and accumulate pertinent system statistics.

Because of its broad-based structure, SIMSCRIPT can be used as a general-purpose language, though its primary area of application is discrete-event simulation (Kiviat, Villanueva, and Markowitz, 1969).

Q-GERT is a predecessor of SLAM which also utilizes a process-oriented network approach to the representation of simulation models. The language permits models to be expressed graphically, with a capability for carrying out direct computer analysis on the graphical representation of the model. The overall characteristics of Q-GERT and SLAM are very similar, though there are some significant differences between the two languages on a more detailed level. The reader who would like more information about Q-GERT is referred to a recent text by Pritsker (1977).

SIMPL/1 is an extension of PL/1 (a general-purpose language similar to ALGOL), which includes a number of specialized, simulation-oriented macro statements. In concept it is thus similar to SIMULA, though the languages differ in detail. Detailed information is available from IBM (SIMPL/1 Program Reference Manual 1972).

8.6 CHOICE OF A SIMULATION LANGUAGE

The decision to use a special-purpose simulation language or a general-purpose language (for example, FORTRAN) is usually a trade-off between the time and the effort required to learn the new language, on the one hand, and the savings that results from the eventual use of that language, on the other. As a rule the special-purpose simulation language is preferable when simulation studies are carried out frequently, or if a particularly large or complex system is being simulated.

The choice of a particular simulation language is often based simply upon availability. Unfortunately there has been no standardization of simulation languages, and no single language has become dominant. The more commonly used languages tend to be the oldest and the least flexible. Thus some careful considerations should be given to the use of one of the newer, more powerful languages (for example, SLAM) if the user is able to have some effect upon this decision.

PROBLEMS

8.1 Determine which special-purpose simulation language is available at your particular school or office. Obtain a programmer's reference manual for that language (if available), a detailed textbook, and any local write-ups that your computer center may provide. Familiarize yourself with the general contents of these documents, in preparation for writing programs with that language.

8.2 Solve the following problems using whatever special-purpose simulation language is available at your particular school or office:
  (a) The facility utilization problem described in Ex. 5.1 (see Prob. 5.1).
  (b) The inventory control problem described in Ex. 5.3 (see Prob. 5.9).
  (c) The PERT problem described in Ex. 5.5 (see Prob. 5.14).
  (d) The risk analysis problem described in Ex. 5.7 (see Prob. 5.15).
  (e) The blood bank problem described in Prob. 5.17.
  (f) The bank problem described in Prob. 7.7.
  (g) The barber shop problem described in Prob. 7.11.
  (h) The job shop problem described in Probs. 7.13 and 7.14.
  (i) The job shop problem described in Prob. 7.20.
  (j) The computer center problem described in Prob. 7.21.
  (k) The barber shop problem described in Prob. 7.24.
  (l) The bank problem described in Probs. 7.25 and 7.26.

BIBLIOGRAPHY

BIRTWISTLE, G. M., O-J. DAHL, B. MYHRHAUG, AND K. NYGAARD, *Simula Begin* (2nd ed.), New York: Van Nostrand Reinhold, 1979.

BOBILLIER, P. A., B. C. KAHAN, AND A. R. PROBST, *Simulation with GPSS and GPSS V*, Englewood Cliffs, N.J.: Prentice-Hall, 1976.

CONWAY, R. W., "Some Tactical Problems in Digital Simulation", *Management Sci.*, Vol. 10, No. 1 (October 1963), pp. 47–61.

COOPER, R. B., *Introduction to Queuing Theory*, New York: Macmillan, 1972.

DEGROOT, M. H., *Probability and Statistics*, Reading, Mass.: Addison–Wesley, 1975.

EMSHOFF, J. R., AND R. L. SISSON, *Design and Use of Computer Simulation Models*, New York: Macmillan, 1970.

FISHMAN, G. S., *Concepts and Methods in Discrete Event Digital Simulation*, New York: John Wiley, 1973.

FISHMAN, G. S., *Principles of Discrete Event Digital Simulation*, New York: John Wiley, 1978.

GORDON, G., *Application of GPSS V to Discrete System Simulation*, Englewood Cliffs, N.J.: Prentice-Hall, 1975.

GOTTFRIED, B. S., *Programming with Fortran IV*, Englewood Cliffs, N.J.: Prentice-Hall, 1982.

GOTTFRIED, B. S., *Programming with Basic*, (2nd ed.), Schaum's Outline Series, New York: McGraw–Hill, 1982.

GOTTFRIED, B. S., *Engineering Calculations for First-Year Students*, Schaum's Outline Series, New York: McGraw–Hill, 1979.

HADLEY, G. J., AND T. M. WHITIN, *Analysis of Inventory Systems*, Englewood Cliffs, N.J.: Prentice-Hall, 1963.

HILLIER, F., AND G. LIEBERMAN, *Introduction to Operations Research*, (3rd ed.), San Francisco: Holden-Day, 1980.

KIVIAT, P. J., R. VILLANUEVA, AND H. MARKOWITZ, *The SIMSCRIPT II Programming Language*, Englewood Cliffs, N.J.: Prentice-Hall, 1969.

KNUTH, D. E., *The Art of Computer Programming* (Semi-numerical Algorithms) (2nd ed.), Chap. 3, Vol. 2, pp. 1–160, Reading, Mass.: Addison-Wesley, 1981.

LAW, A. M., AND W. D. KELTON, *Simulation Modeling and Analysis*, New York: McGraw-Hill, 1982.

MAISEL, H., AND G. GNUGNOLI, *Simulation of Discrete Stochastic Systems*, Chicago: Science Research Associates, 1972.

NAYLOR, T. H., J. L. BALINTFY, D. S. BURDICK, AND K. CHU, *Computer Simulation Techniques*, New York: John Wiley, 1966.

NAYLOR, T. H., *Computer Simulation Experiments with Models of Economic Systems*, New York: John Wiley, 1971.

PRITSKER, A. A. B., *The GASP IV Simulation Language*, New York: John Wiley, 1974.

PRITSKER, A. A. B., *Modeling and Analysis Using Q-GERT Networks*, New York: Halsted Press (John Wiley), 1977.

PRITSKER, A. A. B., AND P. J. KIVIAT, *Simulation with GASP II*, Englewood Cliffs, N.J.: Prentice-Hall, 1969.

PRITSKER, A. A. B., AND C. D. PEGDEN, *Introduction to Simulation and SLAM*, New York: Halsted Press (John Wiley), 1979.

THE RAND CORP., *One Million Random Digits*, Santa Monica, Ca., 1955.

SCHMIDT, J. W., AND R. E. TAYLOR, *Simulation and Analysis of Industrial Systems*, Homewood, Ill.: Richard D. Irwin, 1970.

SCHRIBER, T. J., *Simulation Using GPSS*, New York: John Wiley, 1974.

SHANNON, R. E., *Systems Simulation—The Art and Science*, Englewood Cliffs, N.J.: Prentice-Hall, 1975.

SIMPL/1 Program Reference Manual, SH19-5038-0, New York: IBM Corp., 1972.

WIEST, J. D., AND F. K. LEVY, *A Management Guide to PERT/CPM* (2nd ed.), Englewood Cliffs, N.J.: Prentice-Hall, 1977.

APPENDIX A

SELECTED FORTRAN PROGRAMS AND SUBPROGRAMS

A.1 FACILITY UTILIZATION PROBLEM

```
C**************************************************************************************
C        S A M P L E   F A C I L I T Y   U T I L I Z A T I O N   P R O B L E M   *
C                                                                                  *
C IDENTIFICATION OF VARIABLES:                                                     *
C      A        = ARRIVAL TIME OF THE ITH ORDER                                    *
C      D        = DEPARTURE TIME OF THE ITH ORDER                                  *
C      AT       = TIME INTERVAL BETWEEN ARRIVAL OF SUCCESSIVE ORDERS               *
C                    (EXPONENTIALLY DISTRIBUTED RANDOM VARIATE)                    *
C      PT       = TIME REQUIRED TO PROCESS THE ITH ORDER                          *
C                    (NORMALLY DISTRIBUTED RANDOM VARIATE)                         *
C      IT       = TIME FACILITY IS IDLE BEFORE PROCESSING THE ITH ORDER           *
C      TOTIT    = CUMULATIVE IDLE TIME                                             *
C      FRACT    = FRACTION OF TIME THE FACILITY IS IN USE (OUTPUT ARRAY)          *
C      FAVG     = MEAN VALUE OF COMPUTED FRACT ARRAY (OUTPUT PARAMETER)           *
C      FSTDEV   = STANDARD DEVIATION OF COMPUTED FRACT ARRAY (OUTPUT PARAMETER)   *
C      XBAR     = MEAN VALUE OF TIME INTERVAL BETWEEN ORDERS (INPUT PARAMETER)    *
C      XAVG     = MEAN VALUE OF ORDER PROCESS TIME (INPUT PARAMETER)              *
C      STDEV    = STANDARD DEVIATION OF ORDER PROCESS TIME (INPUT PARAMETER)      *
C      NLOOP    = NUMBER OF CONSECUTIVE SIMULATIONS (INPUT PARAMETER)             *
C      KX       = SEED FOR RANDOM NUMBER GENERATOR (INPUT PARAMETER)              *
C      M        = NUMBER OF INTERVALS IN OUTPUT ARRAY (INPUT PARAMETER)           *
C      N        = NUMBER OF ORDERS PROCESSED PER SIMULATION                       *
C**************************************************************************************
      REAL IT
      DIMENSION FRACT(1000),BOUNDS(26),F(25),Y(25)
  100 FORMAT(3F4.1)
  200 FORMAT(I12,3I4)
  300 FORMAT('1',18X,'S A M P L E   F A C I L I T Y   U T I L I Z A T I
     10 N   P R O B L E M'////,11X,'EXPONENTIALLY DISTRIBUTED TIME INTERV
     2AL BETWEEN SUCCESSIVE ORDERS:   MEAN VALUE = ',F3.1,' HOURS'//,
     311X,'NORMALLY DISTRIBUTED PROCESS TIMES:   MEAN VALUE = ',F3.1,' HO
     4URS'/,48X,'STANDARD DEVIATION = ',F3.1,' HOURS'//)
```

A.1 Continued

```
  400 FORMAT('0',10X,'SEED (RANDOM NUMBER GENERATOR) = ',I12/,11X,'NUMBE
     1R OF CONSECUTIVE SIMULATIONS = ',I4/,11X,'NUMBER OF INTERVALS IN T
     2HE OUTPUT ARRAY = ',I4/,11X,'NUMBER OF ORDERS PROCESSED PER SIMULA
     3TION = ',I4//)
  500 FORMAT('0',10X,'SYSTEM PERFORMANCE CRITERION (FRACTION OF TIME THE
     1 FACILITY IS IN USE):'/)
C READ INPUT DATA
      READ (5,100) XBAR,XAVG,STDEV
      READ (5,200) KX,NLOOP,M,N
C ESTABLISH INTERVAL BOUNDS FOR OUTPUT ARRAY
      BOUNDS(1)=0.
      DO 1 J=1,M
    1 BOUNDS(J+1)=FLOAT(J)/M
C SET UP RANDOM NUMBER GENERATOR
      DUMMY=RAND(KX)
C CARRY OUT M SUCCESSIVE SIMULATIONS
      DO 4 K=1,NLOOP
C INITIALIZE VARIABLES FOR EACH PASS THROUGH THE LOOP
      I=1
      A=0.
      TOTIT=0.
C PROCESS FIRST ORDER (I=1)
      CALL NORMAL(XAVG,STDEV,PT)
      IF (PT.LT.0.) PT=0.
      D=PT
C PROCESS NEXT ORDER (I=2,3,...,N)
    2 I=I+1
    3 CALL EXPON(XBAR,0.,AT)
      A=A+AT
C HAS NEW ORDER ARRIVED BEFORE COMPLETION OF PREVIOUS ORDER?
      IF (A.LT.D) GO TO 3
      DOLD=D
      CALL NORMAL(XAVG,STDEV,PT)
      IF (PT.LT.0.) PT=0.
      D=A+PT
      IT=A-DOLD
C TEST FOR NEGATIVE IDLE TIME (NUMERICAL ERROR)
      IF (IT.LT.0.) IT=0.
      TOTIT=TOTIT+IT
C TEST FOR COMPLETION OF SIMULATION
      IF (I.LT.N) GO TO 2
      FRACT(K)=1.-TOTIT/D
    4 CONTINUE
C GENERATE STATISTICAL OUTPUT DATA
      CALL GROUP(FRACT,BOUNDS,F,Y,M,NLOOP)
      CALL PARAMS(FRACT,FAVG,FSTDEV,NLOOP)
C WRITE OUTPUT DATA
      WRITE (6,300) XBAR,XAVG,STDEV
      WRITE (6,400) KX,NLOOP,M,N
      WRITE (6,500)
      CALL OUTPUT(BOUNDS,F,Y,FAVG,FSTDEV,M,NLOOP)
      STOP
      END
```

A.2 PERT

```
C************************************************************************
C          P E R T   (PROJECT EVALUATION AND REVIEW TECHNIQUE)          *
C                                                                       *
C          THIS PROGRAM IS DESIGNED TO SOLVE THE PERT (PROJECT-         *
C          EVALUATION AND REVIEW TECHNIQUE) PROBLEM, BY DETERMINING ITS *
C          CHARACTERISTIC AS FOLLOWS:                                   *
C               1.) MOST LIKELY CRITICAL PATH                           *
C                                                                       *
C               2.) THE LIKELYHOOD OF AN ACTIVITY FALLING ON A CRITICAL *
C                      PATH.                                            *
C                                                                       *
C               3.) THE EXPECTED VALUE OF THE PROJECT COMPLETION        *
C                      TIME.                                            *
C                                                                       *
C               4.) DISTRIBUTION OF PROJECT COMPLETION TIMES.           *
C                                                                       *
C          VARIABLE LIST:                                               *
C               T(K)      - TIME TO PERFORM THE K-TH ACTIVITY           *
C               I(K)      - ORIGIN OF THE K-TH ACTIVITY                 *
C               J(K)      - DESTINATION OF THE K-TH ACTIVITY            *
C               ES(K)     - EARLIEST STARTING TIME OF THE K-TH ACTIVITY *
C               EF(K)     - EARLIEST FINISH TIME OF THE K-TH ACTIVITY   *
C               ALS(K)    - LATEST STARTING TIME OF THE K-TH ACTIVITY   *
C               ALF(K)    - LATEST FINISH TIME OF THE K-TH ACTIVITY     *
C               AMEAN(K)- THE AVERAGE TIME TO COMPLETE THE K-TH ACTIVITY*
C               STAN(K) - THE STANDARD DEVIATION OF THE K-TH ACTIVITY   *
C               S(K)      - SLACK TIME AT NODE K                        *
C               ATOT(I) - EARLIEST FINISH TIME OF THE SINK FOR CRITICAL *
C                      PATH I                                           *
C               IDATA(I,J)- ARRAY OF CRITICAL PATHS                     *
C               NSTORE(I)- THE NUMBER OF TIMES THAT THE I-TH CRITICAL   *
C                      PATH OCCURS                                      *
C               NACT(I) - THE NUMBER OF. THAT THE I-TH ACTIVITY OCCURS  *
C                      ON THE CRITIVAL PATH                             *
C               N1        - THE NUMBER OF CRITICAL PATHS TO BE GENERATED *
C               M1        - THE NUMBER OF INTERVAL, FOR CUMULATIVE DIST. *
C               M         - THE NUMBER OF ACTIVITIES                    *
C               N         - THE NUMBER OF NODES                         *
C               AMAX      - MAXIMUM VALUE OF THE EF'S FOR A NODE        *
C               TOT       - MAXIMUM VALUE OF THE EF'S FOR THE SINK      *
C               AMIN      - MINIMUM VALUE OF THE ALS' FOR A NODE        *
C               BMAX      - MAXIMUM VALUE OF THE EF TIMES OF ALL        *
C                      CRITICAL PATHS                                   *
C               BMIN      - MINIMUM VALUE OF THE EF TIMES OF ALL        *
C                      CRITICAL PATHS                                   *
C               MAX       - MAXIMUM VALUE OF THE NSTORE ARRAY           *
C                                                                       *
C                                                                       *
C************************************************************************
          DIMENSION T(12),I(12),J(12),ES(12),EF(12),ALS(12),ALF(12)
          DIMENSION STAN(12),S(12),AMEAN(12),ATOT(100),F(25),Y(25)
          DIMENSION IDATA(50,50),NSTORE(50),NACT(20),BOUNDS(26)
          COMMON IDATA,NSTORE,NACT
100       FORMAT(2(I2,1X))
105       FORMAT(2(I1,1X),2(F3.0,1X))
200       FORMAT(2X,I4,3X,I4,3X,I4)
350       FORMAT(2(I3))
400       FORMAT('1',62X,'P E R T',/,63X,7('+'),//,46X,
     1    '(PROJECT EVALUATION AND REVIEW TECHNIQUE)'/)
410       FORMAT(53X,'NUMBER OF SIMULATIONS =',1X,I3/)
420       FORMAT(53X,'NUMBER OF ACTIVITIES =',1X,I2/)
430       FORMAT(53X,'NUMBER OF NODES =',1X,I2/)
440       FORMAT(48X,'ACTIVITY TIMES ARE NORMALLY DISTRIBUTED'/////)
450       FORMAT(54X,'* INPUT DATA SECTION *',/,56X,18('+')/)
460       FORMAT(33X,'ACTIVITY',6X,'ORIGIN',6X,'DESTINATION',
     1    8X,'MEAN',6X,'ST. DEV.'/)
470       FORMAT(36X,I2,T50,I2,T64,I2,T79,F3.0,T91,F3.0/)
480       FORMAT(////132('+')////)
```

A.2 Continued

```
490      FORMAT(43X,'***   S I M U L A T I O N   R E S U L T S   ***'/)
500      FORMAT(52X,'CRITICAL PATH DETERMINATION',/,52X,27('+')/)
510      FORMAT(50X,'FREQUENCY',10X,'CRITICAL PATH'/)
520      FORMAT(53X,I2,12X,5(I2,2X)/)
530      FORMAT(/42X,'MOST LIKELY CRITICAL PATH IS',5(2X,I2)/)
540      FORMAT(///46X,'ACTIVITIES OCCURRING ON THE CRITICAL PATH',/,
     1 46X,41('+')/)
545      FORMAT(51X,'ACTIVITY',12X,'OCCURANCE'/)
550      FORMAT(////43X,'***   S T A T I S T I C A L   A N A L Y S I S
     1   ***'/)
560      FORMAT(52X,'DISTRIBUTION OF ACTIVITY TIMES',/,52X,30('+')/)
565      FORMAT(53X,I2,19X,I3)
         WRITE(6,400)
C    INITIALIZE RANDOM NUMBER GENERATOR
         CALL SETRAN(222233339)
C    READ IN THE NUMBER OF DATA POINTS, NODES, ACTIVITIES AND SIMULATIONS
         READ(5,350) N1,M1
         READ(5,100)M,N
         WRITE(6,410)N1
         WRITE(6,420)M
         WRITE(6,430)N
         WRITE(6,440)
         WRITE(6,450)
         WRITE(6,460)
C    READ IN THE ORIGIN, DESTINATION, MEAN AND STANDARD DEVIATION FOR EACH ACTIVI
         DO 1 K=1,M
         READ(5,105)I(K),J(K),AMEAN(K),STAN(K)
         WRITE(6,470)K,I(K),J(K),AMEAN(K),STAN(K)
C    INITIALIZE NACT ARRAY
         NACT(K)=0
1        CONTINUE
         KOUNT=1
C    INITIALIZE COUNTER AND NSTORE ARRAY
         DO 2 II=1,50
         NSTORE(II)=1
2        CONTINUE
C    WRITE HEADINGS
         WRITE(6,480)
         WRITE(6,490)
         WRITE(6,500)
         WRITE(6,510)
C    START SIMULATION TO GENERATE CRITICAL PATHS
         DO 90 III=1,N1
         KOUNT=KOUNT+1
         LIT=0
         TOT=0.0
C    START FORWARD PASS
         DO 10 K=1,M
         IDATA(KOUNT,K)=0
3        CALL NORMAL(AMEAN(K),STAN(K),X)
         IF(X.LT.0.0) GO TO 3
         T(K)=X
         IF(I(K).EQ.1)GO TO 30
         AMAX=0.0
C    CALCULATE LARGEST EF FOR PRECEEDING ACTIVITY
         DO 20 L=1,K-1
         IF(J(L).NE.I(K))GO TO 20
         IF(EF(L).GT.AMAX)AMAX=EF(L)
20       CONTINUE
         ES(K)=AMAX
         EF(K)=ES(K)+T(K)
         GO TO 40
30       ES(K)=0.0
         EF(K)=T(K)
40       CONTINUE
         IF(J(K).EQ.N)GO TO 5
         GO TO 10
```

A.2 Continued

```
5          IF(EF(K).GT.TOT)TOT=EF(K)
10         CONTINUE
           ATOT(III)=TOT
C    START BACKWARD PASS
           DO 50 K=1,M
           KK=M-K+1
           IF(J(KK).EQ.N)GO TO 60
           AMIN=10.0**5
C    FIND MINIMUM VALUE OF LATEST START AT ORIGIN I(L)
           DO 70 L=KK+1,M
           IF(I(L).NE.J(KK))GO TO 70
           IF(ALS(L).LT.AMIN)AMIN=ALS(L)
70         CONTINUE
           ALF(KK)=AMIN
           ALS(KK)=ALF(KK)-T(KK)
           GO TO 55
60         ALF(KK)=TOT
           ALS(KK)=ALF(KK)-T(KK)
55         S(KK)=ALS(KK)-ES(KK)
           DIFF=ABS(S(KK))
           IF(DIFF.LT..00001) CALL TEST(KK,KOUNT,LIT,1)
50         CONTINUE
           CALL TEST(KK,KOUNT,LIT,0)
90         CONTINUE
C    FIND MIN AND MAX VALUES OF EF AT NETWORK SINK
           BMAX=ATOT(1)
           BMIN=ATOT(1)
           DO 7 I1=2,100
           IF(ATOT(I1).GT.BMAX)BMAX=ATOT(I1)
           IF(ATOT(I1).LT.BMIN)BMIN=ATOT(I1)
7          CONTINUE
C    DEFINE BOUNDS FOR GROUP SUBROUTINE
           BOUNDS(1)=BMIN
           BOUNDS(2)=26.
           BOUNDS(3)=33.
           BOUNDS(4)=BMAX+1
           CALL PARAMS(ATOT,AVG,STDEV,N1)
           CALL GROUP(ATOT,BOUNDS,F,Y,M1,N1)
C    FIND THE MOST LIKELY CRITICAL PATH
           MAX=0
           DO 80 J1=1,KOUNT
           IF(NSTORE(J1).GT.MAX)GO TO 85
           GO TO 80
85         LL=J1
           MAX=NSTORE(J1)
80         CONTINUE
C    OUTPUT CRITICAL PATHYS AND THE NO. OF TIMES THEY OCCUR
           DO 95 II=2,KOUNT
           WRITE(6,520) NSTORE(II),(IDATA(II,JJ),JJ=1,5)
95         CONTINUE
C    OUTPUT MOST LIKELY CRITICAL PATH
           WRITE(6,530)(IDATA(LL,JJ),JJ=1,5)
           WRITE(6,540)
           WRITE(6,545)
C    OUTPUT ACTIVITIES AND THE NO. OF TIMES EACH OCCURS
           DO 97 KL=1,M
           WRITE(6,565)KL,NACT(KL)
97         CONTINUE
           WRITE(6,550)
           WRITE(6,560)
           CALL OUTPUT(BOUNDS,F,Y,AVG,STDEV,M1,N1)
           STOP
           END
C
C
C
```

A.2 Continued

```
      SUBROUTINE NORMAL(XAVG,STDEV,X)
C************************************************************************
C THIS SUBPROGRAM GENERATES A NORMALLY DISTRIBUTED RANDOM VARIATE       *
C     WITH MEAN XAVG AND STANDARD DEVIATION STDEV USING A METHOD BASED   *
C     UPON THE CENTRAL LIMIT THEOREM                                     *
C************************************************************************
      SUM=0.
      DO 1 I=1,12
    1 SUM=SUM+RAN(0)
      Z=SUM-6.
      X=XAVG+STDEV*Z
      RETURN
      END


      SUBROUTINE TEST(KK,KOUNT,LIT,LFLAG)
C************************************************************************
C                                                                      *
C          THIS SUBROUTINE COMPARES THE CRITICAL PATH MOST RECENTLY    *
C     GENERATED TO THOSE WHICH ARE PREVIOUSLY GENERATED. IF IT IS      *
C     IDENTICAL TO ONE ALREADY GENERATED IT WILL INCREASE A            *
C     COUNTER FOR THAT PATH. IF IT IS NOT IDENTICAL TO ANY OF THE      *
C     PATH THEN IT WILL BECOME A NEW PATH FOR COMPARISON BY THE NEXT   *
C     PATH GENERATED.                                                  *
C                                                                      *
C     SEE MAIN PROGRAM FOR VARIABLE LIST.                              *
C************************************************************************
      COMMON IDATA(50,50),NSTORE(50),NACT(20)
      IF(LFLAG.EQ.0) GO TO 1
      NACT(KK)=NACT(KK)+1
      LIT=LIT+1
      IDATA(KOUNT,LIT)=KK
      RETURN
    1 DO 2 I=1,KOUNT-1
      DO 3 J=1,LIT
      IF(IDATA(I,J).EQ.IDATA(KOUNT,J)) GO TO 3
      GO TO 2
    3 CONTINUE
      NSTORE(I)=NSTORE(I)+1
    4 KOUNT=KOUNT-1
      RETURN
    2 CONTINUE
      RETURN
      END
      SUBROUTINE GROUP(DATA,BOUNDS,F,Y,M,N)
C************************************************************************
C THIS SUBPROGRAM GROUPS A SET OF DATA AND THEN CALCULATES THE RELATIVE *
C     FREQUENCIES AND THE CORRESPONDING CUMULATIVE DISTRIBUTION         *
C                                                                      *
C IDENTIFICATION OF VARIABLES:                                         *
C     DATA   = INDIVIDUAL DATA POINTS (INPUT ARRAY)                    *
C     BOUNDS = INTERVAL BOUNDS (INPUT ARRAY)                           *
C     M      = NUMBER OF INTERVALS (INPUT SCALAR)                      *
C     N      = NUMBER OF DATA POINTS (INPUT SCALAR)                    *
C     K      = INTERVAL COUNTERS (INTERNAL ARRAY)                      *
C     F      = RELATIVE FREQUENCIES (OUTPUT ARRAY)                     *
C     Y      = CUMULATIVE DISTRIBUTION (OUTPUT ARRAY)                  *
C************************************************************************
      DIMENSION DATA(100),BOUNDS(26),F(25),Y(25),K(25)
C ZERO THE INTERVAL COUNTERS
      DO 1 J=1,M
    1 K(J)=0
C GROUP THE DATA
      DO 3 I=1,N
      DO 2 J=1,M
      IF (DATA(I).GE.BOUNDS(J+1)) GO TO 2
      K(J)=K(J)+1
      GO TO 3
    2 CONTINUE
    3 CONTINUE
```

A.2 Continued

```
C CALCULATE THE RELATIVE FREQUENCIES AND THE CUMULATIVE DISTRIBUTION
      F(1)=FLOAT(K(1))/N
      Y(1)=F(1)
      DO 4 J=2,M
      F(J)=FLOAT(K(J))/N
    4 Y(J)=Y(J-1)+F(J)
      RETURN
      END
      SUBROUTINE OUTPUT(BOUNDS,F,Y,AVG,STDEV,M,N)
C**********************************************************************
C THIS SUBPROGRAM PRINTS THE MEAN, THE STANDARD DEVIATION, THE RELATIVE *
C      FREQUENCIES AND THE CUMULATIVE DISTRIBUTION FOR A GIVEN DATA SET  *
C                                                                       *
C IDENTIFICATION OF VARIABLES:                                          *
C      BOUNDS = INTERVAL BOUNDS                                         *
C      F      = RELATIVE FREQUENCIES                                   *
C      Y      = CUMULATIVE DISTRIBUTION                                *
C      AVG    = MEAN                                                   *
C      STDEV  = STANDARD DEVIATION                                     *
C      M      = NUMBER OF INTERVALS                                    *
C      N      = NUMBER OF DATA POINTS                                  *
C**********************************************************************
      DIMENSION BOUNDS(26),F(25),Y(25)
  100 FORMAT('0',10X,'MEAN =',E12.5,10X,'STANDARD DEVIATION =',E12.5,
     1 10X,'NUMBER OF DATA POINTS =',I4//)
  200 FORMAT('0',14X,'INTERVAL',14X,'LOWER',14X,'UPPER',14X,'RELATIVE',
     1 14X,'CUMULATIVE'/,16X,'NUMBER',15X,'BOUND',14X,'BOUND',14X,
     2 'FREQUENCY',12X,'DISTRIBUTION'/)
  300 FORMAT(18X,I2,14X,F8.4,11X,F8.4,13X,F8.4,15X,F8.4)
      WRITE (6,100) AVG,STDEV,N
      WRITE (6,200)
      DO 1 J=1,M
      J1=J+1
    1 WRITE (6,300) J,BOUNDS(J),BOUNDS(J1),F(J),Y(J)
      RETURN
      END
      SUBROUTINE PARAMS(DATA,AVG,STDEV,N)
C**********************************************************************
C THIS SUBPROGRAM CALCULATES A MEAN AND A STANDARD DEVIATION FOR A GIVEN *
C      SET OF DATA                                                      *
C                                                                       *
C IDENTIFICATION OF VARIABLES:                                          *
C      DATA   = INDIVIDUAL DATA POINTS (INPUT ARRAY)                    *
C      N      = NUMBER OF DATA POINTS (INPUT SCALAR)                    *
C      AVG    = MEAN (OUTPUT SCALAR)                                    *
C      STDEV  = STANDARD DEVIATION (OUTPUT SCALAR)                      *
C**********************************************************************
      DIMENSION DATA(1000)
      SUM1=0.
      SUM2=0.
      DO 1 I=1,N
      SUM1=SUM1+DATA(I)
    1 SUM2=SUM2+DATA(I)**2
      AVG=SUM1/N
      STDEV=SQRT(SUM2/N-AVG**2)
      RETURN
      END
```

A.3 SINGLE-CHANNEL SINGLE-STATION QUEUE (NEXT-CUSTOMER MODEL)

```
C**********************************************************************************
C      S I N G L E - C H A N N E L   S I N G L E - S T A T I O N   Q U E U E   *
C                                                                              *
C      N E X T - C U S T O M E R    M O D E L                                  *
C                                                                              *
C IDENTIFICATION OF VARIABLES:                                                 *
C      A(I)    = ARRIVAL TIME OF THE ITH CUSTOMER                              *
C      B(I)    = TIME WHEN ITH CUSTOMER ENTERS SERVER                          *
C      D(I)    = DEPARTURE TIME OF THE ITH CUSTOMER                            *
C      T(I)    = TIME OF OCCURRENCE OF ITH EVENT                               *
C      Q(I)    = QUEUE LENGTH RESULTING FROM THE ITH EVENT                     *
C      AT      = TIME INTERVAL BETWEEN ARRIVAL OF SUCCESSIVE CUSTOMERS         *
C                    (EXPONENTIALLY DISTRIBUTED RANDOM VARIATE)                *
C      ST      = TIME REQUIRED TO SERVICE THE ITH CUSTOMER                     *
C                    (NORMALLY DISTRIBUTED RANDOM VARIATE)                     *
C      WT(I)   = WAITING TIME (IN QUEUE) FOR ITH CUSTOMER                      *
C      TOTWT   = CUMULATIVE WAITING TIME                                       *
C      IT      = TIME SERVER IS IDLE BEFORE SERVICING THE ITH CUSTOMER         *
C      TOTIT   = CUMULATIVE IDLE TIME                                          *
C      XBAR    = MEAN VALUE OF TIME INTERVAL BETWEEN ARRIVALS (INPUT PARAMETER) *
C      XAVG    = MEAN VALUE OF CUSTOMER SERVICE TIME (INPUT PARAMETER)         *
C      STDEV   = STANDARD DEVIATION OF CUSTOMER SERVICE TIME (INPUT PARAMETER) *
C      KX      = SEED FOR RANDOM NUMBER GENERATOR (INPUT PARAMETER)            *
C      M       = NUMBER OF INTERVALS IN EACH OUTPUT ARRAY (INPUT PARAMETER)    *
C      N       = NUMBER OF CUSTOMERS PROCESSED PER SIMULATION                  *
C**********************************************************************************
      INTEGER Q
      REAL IT
      DIMENSION A(1000),B(1000),D(1000),WT(1000),Q(2000),T(2000)
      DIMENSION BOUNDS(26),F(25),Y(25)
  100 FORMAT(3F4.1)
  200 FORMAT(I12,3I4)
  300 FORMAT('1',18X,'S I N G L E - C H A N N E L   S I N G L E - S T A
     1T I O N   Q U E U E'////,11X,'EXPONENTIALLY DISTRIBUTED TIME INTERV
     2AL BETWEEN SUCCESSIVE CUSTOMERS:  MEAN VALUE = ',F3.1,' MINUTES'//
     3,11X,'NORMALLY DISTRIBUTED SERVICE TIMES:  MEAN VALUE = ',F3.1,
     4' MINUTES'/,48X,'STANDARD DEVIATION = ',F3.1,' MINUTES'//)
  400 FORMAT('0',10X,'SEED (RANDOM NUMBER GENERATOR) = ',I12/,11X,'NUMBE
     1R OF INTERVALS IN THE OUTPUT ARRAYS = ',I4/,11X,'NUMBER OF CUSTOME
     2RS = ',I4//)
  500 FORMAT('0',10X,'MEAN WAITING TIME = ',F5.2,' MINUTES'/)
  600 FORMAT('0',10X,'MEAN QUEUE LENGTH = ',F5.2,' CUSTOMERS'/)
  700 FORMAT('0',10X,'FRACTION OF TIME SERVER IS IDLE = ',F5.3/)
C READ INPUT DATA
      READ (5,100) XBAR,XAVG,STDEV
      READ (5,200) KX,M,N
C ESTABLISH INTERVAL BOUNDS FOR OUTPUT ARRAYS
      BOUNDS(1)=0.
      DO 1 J=1,M
    1 BOUNDS(J+1)=FLOAT(J)/M
C SET UP RANDOM NUMBER GENERATOR
      DUMMY=RAND(KX)
C INITIALIZE VARIABLES
      I=1
      A(1)=0.
      B(1)=0.
      WT(1)=0.
      TOTWT=0.
      IT=0.
      TOTIT=0.
C PROCESS FIRST CUSTOMER (I=1)
      CALL NORMAL(XAVG,STDEV,ST)
      IF (ST.LT.0.) ST=0.
      D(1)=ST
```

A.3 Continued

```
C PROCESS SUCCESSIVE CUSTOMERS (I=2,3,...,N)
      DO 3 I=2,N
      CALL EXPON(XBAR,0.,AT)
      A(I)=A(I-1)+AT
      CALL NORMAL(XAVG,STDEV,ST)
      IF (ST.LT.0.) ST=0.
      IT=0.
      D(I)=D(I-1)+ST
      WT(I)=D(I)-A(I)-ST
      IF (WT(I).GE.0.) GO TO 2
      WT(I)=0.
      D(I)=A(I)+ST
      IT=D(I)-D(I-1)-ST
    2 B(I)=A(I)+WT(I)
      TOTWT=TOTWT+WT(I)
      TOTIT=TOTIT+IT
    3 CONTINUE
C DETERMINE MEAN WAITING TIME
      WTAVG=TOTWT/N
C DETERMINE MEAN QUEUE LENGTH
      CALL MERGE(A,B,T,Q,N)
      SUM=0.
      N1=2*N-1
      DO 4 I=1,N1
    4 SUM=SUM+(T(I+1)-T(I))*Q(I)
      QAVG=SUM/(T(2*N)-T(1))
C DETERMINE FRACTION OF TIME SERVER IS IDLE
      FRACT=TOTIT/(D(N)-A(1))
C WRITE OUTPUT DATA
      WRITE (6,300) XBAR,XAVG,STDEV
      WRITE (6,400) KX,M,N
      WRITE (6,500) WTAVG
      WRITE (6,600) QAVG
      WRITE (6,700) FRACT
C DETERMINE DISTRIBUTIONS OF WAITING TIME AND QUEUE LENGTH
C
C ******   TO BE ADDED   ******
C
      STOP
      END
```

A.4 SINGLE-CHANNEL SINGLE-STATION QUEUE
(NEXT-EVENT MODEL)

```
C*********************************************************************************
C       S I N G L E - C H A N N E L   S I N G L E - S T A T I O N   Q U E U E    *
C                                                                                *
C       N E X T - E V E N T   M O D E L                                          *
C                                                                                *
C IDENTIFICATION OF PRINCIPAL VARIABLES:                                         *
C       A       = SYSTEM ARRIVAL TIME                                            *
C       B       = TIME WHEN SERVER IS ENTERED                                    *
C       D       = SYSTEM DEPARTURE TIME                                          *
C       TO      = TIME OF LAST EVENT                                             *
C       AT      = SYSTEM INTER-ARRIVAL TIME (RANDOM VARIATE)                     *
C       ST      = SERVICE TIME (RANDOM VARIATE)                                  *
C       DT      = TIME INTERVAL BETWEEN SUCCESSIVE EVENTS                        *
C       I       = ARRIVAL INDEX (I=1,2,...,N)                                    *
C       J       = DEPARTURE INDEX (J=1,2,...,N)                                  *
C       N       = TOTAL NUMBER OF CUSTOMERS PROCESSED PER SIMULATION             *
C                   (INPUT PARAMETER)                                            *
C       KX      = SEED FOR RANDOM NUMBER GENERATOR (INPUT PARAMETER)             *
C       XBAR    = MEAN INTER-ARRIVAL TIME (INPUT PARAMETER)                      *
C       XAVG    = MEAN SERVICE TIME (INPUT PARAMETER)                            *
C       STDEV   = STANDARD DEVIATION OF SERVICE TIME (INPUT PARAMETER)           *
C       Q       = QUEUE LENGTH                                                   *
C       S       = STATUS OF SERVER (0=IDLE, 1=BUSY)                              *
C       NWT     = NUMBER OF CUSTOMERS WHO MUST WAIT BEFORE ENTERING SERVER       *
C       QIN(K)  = ARRIVAL TIMES OF CUSTOMERS IN QUEUE                            *
C       NC(K)   = INTERVAL COUNTERS FOR DISTRIBUTION OF WAITING TIMES            *
C       TQ(K)   = INTERVAL COUNTERS FOR DISTRIBUTION OF QUEUE LENGTHS            *
C                                                                                *
C*********************************************************************************
      INTEGER Q,S
      DIMENSION QIN(20),NC(21)
      COMMON TQ(14)
      REAL IT
  100 FORMAT(3F4.1)
  200 FORMAT(I12,I4)
  300 FORMAT('1',18X,'S I N G L E - C H A N N E L   S I N G L E - S T A
     1T I O N   Q U E U E'///,11X,'EXPONENTIALLY DISTRIBUTED TIME INTERV
     2AL BETWEEN SUCCESSIVE CUSTOMERS:  MEAN VALUE = ',F3.1,' MINUTES'//
     3,11X,'NORMALLY DISTRIBUTED SERVICE TIMES:  MEAN VALUE = ',F3.1,
     4' MINUTES'/,48X,'STANDARD DEVIATION = ',F3.1,' MINUTES'//)
  400 FORMAT('0',10X,'SEED (RANDOM NUMBER GENERATOR) = ',I12/,11X,'NUMBE
     1R OF CUSTOMERS = ',I4//)
  500 FORMAT('0',10X,'MEAN WAITING TIME = ',F5.2,' MINUTES    (BASED UPON
     1 ALL CUSTOMERS)'/)
  505 FORMAT('0',10X,'MEAN WAITING TIME = ',F5.2,' MINUTES    (BASED UPON
     1 THOSE CUSTOMERS THAT MUST WAIT)'/)
  510 FORMAT(' ',10X,'DISTRIBUTION OF WAITING TIMES:'//,
     1 14X,'INTERVAL',8X,'RELATIVE',10X,'CUMULATIVE'/,
     2 15X,'BOUNDS',8X,'FREQUENCY',9X,'DISTRIBUTION')
  520 FORMAT(12X,F4.1,' TO ',F4.1,7X,F5.3,15X,F5.3)
  530 FORMAT(18X,'> ',F4.1,7X,F5.3,15X,F5.3)
  600 FORMAT('1',10X,'MEAN QUEUE LENGTH = ',F5.2,' CUSTOMERS'/)
  610 FORMAT(' ',10X,'DISTRIBUTION OF QUEUE LENGTHS:'//,
     1 15X,'QUEUE',10X,'RELATIVE',10X,'CUMULATIVE'/,
     2 14X,'LENGTH',9X,'FREQUENCY',9X,'DISTRIBUTION')
  620 FORMAT(16X,I2,13X,F5.3,15X,F5.3)
  630 FORMAT(15X,'>',I2,13X,F5.3,15X,F5.3//)
  700 FORMAT('0',10X,'FRACTION OF TIME SERVER IS IDLE = ',F5.3/)
C READ INPUT DATA
      READ (5,100) XBAR,XAVG,STDEV
      A=A+AT
      IF (I.GT.N) A=1.E+38
C TEST FOR SIMULTANEOUS ARRIVAL AND DEPARTURE
      IF (TO.NE.D) GO TO 1
```

A.4 Continued

```
C *****************************************************************************
C DEPARTURE IS NEXT                                                          *
C *****************************************************************************
    5 DT=D-TO
      TO=D
C GROUP DT VS Q
      QSUM=QSUM+Q*DT
      CALL QDIST(DT,Q)
C TEST FOR LAST DEPARTURE
      IF (J.EQ.N) GO TO 8
      IF (Q.GT.0) GO TO 6
C QUEUE IS EMPTY - SERVER BECOMES IDLE
      S=0
      GO TO 2
C QUEUE IS NOT EMPTY - SERVER REMAINS BUSY (NEXT CUSTOMER ENTERS)
    6 B=D
      Q=Q-1
C LEAVE QUEUE
      WT=B-QIN(POINTR)
      WTSUM=WTSUM+WT
      DO 60 K=1,20
      IF (WT.GT.0.5*K) GO TO 60
      NC(K)=NC(K)+1
      GO TO 61
   60 CONTINUE
      NC(21)=NC(21)+1
C COMPRESS WAITING LINE
   61 DO 62 K=1,Q
   62 QIN(K)=QIN(K+1)
C GENERATE TIME OF NEXT DEPARTURE
    7 J=J+1
      CALL NORMAL(XAVG,STDEV,ST)
      IF (ST.LT.0.) ST=0.
      D=B+ST
      GO TO 1
C *****************************************************************************
C CALCULATE STATISTICAL INFORMATION AND                                      *
C WRITE OUTPUT DATA                                                          *
C*****************************************************************************
    8 WTAVG=WTSUM/N
      WTAVG1=WTSUM/NWT
      QAVG=QSUM/B
      FRACT=IT/D
      WRITE (6,300) XBAR,XAVG,STDEV
      WRITE (6,400) KX,N
C OBTAIN DISTRIBUTION OF WAITING TIMES
      WRITE (6,500) WTAVG
      WRITE (6,505) WTAVG1
      WRITE (6,510)
      YWT=0.
      DO 9 K=1,20
      B1=.5*(K-1)
      B2=.5*K
      FWT=FLOAT(NC(K))/NWT
      YWT=YWT+FWT
    9 WRITE (6,520) B1,B2,FWT,YWT
      READ (5,200) KX,N
C SET UP RANDOM NUMBER GENERATOR
      DUMMY=RAND(KX)
C INITIALIZE VARIABLES
      TO=0.
      IT=0.
      WTSUM=0.
      QSUM=0.
      NWT=0
      Q=0
      S=1
      DO 50 I=1,14
   50 TQ(I)=0.
      DO 55 K=1,21
   55 NC(K)=0
```

A.4 Continued

```
C GENERATE TIME OF FIRST DEPARTURE
      J=1
      CALL NORMAL(XAVG,STDEV,ST)
      IF (ST.LT.0.) ST=0.
      D=ST
C GENERATE TIME OF SECOND ARRIVAL
      I=2
      CALL EXPON(XBAR,0.,AT)
      A=AT
C **********************************************************************************
C BEGIN LOOP                                                                       *
C **********************************************************************************
    1 IF (S.EQ.0) GO TO 2
C COMPARE TIME OF NEXT ARRIVAL WITH TIME OF NEXT DEPARTURE
      IF (A.GT.D) GO TO 5
C **********************************************************************************
C ARRIVAL IS NEXT                                                                  *
C **********************************************************************************
    2 DT=A-TO
      TO=A
C GROUP DT VS Q
      QSUM=QSUM+Q*DT
      CALL QDIST(DT,Q)
      IF (S.EQ.1) GO TO 3
C SERVER IS IDLE - ENTER DIRECTLY
      S=1
      B=A
      IT=IT+DT
C GENERATE DEPARTURE TIME FOR CURRENT ARRIVAL
      J=J+1
      CALL NORMAL(XAVG,STDEV,ST)
      IF (ST.LT.0.) ST=0.
      D=B+ST
      GO TO 4
C SERVER IS BUSY - WAIT IN LINE
    3 Q=Q+1
C ENTER QUEUE
      NWT=NWT+1
      QIN(Q)=A
C DETERMINE NEXT DEPARTURE FROM QUEUE
C (PRIORITY RULE IS FIRST-IN, FIRST-OUT)
      POINTR=1
C GENERATE TIME OF NEXT ARRIVAL
    4 I=I+1
      CALL EXPON(XBAR,0.,AT)
      FWT=FLOAT(NC(21))/NWT
      YWT=YWT+FWT
      WRITE (6,530) B2,FWT,YWT
C OBTAIN DISTRIBUTION OF QUEUE LENGTHS
      WRITE (6,600) QAVG
      WRITE (6,610)
      YQ=0.
      DO 10 I=1,13
      IQ=I-1
      FQ=TQ(I)/D
      YQ=YQ+FQ
   10 WRITE (6,620) IQ,FQ,YQ
      FQ=TQ(14)/D
      YQ=YQ+FQ
      WRITE (6,630) IQ,FQ,YQ
      WRITE (6,700) FRACT
      STOP
      END
```

A.5 SINGLE-CHANNEL MULTISTATION QUEUE

```
C*********************************************************************************
C       S I N G L E - C H A N N E L   M U L T I - S T A T I O N  Q U E U E     Q
C                                                                              *
C       N E X T - C U S T O M E R    M O D E L                                 *
C                                                                              *
C IDENTIFICATION OF VARIABLES:                                                 *
C     A(I,J) = ARRIVAL TIME OF ITH CUSTOMER TO JTH STATION                     *
C     B(I,J) = TIME WHEN ITH CUSTOMER ENTERS JTH SERVER                        Q
C     D(I,J) = DEPARTURE TIME OF ITH CUSTOMER FROM JTH SERVER                  *
C     T(I)   = TIME OF OCCURRENCE OF ITH EVENT                                 *
C     Q(I)   = QUEUE LENGTH RESULTING FROM THE ITH EVENT                       *
C     AT     = TIME INTERVAL BETWEEN ARRIVAL OF SUCCESSIVE CUSTOMERS           *
C                 (EXPONENTIALLY DISTRIBUTED RANDOM VARIATE)                    *
C     ST     = TIME REQUIRED TO SERVICE THE ITH CUSTOMER IN THE JTH SERVER     *
C                 (NORMALLY DISTRIBUTED RANDOM VARIATE)                         *
C     WT(I,J)= WAITING TIME OF ITH CUSTOMER IN JTH QUEUE                       *
C     TOTWT  = CUMULATIVE WAITING TIME FOR THE JTH STATION                     *
C     IT     = TIME JTH SERVER IS IDLE BEFORE SERVICING THE ITH CUSTOMER       *
C     TOTIT  = CUMULATIVE IDLE TIME FOR THE JTH STATION                        *
C     TOTTS  = CUMULATIVE TOTAL TIME IN THE SYSTEM                             *
C     XBAR   = MEAN VALUE OF TIME INTERVAL BETWEEN ARRIVALS (INPUT PARAMETER)  *
C     XAVG   = MEAN VALUE OF CUSTOMER SERVICE TIME (INPUT PARAMETER)           *
C     STDEV  = STANDARD DEVIATION OF CUSTOMER SERVICE TIME (INPUT PARAMETER)   *
C     KX     = SEED FOR RANDOM NUMBER GENERATOR (INPUT PARAMETER)              *
C     M      = NUMBER OF INTERVALS IN EACH OUTPUT ARRAY (INPUT PARAMETER)      *
C     N      = NUMBER OF CUSTOMERS PROCESSED PER SIMULATION (MAXIMUM IS 1000)  *
C     NUNITS = NUMBER OF STATIONS (MAXIMUM IS 6)                               *
C*********************************************************************************
      INTEGER Q
      REAL IT
      DIMENSION X1(1000),X2(1000)
      DIMENSION A(1000,6),B(1000,6),D(1000,6),WT(1000,6),Q(2000),T(2000)
      DIMENSION XAVG(6),STDEV(6),TOTWT(6),TOTIT(6)
      DIMENSION BOUNDS(26),F(25),Y(25)
  100 FORMAT(I12,4I4)
  200 FORMAT(13F5.1)
  300 FORMAT('1',18X,'S I N G L E - C H A N N E L    M U L T I - S T A T
     1I O N    Q U E U E'///)
  400 FORMAT('0',10X,'SEED (RANDOM NUMBER GENERATOR) = ',I12/,11X,'NUMBE
     1R OF INTERVALS IN THE OUTPUT ARRAYS = ',I4/,11X,'NUMBER OF CUSTOME
     2RS = ',I4/,11X,'NUMBER OF STATIONS = ',I4//)
  410 FORMAT(11X,'EXPONENTIALLY DISTRIBUTED TIME INTERVAL BETWEEN SUCCES
     1SIVE CUSTOMERS:  MEAN VALUE = ',F3.1,' MINUTES'//,
     2 11X,'NORMALLY DISTRIBUTED SERVICE TIMES:'/)
  420 FORMAT(15X,'STATION NO:',I2,5X,'MEAN = ',F3.1,5X,'STANDARD DEVIATI
     1ON = ',F3.1)
  500 FORMAT('0',10X,'MEAN TOTAL TIME IN SYSTEM = ',F5.2,' MINUTES'/)
  600 FORMAT('1',10X,'STATION NUMBER:',I2/)
  610 FORMAT('0',10X,'MEAN WAITING TIME = ',F5.2,' MINUTES'/)
  620 FORMAT('0',10X,'MEAN QUEUE LENGTH = ',F5.2,' CUSTOMERS'/)
  630 FORMAT('0',10X,'FRACTION OF TIME SERVER IS IDLE = ',F5.3/)
C READ INPUT DATA
      READ (5,100) KX,M,N,NUNITS
      READ (5,200) XBAR,((XAVG(J),STDEV(J)),J=1,NUNITS)
C WRITE INPUT DATA
      WRITE (6,300)
      WRITE (6,400) KX,M,N,NUNITS
      WRITE (6,410) XBAR
      DO 10 J=1,NUNITS
   10 WRITE (6,420) J,XAVG(J),STDEV(J)
C ESTABLISH INTERVAL BOUNDS FOR OUTPUT ARRAYS
      BOUNDS(1)=0.
      DO 1 J=1,M
    1 BOUNDS(J+1)=FLOAT(J)/M
C SET UP RANDOM NUMBER GENERATOR
      DUMMY=RAND(KX)
```

A.5 Continued

```
C PROCESS FIRST CUSTOMER (I=1)
      TOTTS=0.
      I=1
      A(1,1)=0.
      DO 2 J=1,NUNITS
      B(1,J)=A(1,J)
      WT(1,J)=0.
      TOTWT(J)=0.
      TOTIT(J)=A(1,J)
      CALL NORMAL(XAVG(J),STDEV(J),ST)
      IF (ST.LT.0.) ST=0.
      D(1,J)=A(1,J)+ST
      IF (J.LT.NUNITS) A(1,J+1)=D(1,J)
    2 TOTTS=TOTTS+ST
C PROCESS SUCCESSIVE CUSTOMERS (I=2,3,...,N)
      DO 4 I=2,N
      CALL EXPON(XBAR,0.,AT)
      A(I,1)=A(I-1,1)+AT
      DO 4 J=1,NUNITS
      CALL NORMAL(XAVG(J),STDEV(J),ST)
      IF (ST.LT.0.) ST=0.
      IT=0.
      D(I,J)=D(I-1,J)+ST
      WT(I,J)=D(I,J)-A(I,J)-ST
      IF (WT(I,J).GE.0.) GO TO 3
      WT(I,J)=0.
      D(I,J)=A(I,J)+ST
      IT=D(I,J)-D(I-1,J)-ST
    3 B(I,J)=A(I,J)+WT(I,J)
      TOTWT(J)=TOTWT(J)+WT(I,J)
      TOTIT(J)=TOTIT(J)+IT
      IF (J.LT.NUNITS) A(I,J+1)=D(I,J)
    4 TOTTS=TOTTS+WT(I,J)+ST
C DETERMINE TOTAL TIME IN SYSTEM
      TSAVG=TOTTS/N
      WRITE (6,500) TSAVG
C DETERMINE STATISTICS FOR EACH STATION
      DO 7 J=1,NUNITS
      WRITE (6,600) J
      WTAVG=TOTWT(J)/N
      WRITE (6,610) WTAVG
      DO 5 I=1,N
      X1(I)=A(I,J)
    5 X2(I)=B(I,J)
      CALL MERGE(X1,X2,T,Q,N)
      SUM=0.
      N1=2*N-1
      DO 6 I=1,N1
    6 SUM=SUM+(T(I+1)-T(I))*Q(I)
      QAVG=SUM/(T(2*N)-T(1))
      WRITE (6,620) QAVG
      FRACT=TOTIT(J)/(D(N,J)-A(1,J))
      WRITE (6,630) FRACT
C DETERMINE DISTRIBUTIONS OF WAITING TIME AND QUEUE LENGTH
C
C ******   TO BE ADDED   ******
C
    7 CONTINUE
      STOP
      END
```

A.6 MULTICHANNEL SINGLE-STATION QUEUE

```
C IDENTIFICATION OF VARIABLES:                                                    *
C      BOUNDS = INTERVAL BOUNDS                                                    *
C      F      = RELATIVE FREQUENCIES                                               *
C      Y      = CUMULATIVE DISTRIBUTION                                            *
C      AVG    = MEAN                                                               *
C      STDEV  = STANDARD DEVIATION                                                 *
C      M      = NUMBER OF INTERVALS                                                *
C      N      = NUMBER OF DATA POINTS                                              *
C**********************************************************************************
      DIMENSION BOUNDS(26),F(25),Y(25)
  100 FORMAT('0',10X,'MEAN =',E12.5,10X,'STANDARD DEVIATION =',E12.5,
     1 10X,'NUMBER OF DATA POINTS =',I4//)
  200 FORMAT('0',14X,'INTERVAL',14X,'LOWER',14X,'UPPER',14X,'RELATIVE',
     1 14X,'CUMULATIVE'/,16X,'NUMBER',15X,'BOUND',14X,'BOUND',14X,
     2 'FREQUENCY',12X,'DISTRIBUTION'/)
  300 FORMAT(18X,I2,14X,F6.4,11X,F8.4,13X,F8.4,15X,F8.4)
      WRITE (6,100) AVG,STDEV,N
      WRITE (6,200)
      DO 1 J=1,M
      J1=J+1
    1 WRITE (6,300) J,BOUNDS(J),BOUNDS(J1),F(J),Y(J)
      RETURN
      END



      SUBROUTINE QOUT(QMAX,QAVG,WSTAN)
      COMMON QSTORE(30),FREQ(30)
      WRITE(6,400) (QAVG,WSTAN)
      WRITE(6,100)
      WRITE(6,200)
      IQ=QMAX
      SUM=0.0
      DO 10 I=1,IQ+1
      L=I-1
      SUM=SUM+FREQ(I)
      WRITE(6,300) I,L,FREQ(I),SUM
10        CONTINUE
100       FORMAT(26X,'INTERVAL',10X,'QUEUE LENGTH',10X,'RELATIVE',10X,
     1 'CUMULATIVE')
200       FORMAT(66X,'FREQUENCY',9X,'DISTRIBUTION'/)
300       FORMAT(29X,I2,18X,I2,15X,F6.4,13X,F6.4)
400       FORMAT(/36X,'AVG = ',F7.3,20X,'STAN. DEV. = ',F7.3///)
      RETURN
      END
      SUBROUTINE PARAMS(DATA,AVG,STDEV,N)
C**********************************************************************************
C THIS SUBPROGRAM CALCULATES A MEAN AND A STANDARD DEVIATION FOR A GIVEN          *
C      SET OF DATA                                                                 *
C                                                                                  *
C IDENTIFICATION OF VARIABLES:                                                     *
C      DATA    = INDIVIDUAL DATA POINTS (INPUT ARRAY)                              *
C      N       = NUMBER OF DATA POINTS (INPUT SCALAR)                              *
C      AVG     = MEAN (OUTPUT SCALAR)                                              *
C      STDEV   = STANDARD DEVIATION (OUTPUT SCALAR)                               *
C**********************************************************************************
      DIMENSION DATA(1000)
      SUM1=0.
      SUM2=0.
      DO 1 I=1,N
      SUM1=SUM1+DATA(I)
    1 SUM2=SUM2+DATA(I)**2
      AVG=SUM1/N
      STDEV=SQRT(SUM2/N-AVG**2)
      RETURN
      END
```

A.6 Continued

```
      SUBROUTINE GROUP(DATA,BOUNDS,F,Y,M,N)
C**********************************************************************************
C THIS SUBPROGRAM GROUPS A SET OF DATA AND THEN CALCULATES THE RELATIVE          *
C     FREQUENCIES AND THE CORRESPONDING CUMULATIVE DISTRIBUTION                  *
C                                                                               *
C IDENTIFICATION OF VARIABLES:                                                  *
C     DATA    = INDIVIDUAL DATA POINTS (INPUT ARRAY)                            *
C     BOUNDS  = INTERVAL BOUNDS (INPUT ARRAY)                                   *
C     M       = NUMBER OF INTERVALS (INPUT SCALAR)                              *
C     N       = NUMBER OF DATA POINTS (INPUT SCALAR)                            *
C     K       = INTERVAL COUNTERS (INTERNAL ARRAY)                             *
C     F       = RELATIVE FREQUENCIES (OUTPUT ARRAY)                            *
C     Y       = CUMULATIVE DISTRIBUTION (OUTPUT ARRAY)                         *
C**********************************************************************************
      DIMENSION DATA(1000),BOUNDS(26),F(25),Y(25),K(25)
C ZERO THE INTERVAL COUNTERS
      DO 1 J=1,M
    1 K(J)=0
C GROUP THE DATA
      DO 3 I=1,N
      DO 2 J=1,M
      IF (DATA(I).GE.BOUNDS(J+1)) GO TO 2
      K(J)=K(J)+1
      GO TO 3
    2 CONTINUE
    3 CONTINUE
C CALCULATE THE RELATIVE FREQUENCIES AND THE CUMULATIVE DISTRIBUTION
      F(1)=FLOAT(K(1))/N
      Y(1)=F(1)
      DO 4 J=2,M
      F(J)=FLOAT(K(J))/N
    4 Y(J)=Y(J-1)+F(J)
      RETURN
      END
      SUBROUTINE OUTPUT(BOUNDS,F,Y,AVG,STDEV,M,N)
C**********************************************************************************
C THIS SUBPROGRAM PRINTS THE MEAN, THE STANDARD DEVIATION, THE RELATIVE          *
C     FREQUENCIES AND THE CUMULATIVE DISTRIBUTION FOR A GIVEN DATA SET           *
C                                                                               *
C                                                                               *
C             SUBROUTINES                                                       *
C         3) CALCULATES THE MEAN AND STANDARD DEVIATION OF THE                  *
C            DATA USING THE PARAMS SUBROUTINE                                   *
C         4) GROUPS DATA USING THE GROUP SUBROUTINE                             *
C         5) OUTPUTS DATA USING THE OUTPUT SUBROUTINE                           *
C                                                                               *
C     VARIABLE LIST:                                                           *
C         DATA(J)     = INPUT ARRAY                                            *
C         CMIN        = MINIMUM VALUE OF DATA                                  *
C         CMAX        = MAXIMUM VALUE OF DATA ARRAY                            *
C         ADD         = INTERVAL BETWEEN BOUNDS                                *
C         BOUNDS(J) = INTERVAL BOUNDARIES                                      *
C                                                                               *
C**********************************************************************************
      DIMENSION DATA(500),BOUNDS(26),F(25),Y(25)
      DO 1 J=1,10
    1 BOUNDS(J)=0.0
C     CALCULATE MAX AND MIN OF INPUT ARRAY
      CMIN=DATA(1)
      CMAX=DATA(1)
      DO 10 J=2,I
      IF(DATA(J).LT.CMIN)CMIN=DATA(J)
      IF(DATA(J).GT.CMAX)CMAX=DATA(J)
   10 CONTINUE
      ADD=(CMAX-CMIN)/10.
      BOUNDS(1)=CMIN
      DO 20 J=2,10
      L=J-1
```

A.6 Continued

```
           BOUNDS(J)=BOUNDS(L)+ADD
20         CONTINUE
           BOUNDS(11)=CMAX+1.
           CALL PARAMS(DATA,AVG,STDEV,I)
           CALL GROUP(DATA,BOUNDS,F,Y,10,I)
           CALL OUTPUT(BOUNDS,F,Y,AVG,STDEV,10,I)
           RETURN
           END

           SUBROUTINE EXPON(XAVG,X0,X)
C******************************************************************************
C THIS SUBPROGRAM GENERATES AN EXPONENTIALLY DISTRIBUTED RANDOM VARIATE       *
C      WITH MEAN XAVG AND MINIMUM X0 USING THE INVERSE TRANSFORMATION METHOD   *
C******************************************************************************
           ALPHA=1./(XAVG-X0)
           X=X0-ALOG(RAN(0))/ALPHA
           RETURN
           END
           SUBROUTINE NORMAL(XAVG,STDEV,X)
C******************************************************************************
C THIS SUBPROGRAM GENERATES A NORMALLY DISTRIBUTED RANDOM VARIATE             *
C      WITH MEAN XAVG AND STANDARD DEVIATION STDEV USING A METHOD BASED        *
C      UPON THE CENTRAL LIMIT THEOREM                                          *
C******************************************************************************
           SUM=0.
           DO 1 I=1,12
         1 SUM=SUM+RAN(0)
           Z=SUM-6.
           X=XAVG+STDEV*Z
           RETURN
           END
           IF(L.EQ.1) GO TO 7
           DIFF=T(L)-T(L1)
C    ITERATION FOR CALCULATING QAVG
           QAVG=QAVG+(Q*DIFF)
           JJ=Q+1.
C    SUM TIME FOR JJ CUSTOMERS IN QUEUE
           QSTORE(JJ)=QSTORE(JJ)+DIFF
           L=L+1
           GO TO 3
C    SPECIAL CASE WHEN STARTING MERGE (L=1)
7          QAVG=QAVG+(T(L)*Q)
           L=L+1
           GO TO 3
C    FINDS T(L) WHEN B > SAVE
5          T(L)=SAVE(K)
           K=K+1
           Q=Q-1.
           DIFF=T(L)-T(L1)
           QAVG=QAVG+(Q*DIFF)
           JJ=Q+1.
           QSTORE(JJ)=QSTORE(JJ)+DIFF
           L=L+1
           GO TO 3
C    FINDS T(L) WHEN B=SAVE
6          T(L)=B(J)
           Q=Q+1.
           M=L+1
           T(M)=T(L)
           Q=Q-1.
           DIFF=T(L)-T(L1)
           QAVG=QAVG+(Q*DIFF)
           JJ=Q+1.
           QSTORE(JJ)=QSTORE(JJ)+DIFF
           L=L+2
           K=K+1
           J=J+1
           GO TO 3
C    DETERMINE Q-AVERAGE
```

A.6 Continued

```
30          TOT=T(L1)-B(1)
            QAVG=QAVG/TOT
            TL1=T(L1)
            IQ=QMAX
C   DETERMINE QUEUE FREQUENCY
            DO 60 IFR=1,IQ+1
            FREQ(IFR)=QSTORE(IFR)/TOT
60          CONTINUE
            SVAR=0.0
            DO 70 LL=1,IQ+1
70          SVAR=FREQ(LL)*((FLOAT(LL-1)-QAVG)**2)+SVAR
            WSTAN=SQRT(SVAR)
            RETURN
            END


            SUBROUTINE SETUP(DATA,I)
C**********************************************************************************
C                                                                                *
C       THIS SUBROUTINE ANALIZES THE DATA AS FOLLOWS:                            *
C            1) CALCULATES THE MIN AND MAX VALUES OF DATA ARRAY                   *
C            2) CALCULATES THE BOUNDS FOR THE GROUP AND OUTPUT                    *
C       VARIABLE LIST:                                                           *
C          B(I)          = CORRESPONDS TO A(I) IN MAIN                           *
C          C(K,J)        = CORRESPONDS TO B(I,J) IN MAIN                         *
C          QSTORE(JJ)= TOTAL TIME A QUEUE HAS JJ-1 CUSTOMERS                     *
C                        IN IT                                                   *
C          SAVE(K)       = SUMMARIZES THE C(K,J) ARRAY INTO ITS                  *
C                          CORRESPONDING ONE DIMENSIONAL ARRAY                   *
C          QAVG          = AVERAGE NO. OF CUSTOMERS IN THE QUEUE                 *
C          T(L)          = THE SEQUENCED ARRAY OF ARRIVALS AND DEPARTURS         *
C                          FROM THE QUEUE                                        *
C          DIFF          = TIME BETWEEN SEQUENCED EVENTS                         *
C          TOT           = TOTAL TIME OF QUEUE OPERATION                         *
C          FREQ(I)       = PERCENT OF TOTAL QUEUE TIME IN WHICH                  *
C                          THERE ARE I-1 CUSTOMERS IN THE QUEUE                  *
C                                                                                *
C**********************************************************************************
            DIMENSION B(500),C(500,3),T(1050),SAVE(500)
            COMMON QSTORE(30),FREQ(30)
C   INITIALIZE ARRAYS
            DO 1 K=1,500
1           SAVE(K)=0.0
            DO 99 K=1,30
            QSTORE(K)=0.0
99          FREQ(K)=0.0
C   RENAME AND ELIMINATE ZEROS FROM C(K,J)
            DO 10 K=1,I
            DO 20 J=1,3
            IF(C(K,J).GT.0.0) SAVE(K)=C(K,J)
20          CONTINUE
10          CONTINUE
C   SORT SAVE(K)
            DO 40 LL=1,I
            DO 50 K=1,I-1
            M=K+1
            IF(SAVE(K).LT.SAVE(M)) GO TO 50
            TEMP=SAVE(K)
            SAVE(K)=SAVE(M)
            SAVE(M)=TEMP
50          CONTINUE
            SAVE(I)=SAVE(M)
40          CONTINUE
```

A.6 Continued

```
C     START MERGING OF B AND SAVE ARRAYS
          L=1
          J=1
          K=1
          Q=0.
          QAVG=0.0
          QMAX=0.0
          IF(L.EQ.1) GO TO 4
3         L1=L-1
C     FIND QMAX
          IF(Q.GT.QMAX) QMAX=Q
C     DETERMINE PROPER MERGE PROCEDURE
          IF(T(L1).EQ.SAVE(I)) GO TO 30
4         IF(B(J).GT.SAVE(K)) GO TO 5
          IF(B(J).EQ.SAVE(K)) GO TO 6
C     FINDS T(L) WHEN B < SAVE
          T(L)=B(J)
          J=J+1
          Q=Q+1.
C     FIND THE MAXIMUM WAITING TIME
          IF(WT(I).GT.WTHIGH)WTHIGH=WT(I)
C     GENERATE SERVICE TIME AND TEST
8         CALL NORMAL(AMEAN,STAN,X)
          ST(I,J)=X
          IF(ST(I,J).LT.0.0)GO TO 8
C     FIND THE MAX AND MIN SERVICE TIMES
          IF(ST(I,J).GT.STHIGH)STHIGH=ST(I,J)
          IF(ST(I,J).LT.STLOW)STLOW=ST(I,J)
          IT(I,J)=B(I,J)-AMAX(J)
          SIT(J)=SIT(J)+IT(I,J)
          IF(IT(I,J).LT.0.0)IT(I,J)=0.0
          IF(WT(I).GT.0.0)IT(I,J)=0.0
          IF(IT(I,J).GT.0.0) WT(I)=0.0
          D(I,J)=AMAX(J)+IT(I,J)+ST(I,J)
C     FIND THE LAST DEPARTURE TIME
          IF(D(I,J).GT.DTHIGH) DTHIGH=D(I,J)
          IF(AMAX(J).LT.D(I,J))AMAX(J)=D(I,J)
C     FIND MINIMUM VALUE OF AMAX(J)'S
          BMIN=AMAX(1)
          DO 6 K=2,3
          IF(AMAX(K).LE.BMIN) BMIN=AMAX(K)
6         CONTINUE
          STORE(I)=D(I,J)-A(I)
          GO TO 101
C     ZERO UNUSED B(I,J)
5         B(I,J)=0.0
          GO TO 3
101       CONTINUE
          A(I1)=1000.
C     MERGE ARRIVALS AND DEPARTURES FROM QUEUE
          CALL MERGE(A,B,I,QAVG,L1,TL1,QMAX,WSTAN)
          WRITE(6,105)JRE
          WRITE(6,110)
          WRITE(6,115)JRE
          WRITE(6,120)
          WRITE(6,125)ALPHA,AMEAN,STAN
          WRITE(6,235)
          WRITE(6,230)
          WRITE(6,235)
          WRITE(6,130)
          WRITE(6,135)
```

A.6 Continued

```
          C    SUMMARIZE AND OUTPUT RESULTS
                 CALL SETUP(STORE,I)
                 WRITE(6,235)
                 WRITE(6,140)
                 CALL SETUP(WT,I)
                 WRITE(6,235)
                 WRITE(6,145)
                 CALL QOUT(QMAX,QAVG,WSTAN)
                 WRITE(6,235)
                 WRITE(6,150)
                 WRITE(6,155)
          C    CALCULATE PERCENT  IDLE TIME
                 DO 10 J=1,3
                 SIDLE=SIT(J)/TL1
                 WRITE(6,160)J,SIDLE
          10       CONTINUE
                 WRITE(6,235)
                 WRITE(6,230)
                 WRITE(6,235)
                 WRITE(6,165)
                 WRITE(6,170)
                 WRITE(6,175)
                 DO 11 NN=1,3
          11       WRITE(6,180) NN,ICOUNT(NN)
                 WRITE(6,195) WTHIGH
                 WRITE(6,205) STLOW
                 WRITE(6,210) STHIGH
                 WRITE(6,220) QMAX
                 TOTAL=DTHIGH-A(1)
                 WRITE(6,225) TOTAL
                 STOP
                 END


C*********************************************************************************
C          THIS SUBROUTINE RE-INITIALIZES THE VARIABLES AFTER            *
C    A WARM-UP PERIOD OF 100 CUSTOMERS.                                  *
C                                                                        *
C    VARIABLE LIST:                                                      *
C        AS = SAVES VALUE OF LAST ARRIVAL TIME                           *
C        ALL OTHERS - SEE MAIN PROGRAM                                   *
C                                                                        *
C*********************************************************************************
C
      SUBROUTINE RESTRT(I1,JRE,A,D,IT,ST,B,SIT,DTLOW,ICOUNT)
      DIMENSION A(500),D(500,3),AMAX(3),WT(500),IT(500,3),B(500,3)
      DIMENSION ICOUNT(3),ST(500,3),SIT(3)
C    RESTART COUNTER
      I1=1
C    INITIALIZE REMAINING VARIABLES
      AS=A(100)
      DTLOW=1000.00
      JRE=JRE-100
      DO 1 I=1,JRE
      A(I)=1000.
      DO 2 J=1,3
      D(I,J)=0.0
      IT(I,J)=0.0
      ST(I,J)=0.0
      B(I,J)=0.0
   2    CONTINUE
   1    CONTINUE
      DO 3 J=1,3
      ICOUNT(J)=0.0
   3    SIT(J)=0.0
      A(1)=AS
      RETURN
      END
```

A.6 Continued

```
      SUBROUTINE MERGE(B,C,I,QAVG,L1,TL1,QMAX,WSTAN)
C********************************************************************
C                                                                  *
C     THIS SUBROUTINE SEQUENCES THE ARRIVALS AND DEPARTURES FROM   *
C     THE QUEUE,ACCORDING TO CLOCK TIMES.                          *
C                                                                  *
C********************************************************************
C     M U L T I - C H A N N E L   S I N G L E - S T A T   Q U E U E *
C                                                                  *
C     N E X T - C U S T O M E R   M O D E L                        *
C                                                                  *
C        THIS PROGRAM IS DESIGNED TO SIMULATE A MULTI-CHANNEL      *
C     QUEUEING SYSTEM USING THE NEXT CUSTOMER MODEL, WITH THREE    *
C     SERVICE FACILITIES. HOWEVER, BY CHANGING ONLY THE SYSTEM     *
C     PARAMETER J (THE NO. OF SERVICE FACILITIES) AND THE DIMENSION*
C     STATEMENTS IT WILL SIMULATE ANY NUMBER OF SERVICE FACILITIES.*
C     (SEE SUBROUTINES FOR DESCRIPTION OF OTHER PROGRMMING FEATURES)*
C                                                                  *
C     VARIABLE LIST:                                               *
C        A(I)      = ARRIVAL TIME OF THE I-TH CUSTOMERS (CLOCK TIME)*
C        AT        = TIME BETWEEN ARRIVALS (RANDOM VARIATE)        *
C        J         = RANDOMLY GENERATED SERVICE FACILITY, FOR      *
C                    I-TH  CUSTOMER TO ENTER                       *
C        WT(I)     = WAITING TIME OF THE I-TH CUSTOMER(TIME INTERVAL)*
C        B(I,J)    = TIME THE I-TH CUSTOMER ENTERS THE J-TH SERVICE*
C                    FACILITY                                      *
C        ST(I,J)   = SERICE TIME OF THE I-TH CUSTOMER AT THE J-TH  *
C                    SERVICE FACILITY                              *
C        IT(I,J)   = IDLE TIME OF THE J-TH SERVICE FACILITY        *
C                    ASSOCIATED WITH THE I-T CUSTOMER              *
C        SIT(J)    = TOTAL TIME THE J-TH SERVICE FACILITY IS IDLE  *
C        AMAX(J)   = VALUE OF THE PRECEEDING D(I,J) FOR THE J-TH   *
C                    SERVICE FACILITY                              *
C        BMIN      = MINIMUM VALUE OF THE AMAX(J)'S                *
C        STORE(I)  = TIME THE I-TH CUSTOMER HAS SPENT IN THE       *
C                    SYSTEM                                        *
C        SIDLE     = THE FRACTION OF TIME THE J-TH SERVICE FACILITY*
C                    IS IDLE                                       *
C                                                                  *
C                                                                  *
C********************************************************************
      DIMENSION A(500),D(500,3),AMAX(3),WT(500),IT(500,3),B(500,3)
      DIMENSION QSTORE(30),ST(500,3),SIT(3),STORE(1000),ICOUNT(3)
      DIMENSION FREQ(30)
      COMMON QSTORE,FREQ
100   FORMAT('1',18X,'***   S I M U L A T I O N   O F   A   M U L T I -
     1 C H A N N E L   Q U E U E I N G   S Y S T E M
     2   ***',/,25X,84('*')/)
105   FORMAT(50X,'THE NUMBER OF SIMULATIONS TO PRODUCE A',/,50X,
     1 '95% CONFIDENCE INTERVAL IS',1X,I4/)
110   FORMAT(50X,'NUMBER OF SERVICE FACILITIES = 3 '/)
115   FORMAT(50X,'NUMBER OF TRANSACTIONS= ',I4/)
120   FORMAT(50X,'WARMUP PERIOD FOR SIMULATION IS 100 RUNS'/)
125   FORMAT(18X,'ASSUMPTIONS:',/,22X,'1.)INTER-ARRIVAL TIMES
     1 ARE EXPONENTIALLY DISTRIBUTED WITH ALPHA =',1X,F2.0,/,
     2 22X,'2.)SERVICE TIMES ARE NORMALLY DISTRIBUTED WITH A MEAN
     3 OF',1X,F2.0,1X,'AND A STANDARD DEVIATION OF',1X,F2.0/)
130   FORMAT(22X,'**  S I M U L A T I O N   A N D   S T A T I S T I C A
     1 L   A N A L Y S I S   S E C T I O N   **',/,26X,81('*')/)
135   FORMAT(35X,'DISTRIBUTION OF TOTAL TIME OF EACH CUSTOMER
     1 SPENT IN THE SYSTEM'/)
140   FORMAT(47X,'DISTRIBUTION OF CUSTOMER WAITING TIMES'/)
145   FORMAT(51X,'DISTRIBUTION OF QUEUE LENGTH'/)
150   FORMAT(43X,'FRACTION OF TIME EACH SERVICE FACILITY
     1 IS IDLE',/,43X,46('*')/)
155   FORMAT(44X,'SERVICE FACILITY',6X,'FRACTION OF IDLE TIME'/)
160   FORMAT(51X,I1,T73,F6.4/)
165   FORMAT(28X,'***  S I M U L A T I O N   A N D
     1 S T A T I S T I C A L   S U M M A R Y   ***',/,33X,64('*')/)
```

A.6 Continued

```
      170        FORMAT(45X,'NUMBER OF CUSTOMERS ENTERING SERVICE FACILITIES'/)
      175        FORMAT(49X,'SERVICE FACILITY',5X,'CUSTOMERS SERVICED'/)
      180        FORMAT(56X,I1,T77,I3/)
      195        FORMAT(///51X,'MAXIMUM WAITING TIME = ',F6.2/)
      205        FORMAT(51X,'MINIMUM SERVICE TIME = 'F6.2/)
      210        FORMAT(51X,'MAXIMUM SERVICE TIME = ',F6.2/)
      220        FORMAT(51X,'MAXIMUM QUEUE LENGTH = ',F3.0/)
      225        FORMAT(51X,'TOTAL TIME OF OPERATION = ',F6.2/)
      230        FORMAT(132('+'))
      235        FORMAT(4(/))
               REAL IT
      C     INITIALIZE THE RANDOM NUMBER NUMBER GENERATOR
               CALL SETRAN(12345)
      C     WRITE OUT THE HEADINGS
               WRITE(6,100)
      C     INITIALIZE ALL VARIABLES
               DATA WTHIGH,STHIGH,STLOW,I1,JJJ/0.0,0.0,100.0,1,245/
               DATA DTHIGH/0.0/
               DATA ALPHA,AMEAN,STAN/2.0,5.0,2.0/
               DO 1 I=1,JJJ
               A(I)=1000.
               DO 2 J=1,3
               D(I,J)=0.0
               IT(I,J)=0.0
               ST(I,J)=0.0
               B(I,J)=0.0
      2        CONTINUE
      1        CONTINUE
               DO 4 J=1,3
               ICOUNT(J)=0
               SIT(J)=0.0
               WT(J)=0.0
      4        AMAX(J)=0.0
               A(1)=0.0
               BMIN=0.0
               JRE=JJJ
      C     START PROGRAM
               DO 101 LL=1,JJJ
      C     RUN PROGRAM FOR WARM UP PERIOD, THEN RE-INITIALIZE
               IF(LL.EQ.101) CALL RESTRT(I1,JRE,A,D,IT,ST,B,SIT,DTLOW,ICOUNT)
               I1=I1+1
      C     GENERATE RANDOM VARIATE (AT) AND TEST FOR NEGATIVE VALUE
      7        CALL EXPON(ALPHA,0.,X)
               AT=X
               IF(AT.LT.0.0)GO TO 7
               I=I1-1
      C     CALCULATE THE ARRIVAL TIME OF THE I+1 CUSTOMER
               A(I1)=A(I)+AT
      C     GENERATE A UNIFORM RANDOM NO.(SERVICE FACILITY)
      3        J=1+INT(RAN(0)*3)
               WT(I)=BMIN-A(I)
               IF(WT(I).LT.0.0) WT(I)=0.0
               B(I,J)=A(I)+WT(I)
      C     TEST FOR ABILITY TO USE CALCULATED B(I,J)
               IF(AMAX(J).GT.B(I,J))GO TO 5
      C     COUNT THE NUMBER OF CUSTOMERS GOING INTO EACH SERVICE AREA
               ICOUNT(J)=ICOUNT(J)+1
```

A.7 BETA DISTRIBUTION

```
      SUBROUTINE BETA(IBETA1,IBETA2,X)
C*********************************************************************************
C     THIS SUBPROGRAM GENERATES A BETA DISTRIBUTED RANDOM VARIATE WITH
C     PARAMETERS  IBETA1, IBETA2 (INTEGER VALUES ONLY) BY DIRECT
C     SIMULATION, USING THE EQUATION  X=X1/(X1+X2) WHERE X1 AND X2
C     ARE GAMMA DISTRIBUTED RANDOM VARIATES WITH PARAMETERS ALFA, IBETA1,
C     AND IBETA2 RESPECTIVELY.
C
C IDENTIFICATION OF VARIABLES
C     IBETA1   =A POSITIVE INTEGER PARAMETER(INPUT SCALAR)
C     IBETA2   =A POSITIVE INTEGER PARAMETER (INPUT SCALAR)
C     X        =THE GENERATED RANDOM VARIATE (OUTPUT SCALAR)
C*********************************************************************************
      LL=IBETA1+1
      LU=IBETA1+IBETA2
      CALL GAMMA(IBETA1,1.,1,X1)
      CALL GAMMA(LU,1.,LL,X2)
      X= X1/(X1+X2)
      RETURN
      END
```

A.8 CONTINUOUS DISTRIBUTION

```
      SUBROUTINE CONTR(X,A,B)
C*********************************************************************************
C     THIS SUBPROGAM GENERATES A CONTINUOUS RANDOM VARIATE, UNIFORMLY
C     DISTIBUTED WITHIN THE INTERVAL (A,B), WHERE A < B.
C
C IDENTIFICATION OF VARIABLES :
C     A=INTERVAL LOWER BOUND (INPUT SCALAR)
C     B=INTERVAL UPPER BOUND (INPUT SCALAR)
C     X=CONTINUOUS RANDOM VARIATE (OUTPUT SCALAR)
C*********************************************************************************
      U=RAND(0)
      X=A+(B-A)*U
      RETURN
      END
```

A.9 DISCRETE UNIFORM DISTRIBUTION

```
      SUBROUTINE DISCR(LX,A,B)
C*********************************************************************************
C     THIS SUBPROGRAM GENERATES A DISCRETE INTEGER-VALUED RANDOM
C     VARIATE, UNIFORMLY DISTRIBUTED WITHIN THE INTERVAL (A,B)
C     WHERE A < B.
C
CIDENTIFICATION OF VARIABLES :
C     A=INTERVAL LOWER BOUND (INPUT SCALAR)
C     B=INTERVAL UPPER BOUND (INPUT SCALAR)
3&    LX=DISCRETE RANDOM VARIATE (OUTPUT SCALAR)
C*********************************************************************************
      U=RAND(0)
      LX=A+INT((B-A+1)*U)
      RETURN
      END
```

A.10 EMPIRICAL DISTRIBUTION

```fortran
      SUBROUTINE EMPI(BOUNDS,Y,M,X)
C***********************************************************************************
C THIS SUBPROGRAM GENERATES VALUES OF A RANDOM VARIABLE X
C     BASED ON AN EMPIRICAL DISTRIBUTION .
C IDENTIFICATION OF VARIABLES:
C     BOUNDS = INTERVAL BOUNDS (INPUT ARRAY)
C     Y      = CUMULATIVE DISTRIBUTION ( INPUT ARRAY)
C     M      = NUMBER OF INTERVALS (INPUT SCALAR)
C     X      = CALCULATED VALUE OF THE RANDOM VARIABLE (OUTPUT SCALAR)
C***********************************************************************************
      DIMENSION  BOUNDS(26),Y(25)
      U=RAND(0)
      DO 1 J=1,M
      IF (U.LE.Y(J)) GO TO 2
    1 CONTINUE
    2 Y1=0.
      IF (J.GT.1) Y1=Y(J-1)
      X=BOUNDS(J)+((U-Y1)/(Y(J)-Y1))*(BOUNDS(J+1)-BOUNDS(J))
      RETURN
      END
```

A.11 EXPONENTIAL DISTRIBUTION

```fortran
      SUBROUTINE EXPON(XAVG,X0,X)
C***********************************************************************************
C THIS SUBPROGRAM GENERATES AN EXPONENTIALLY DISTRIBUTED RANDOM VARIATE          *
C     WITH MEAN XAVG AND MINIMUM X0 USING THE INVERSE TRANSFORMATION METHOD      *
C***********************************************************************************
      ALPHA=1./(XAVG-X0)
      X=X0-ALOG(RAND(0))/ALPHA
      RETURN
      END
```

A.12 GAMMA DISTRIBUTION

```fortran
      SUBROUTINE GAMMA(IBETA,ALFA,L,X)
C***********************************************************************************
C     THIS SUBPROGRAM GENERATES A GAMMA DISTRIBUTED RANDOM VARIATE WITH
C     PARAMETERS ALFA AND IBETA (INTEGER VALUES ONLY), USING
C     DIRECT SIMULATION.
C
C IDENTIFICATION OF VARIABLES:
C     IBETA  = A POSITIVE INTEGER PARAMETER (INPUT SCALAR)
C     ALFA   = A POSITIVE REAL PARAMETER (INPUT SCALAR)
C     L      = THE LOWER LIMIT OF THE PRODUCT (USUALLY THIS VALUE IS 1)
C              (INPUT SCALAR)
C     X      = THE CALCULATED RANDOM VARIATE (OUTPUT SCALAR)
C***********************************************************************************
      P=1.
      DO 1 I=L,IBETA
      U=RAND(0)
      P=P*U
      CONTINUE
      X= -(1./ALFA)*ALOG(P)
      RETURN
      END
```

A.13 GEOMETRIC DISTRIBUTION

```
      SUBROUTINE GEOM(Q,LX)
C***********************************************************************
C     THIS SUBPROGAM GENERATES A GEOMETRICALLY DISTRIBUTED RANDOM VARIATE
C     WITH PARAMETER Q. USING THE INVERSE TRANSFORMATION METHOD.
C
C IDENTIFICATION OF VARIABLES:
C     Q       = PROBABILITY OF FAILURE FOR EACH EXPERIMENT (INPUT SCALAR)
C     LX      = THE GENERATED RANDOM VARIATE (OUTPUT SCALAR)
C***********************************************************************
      U=RAND(0)
      W=ALOG(U)/ALOG(Q)
      LX=INT(W)
      RETURN
      END
```

A.14 DATA GROUPING ROUTINE

```
      SUBROUTINE GROUP(DATA,BOUNDS,F,Y,M,N)
C***********************************************************************
C THIS SUBPROGRAM GROUPS A SET OF DATA AND THEN CALCULATES THE RELATIVE   *
C     FREQUENCIES AND THE CORRESPONDING CUMULATIVE DISTRIBUTION           *
C                                                                         *
C IDENTIFICATION OF VARIABLES:                                            *
C     DATA    = INDIVIDUAL DATA POINTS (INPUT ARRAY)                      *
C     BOUNDS  = INTERVAL BOUNDS (INPUT ARRAY)                             *
C     M       = NUMBER OF INTERVALS (INPUT SCALAR)                        *
C     N       = NUMBER OF DATA POINTS (INPUT SCALAR)                      *
C     K       = INTERVAL COUNTERS (INTERNAL ARRAY)                        *
C     F       = RELATIVE FREQUENCIES (OUTPUT ARRAY)                       *
C     Y       = CUMULATIVE DISTRIBUTION (OUTPUT ARRAY)                    *
C***********************************************************************
      DIMENSION DATA(1000),BOUNDS(26),F(25),Y(25),K(25)
C ZERO THE INTERVAL COUNTERS
      DO 1 J=1,M
    1 K(J)=0
C GROUP THE DATA
      DO 3 I=1,N
      DO 2 J=1,M
      IF (DATA(I).GE.BOUNDS(J+1)) GO TO 2
      K(J)=K(J)+1
      GO TO 3
    2 CONTINUE
    3 CONTINUE
C CALCULATE THE RELATIVE FREQUENCIES AND THE CUMULATIVE DISTRIBUTION
      F(1)=FLOAT(K(1))/N
      Y(1)=F(1)
      DO 4 J=2,M
      F(J)=FLOAT(K(J))/N
    4 Y(J)=Y(J-1)+F(J)
      RETURN
      END
```

A.15 MERGE ROUTINE

```
      SUBROUTINE MERGE(A,B,T,Q,N)
C ***************************************************************************
C THIS SUBPROGRAM MERGES QUEUE ARRIVAL TIMES WITH QUEUE DEPARTURE TIMES     *
C     AND DETERMINES QUEUE LENGTH AS A FUNCTION OF TIME                     *
C                                                                           *
C IDENTIFICATION OF VARIABLES:                                              *
C     A(J)   = ARRIVAL TIME OF JTH CUSTOMER, J=1,2,....N                    *
C     B(K)   = DEPARTURE TIME OF KTH CUSTOMER, K=1,2,....N                  *
C     T(I)   = TIME OF ITH EVENT (ARRIVAL OR DEPARTURE), I=1,2,....,2*N     *
C     Q(I)   = QUEUE LENGTH RESULTING FROM THE ITH EVENT                    *
C     N      = NUMBER OF ARRIVALS (OR DEPARTURES)                           *
C ***************************************************************************
      INTEGER Q
      DIMENSION A(1000),B(1000),T(2000),Q(2000)
      T(1)=A(1)
      Q(1)=1
      I=1
      J=2
      K=1
C BEGIN LOOP
    1 IF (J.GT.N) GO TO 3
    2 IF (A(J).GT.B(K)) GO TO 3
C ARRIVAL IS NEXT
      T(I+1)=A(J)
      Q(I+1)=Q(I)+1
      I=I+1
      J=J+1
      IF (A(J-1).NE.B(K)) GO TO 1
C DEPARTURE IS NEXT
    3 IF (K.GT.N) RETURN
      T(I+1)=B(K)
      Q(I+1)=Q(I)-1
      I=I+1
      K=K+1
      GO TO 1
      END
```

A.16 NORMAL DISTRIBUTION

```
      SUBROUTINE NORMAL(XAVG,STDEV,X)
C***************************************************************************
C THIS SUBPROGRAM GENERATES A NORMALLY DISTRIBUTED RANDOM VARIATE          *
C     WITH MEAN XAVG AND STANDARD DEVIATION STDEV USING A METHOD BASED     *
C     UPON THE CENTRAL LIMIT THEOREM                                       *
C***************************************************************************
      SUM=0.
      DO 1 I=1,12
    1 SUM=SUM+RAND(0)
      Z=SUM-6.
      X=XAVG+STDEV*Z
      RETURN
      END
```

A.17 OUTPUT ROUTINE

```
      SUBROUTINE OUTPUT(BOUNDS,F,Y,AVG,STDEV,M,N)
C**********************************************************************************
C THIS SUBPROGRAM PRINTS THE MEAN, THE STANDARD DEVIATION, THE RELATIVE          *
C     FREQUENCIES AND THE CUMULATIVE DISTRIBUTION FOR A GIVEN DATA SET           *
C                                                                                *
C IDENTIFICATION OF VARIABLES:                                                   *
C     BOUNDS = INTERVAL BOUNDS                                                   *
C     F      = RELATIVE FREQUENCIES                                             *
C     Y      = CUMULATIVE DISTRIBUTION                                          *
C     AVG    = MEAN                                                             *
C     STDEV  = STANDARD DEVIATION                                              *
C     M      = NUMBER OF INTERVALS                                             *
C     N      = NUMBER OF DATA POINTS                                           *
C**********************************************************************************
      DIMENSION BOUNDS(26),F(25),Y(25)
  100 FORMAT('0',10X,'MEAN =',E12.5,10X,'STANDARD DEVIATION =',E12.5,
     1 10X,'NUMBER OF DATA POINTS =',I4//)
  200 FORMAT('0',14X,'INTERVAL',14X,'LOWER',14X,'UPPER',14X,'RELATIVE',
     1 14X,'CUMULATIVE'/,16X,'NUMBER',15X,'BOUND',14X,'BOUND',14X,
     2 'FREQUENCY',12X,'DISTRIBUTION'/)
  300 FORMAT(18X,I2,14X,F8.4,11X,F8.4,13X,F8.4,15X,F8.4)
      WRITE (6,100) AVG,STDEV,N
      WRITE (6,200)
      DO 1 J=1,M
      J1=J+1
    1 WRITE (6,300) J,BOUNDS(J),BOUNDS(J1),F(J),Y(J)
      RETURN
      END
```

A.18 PARAMETER ROUTINE (MEAN AND STANDARD DEVIATION)

```
      SUBROUTINE PARAMS(DATA,AVG,STDEV,N)
C**********************************************************************************
C THIS SUBPROGRAM CALCULATES A MEAN AND A STANDARD DEVIATION FOR A GIVEN         *
C     SET OF DATA                                                                *
C                                                                                *
C IDENTIFICATION OF VARIABLES:                                                   *
C     DATA   = INDIVIDUAL DATA POINTS (INPUT ARRAY)                             *
C     N      = NUMBER OF DATA POINTS (INPUT SCALAR)                            *
C     AVG    = MEAN (OUTPUT SCALAR)                                            *
C     STDEV  = STANDARD DEVIATION (OUTPUT SCALAR)                             *
C**********************************************************************************
      DIMENSION DATA(1000)
      SUM1=0.
      SUM2=0.
      DO 1 I=1,N
      SUM1=SUM1+DATA(I)
    1 SUM2=SUM2+DATA(I)**2
      AVG=SUM1/N
      STDEV=SQRT(SUM2/N-AVG**2)
      RETURN
      END
```

A.19 POISSON DISTRIBUTION

```
      SUBROUTINE PCISSC (RLANDA,LX)
C***********************************************************************************
C THIS SUBPROGRAM GENERATES A POISSON DISTRIBUTED RANDOM VARIATE
C     WITH PARAMETER RLANDA
C
C IDENTIFICATION OF VARIABLES:
C     RLANDA = NUMBER OF EVENTS THAT OCCUR IN SOME FINITE TIME INTERVAL
C     LX     = THE GENERATED RANDOM VARIATE (OUTPUT SCALAR)
C***********************************************************************************
      F=EXP(-RLANDA)
      P=1.
      I=1
2     U=RAND(0)
      P=P*U
      IF(P.LT.F) GO TO 1
      I=I+1
      GO TO 2
1     LX=I-1
      RETURN
      END
```

A.20 GROUPING ROUTINE

```
      SUBROUTINE QDIST(DT,Q)
C ***********************************************************************************
C THIS SUBPROGRAM GROUPS TIME INTERVAL (DT) VS QUEUE LENGTH (Q)              *
C ***********************************************************************************
      INTEGER Q
      COMMON TQ(14)
      DO 1 I=1,13
      IF ((Q+1).GT.I) GO TO 1
      TQ(I)=TQ(I)+DT
      RETURN
    1 CONTINUE
      TQ(14)=TQ(14)+DT
      RETURN
      END
```

A.21 UNIFORM (0, 1) RANDOM NUMBER
GENERATOR FOR A 36-BIT COMPUTER

```
      FUNCTION RAND(KX)
C***********************************************************************************
C THIS FUNCTION GENERATES A UNIFORMLY DISTRIBUTED RANDOM NUMBER             *
C     BETWEEN ZERO AND ONE, USING THE POWER RESIDUE METHOD.                 *
C                                                                           *
C KX SHOULD BE ASSIGNED A POSITIVE, ODD, INTEGER VALUE, NOT EXCEEDING       *
C     34359738367, THE FIRST TIME THE FUNCTION IS CALLED.                   *
C     THEREAFTER, KX SHOULD BE ASSIGNED A VALUE OF ZERO.                    *
C                                                                           *
C NOTE: THIS FUNCTION IS VALID ONLY FOR A COMPUTER HAVING A 36-BIT WORD.    *
C*************************************************************************QQQQ****
      IF (KX.GT.0) IX=KX
      IY=262147*IX
      IF (IY.LT.0) IY=IY+34359738367+1
      RAND=FLOAT(IY)/34359738367
      IX=IY
      RETURN
      END
```

APPENDIX B

ANSWERS TO

SELECTED PROBLEMS

CHAPTER 1

(1.1) *System parameters:* sales volumes and their associated probabilities; manufacturing and distribution costs and their associated probabilities.
Decision variable: selling price
State variables: yearly sales volume; cost per unit.
Cause-and-effect relationships: equations relating sales volume parameters, manufacturing and distribution cost parameters, and selling price to the yearly sales volume and the cost per unit.
System performance criterion: Yearly profit corresponding to a specified selling price.

(1.2) *System parameters:* numbering, color, and arrangement of the squares.
Decision variables: parameters associated with each bet (choice of square, color, etc.) and the amount of money associated with each bet.
State variables: the final square where the marble comes to rest (for each game)
Cause-and-effect relationships: equations relating a randomly generated number of the particular square where the marble comes to rest.
System performance criteria: the outcome of each single game (win-lose), the likelihood of winning in a large number of games (number of wins per total number of games), and the amount of money won using specified types of bets.

(1.3) *System parameters:* demand for the fourth press, service times, setup time, cost of the fourth press, hourly (or daily) profit.
Decision variable: mathematical representation of the decision to purchase a fourth press or to subcontract.

State variables: the times that orders arrive, begin to be processed, and are completed.

Cause-and-effect relationships: equations relating arrival times, processing start times, and processing completion times to the total time of operation of the fourth press.

Performance criteria: The fraction of time that the fourth press is in operation, and the yearly profit associated with its operation.

(1.4) *System parameters:* initial investment, sales price distribution, sales volume distribution, annual costs, tax rate, interest rate, and method of depreciation.

Decision variables: as the problem is stated, there are none. However the tax rate, interest rate, and method of depreciation could be treated as decision variables for purposes of analysis.

State variables: annual sales volume, annual revenue and annual costs.

Cause-and-effect relationships: equations relating the system parameters to the annual sales volume, the annual revenue, the annual costs, and the annual net revenue (profit).

System performance criterion: annual net profit.

(1.5) *System parameters:* customer inter-arrival time and customer service times.

Decision variables: as the problem is stated, there are none. However the inter-arrival time, service times, and number of checkout counters could be treated as decision variables for purposes of analysis.

State variables: the time that each customer arrives, enters the checkout counter, and leaves the checkout counter.

Cause-and-effect relationships: equations relating the system parameters to the customers' arrival times, the time customers enter the checkout counter, and the time customers leave.

System performance criteria: customer waiting time, total checkout time (waiting time plus service time), length of the waiting line, and fraction of time that the checkout counter is not in use.

(1.6) *System parameters:* passenger inter-arrival times and passenger service (check-in) times.

Decision variable: the number of check-in counters that are open during different times of the day.

State variables: the time that each passenger arrives, checks in, and leaves the check-in counter.

Cause-and-effect relationships: equations relating the system parameters and the decision variable to the various passenger event times.

System performance criteria: passenger waiting time, total check-in time, length of each waiting line, and fraction of time that each check-in window is not in use.

(1.8) The yearly profit should be expressed as an empirical cumulative distribution, showing the percent of all simulated results that are less than or equal to various profit levels. This information indicates the likelihood of obtaining these various yearly profit figures.

CHAPTER 2:

(2.1) $n_1 = 7308$ $n_7 = 2917$
$n_2 = 4068$ $n_8 = 5088$
$n_3 = 5486$ $n_9 = 8877$
$n_4 = 0961$ $n_{10} = 8011$
$n_5 = 2352$ $n_{11} = 1761$
$n_6 = 5319$ $n_{12} = 1011$

(2.3) $\bar{x} \cong 0.5$ $s \cong 0.2887$

(2.4) $n_1 = 2520$ $n_7 = 6639$
$n_2 = 5473$ $n_8 = 2829$
$n_3 = 7919$ $n_9 = 7817$
$n_4 = 3406$ $n_{10} = 1142$
$n_5 = 9721$ $n_{11} = 9270$
$n_6 = 1097$ $n_{12} = 5863$

(2.6) $\bar{x} \cong 0.5$ $s \cong 0.2887$

(2.7) 256 values
0, 1, 2, . . . , 255

(2.8) $p_1 = 8$ There are only 2 power residues.
$p_2 = 0$

(2.9) $p_1 = 9$
$p_2 = 17$
$p_3 = 25$
$p_4 = 1$
$p_5 = 9$
$p_6 = 17$

(2.10) $p_1 = 19$ $p_5 = 3$
$p_2 = 9$ $p_6 = 25$
$p_3 = 11$ $p_7 = 0$
$p_4 = 17$ $p_8 = 1$

(2.12) a. period $= 512$
b. 373
c. No

(2.13) $x \cong 0.5$ $s \cong 0.2887$

(2.17) The mean values are normally distributed.

(2.20) a. ~ 0.216
b. ~ 0.1296
c. ~ 0.064
d. ~ 0.03456

(2.21) a. 0.0833
b. 0.0005787
c. 0.167

(2.22) a. $10.46 > 9.037$; Reject the hypothesis.
b. $10.46 < 14.07$; Accept the hypothesis.
c. $10.46 < 18.48$; Accept the hypothesis.

(2.23) Chi-square statistic = 3.33
Tabulated chi-square = 16.92 for 9 degrees of freedom.
The test is successful.

(2.24) Chi-square statistic = 3.93
Tabulated chi-square = 11.07; $\nu = 5$; p = 5%. Since $3.93 < 11.07$, accept the hypothesis: the sequence is randomly ordered.

(2.25) 32 Runs $E_{tot} = 33$
Regrouping the data:

i	n	O	E
1	1	20	20.92
2	2–49	11	12.08

$$\chi^2 = 0.1370$$

Tabulated chi-square = 3.841; $\nu = 1$; p = 5%.
Since $3.841 > 0.1370$,
accept the hypothesis: the numbers are sequenced randomly.

(2.26) $E_{tot} = 25.5$
$\chi^2 = 2.34$
Tabulated chi-square = 5.991; $\nu = 2$; p = 5%. Since $5.991 > 2.34$,
accept the hypothesis: the numbers are sequenced randomly.

CHAPTER 3:

The answers given below were obtained from actual simulations and are therefore subject to some random variation.

(3.4) a.

Player	# Games Won	Score	Freq
1	22	≤ 12	0
2	22	13	3
3	23	14	4
4	25	15	9
5	22	16	20
		17	24
		18	22
		19	21
		20	94
		21	89
		≥ 22	214

(3.5) **a.**

Score	Freq
≤11	105
12	41
13	48
14	53
15	54
16	37
17	38
18	36
19	33
20	42
21	13
≥22	0

b.

Score	Freq
12	0
13	1
14	12
15	16
16	25
17	44
18	81
19	48
20	82
21	65
22	126

(3.7)

Score	Freq
0	31
1	28
2	31
3	30
4	29
5	42
6	73
7	42
8	41
9	53

(3.11) This is a well-known, old problem. If $N \geq 23$, there is at least a 50% probability that two or more people will have their birthday on the same day.

(3.16) $S = 0.9022$

(3.18)

profit	.38	.43	.48	.64	.80	.90	.95	1.16	1.42	1.52	1.89
freq	.03	.05	.04	.14	.15	.19	.06	.18	.08	.04	.04

(3.19) $S = 0.410$

(3.20) $I \cong 0.4773$

(3.21) $I \cong 9$

CHAPTER 4:

(4.1) **a.** $x = 4\,U + 1$
     b. $x = (63\,U + 1)^{1/3}$
     c. $x = (6\,U)^{1/2} + 1$ $0 \leq U \leq 0.67$
       $x = [3(1 - U)]^{1/2} + 4$ $0.67 \leq U \leq 1$

(4.3) **a.** see appendix A.10
     b. see appendix A.11

    c. see appendix A.13
    d. see appendix A.12
    e. see appendix A.19
    f. see appendix A.16

(4.4)   see appendix A.14, A.17, A.18

(4.12)   see appendix A.7

CHAPTER 5:

The numerical solutions given below were obtained from actual simulations and are therefore subject to some random variation.

(5.1) see appendix A.1

(5.3) a. 3.40
      b. 0.04
      c. 0

(5.5) $x_0 = 0$: a) 2.83 years $x_0 = 1$: a) 2.16 years
             b) 0.11   b) 0.04
             c) 0   c) 0

(5.6) *Mean of yearly costs*   Maintenance cost: $50/machine
    a. $19,000   Repair cost: $100/day downtime
    b. $37,500
    c. $27,100
    d. $23,400
    Mean of yearly costs   Maintenance cost: $65/machine
    a. $ 6,310   Repair cost: $25/breakdown +
    b. $40,900 $40/hr. downtime
    c. $23,200
    d. $14,800

(5.7) *Mean of yearly costs*   Maintenance cost: $65;
    a. $69,600   Repair cost: $25/breakdown +
    b. $42,500 $40/hr. downtime
    c. $45,200
    d. $59,600

(5.9)

ROP	Q	Mean of yearly costs
35	40	$48,400
45	45	$ 4,210
55	50	$ 4,280
65	55	$ 5,800

(5.10) *Mean of yearly costs ($)*

ROP	Q = 40	45	50	55	60	75	85	95	105
25			13000						
30			13500						
35			11300						
40	41600	19000	11600	8400	7500	7000	6800	7100	7200
45			10700						
50			9810						
55			9840						
60			9520						
65			10300						
70			10100						
75			8800						
80			9630						
90			9790						

(5.12)

Lead time	ROP	Q	Mean
0	0	46	3310
0	0	45	3300
0	0	44	3290
0	0	43	3300
0	0	42	3300
0	0	41	3310
5	39	50	3610
5	39	49	3600
5	39	48	3570
5	40	50	3660
5	40	49	3620
5	40	48	3610
5	35	55	3620
5	35	54	3600
5	35	53	3590
5	34	55	3610
5	34	54	3600

(5.14)   **a.** 2-5-10-12

     b. 0.38

     c. 32.98

     d. 0.043

     e. 0.47

     (see appendix A.2)

(5.15)

Interest rate	a	b	c
12%	1,620,000	0.347	0.435
15%	−231,000	0.747	0.090

(5.16)	Interest rate	a	b	c
	12%	665,000	0.587	0.265
	15%	−1,090,000	0.859	0.069

CHAPTER 6:

(6.2) a. 0.608, 0.644
b. 0.608, 0.644

(6.3) a. 318
b. 777

(6.4) a. 13.17, 20.24
b. 55%
c. 1242

(6.5) a. 4.645, 4.935
b. 1807

(6.6) a. 4.6265, 4.9475
b. 267

CHAPTER 7:

The numerical solutions given below were obtained from actual simulations and are therefore subject to some random variation.

(7.3) See appendix A.15

(7.4) See appendix A.3

(7.8)

	a	b	c	d	e
Number of customers	1000	1000	1000	200	200
Mean queue length	0.155	1.583	6.993	6.057	7.37
Mean waiting time, all customers	0.719	4.549	17.44	16.17	19.09
Fraction idle time	.67	0.31	0.13	0.06	0.03
Total simulation time	4648	2872	2513	532	517

(see appendix A.4)

(7.9)

	Day→1	2	3	4	5	6	7	8	9	10
Number of customers	121	113	126	122	116	104	119	130	114	104
Mean queue length	8.6	4.8	4.1	12.4	9.6	2.0	7.15	5.6	10.1	10.3
Mean waiting time	30.1	18.0	13.9	43.3	35.1	8.1	25.3	18.0	39.2	40.4
Mean time spent within the system	33.2	21.2	17.1	46.5	38.4	11.2	28.4	21.0	40.7	44.0
Fraction idle time	0.11	0.16	0.07	0.10	0.11	0.25	0.11	0.11	0.06	0.12

(7.11)

	a	b	c
Queue length	2.31	0.8	0.59
Mean waiting time (all customers)	46.4	24.2	16.7
Time in system	64.6	36.6	29.9
Fraction idle time	0.09	0.11	0.13

(7.12) *Schedule Policy:*

Time	Number of windows open
6.00	1
7.50	2
8.00	3
11.50	2
12.00	1
15.00	3
16.00	5
17.00	6
19.25	4
20.00	3
20.50	2
21.00	1
23.00	0

Average queue length = 1.214
Average waiting time = 3.147
Fraction idle time:

Window	
1	0.38
2	0.36
3	0.33
4	0.20
5	0.11
6	0.05

(7.13) See appendix A.4

(7.14)

Warm-up period:	351.7
Total time:	3202
Average queue length:	1.278
Mean waiting time:	25.25
Fraction idle time:	0.272

(7.16)

Cutoff point:	4.5 days
Total time in system:	14.70

Average waiting time:	11.34
Average service time:	3.36
Maximum queue length:	0.13

(7.19) See appendix A.5

(7.20) Number of jobs simulated: 50
Mean total manufacturing time: 18.97

	station 1	station 2	station 3	station 4
Mean waiting time	3.06	2.27	4.87	0.02
Mean queue length	0.91	0.64	1.43	0.005
Fraction idle time	0.17	0.32	0.17	0.60

(7.23) See appendix A.6

(7.24)

	a	b	c
Mean total time spent in system	45.4	36.8	68.5
Mean waiting time	19.8	19.8	48.8
Mean queue length	4.26	3.13	7.21
Fraction idle time			
Barber 1	0.0923	0.08	0.055
Barber 2	0.0423	0.087	0.046
Barber 3	0.0701	0.089	0.062

(7.25) Number of customers   50
Mean total time in system   14.7 min
Mean waiting time   5.7
Mean queue length   2.2
Fraction idle time   0.11
Maximum time in system   28.7

INDEX